Between Land and Sea

Between Land and Sea

The Atlantic Coast and the Transformation of New England

CHRISTOPHER L. PASTORE

Harvard University Press

Cambridge, Massachusetts
London, England
2014

Copyright © 2014 by the President and Fellows of Harvard College
All rights reserved
Printed in the United States of America

First Printing

Library of Congress Cataloging-in-Publication Data
Pastore, Christopher L.
Between land and sea : the Atlantic Coast and the transformation
of New England / Christopher L. Pastore.
pages cm.
Includes bibliographical references and index.
ISBN 978-0-674-28141-7 (alk. paper)
1. Coast changes—New England—History. 2. Atlantic Coast
(New England)—History. I. Title.
GB459.4.P37 2014
551.45'70974—dc23
2014008171

For Susie, Rosie, and Abe

Contents

Preface, ix

Prologue: From Sweetwater to Seawater, 1

1. Clams, Dams, and the Desiccation of New England, 11

2. Shoveling Dung against the Tide, 50

3. The Geographic Quicksilver of Narragansett Bay, 82

4. Natural Knowledge and a Bay in Transition, 130

5. Improving Coastal Space during a Century of War, 161

6. Carving the Industrial Coastline, 196

Epilogue: Between Progress and the Pull of the Sea, 228

Notes, 239

Acknowledgments, 287

Index, 293

Preface

I grew up at the northern end of Narragansett Bay, a few blocks from the tidal Providence River. When I was an infant, so I am told, my parents lulled me to sleep to the hum of an outboard engine. As a child I spent summer days catching crabs and minnows in the tidal pools down the road and countless weekend afternoons in a bathing suit, digging quahogs in the passage between Prudence and Patience Islands. On other days, we left the house early with piles of books and newspapers, coolers, umbrellas, blankets, and buckets and drove to Newport, where we arrived on the beach before the lifeguards and left when the sun hung low on the horizon. I grew up sailing and fishing on the Bay and sitting by bonfires on its beaches. So my interest in the coast of southern New England grew not from some burning historical question or exotic archival discovery but from many years of living with the Bay as an important part of my everyday existence.

As an environmental historian of early America, I wanted to know how, over the long sweep of time, the people of Narragansett Bay interacted with and understood the estuary that sustained them. I discovered that for centuries Rhode Islanders (as well as the people of southeastern Massachusetts) have been deeply connected to the place, so much so that they felt compelled to change it. But as they did, it shaped them in return. So here I have endeavored to show that process. Although I have not lived by the Bay for some time, it has molded my imagination. It has surely conditioned the way I have read my evidence and the patterns

Preface

by which I have constructed this narrative. Indeed, by design, this book undulates in its early chapters and grows increasingly structured, thereby mirroring the changing estuary it describes. Ultimately, I make a case for cultivating a more dynamic relationship between people and the sea. And to do that, we must let the ocean in. So if these pages evoke the smell of mud and salt or the feeling of standing alongshore before the breeze, well, that is all I can hope.

PROLOGUE

From Sweetwater to Seawater

"But for their later Descent, and whence they came . . . ," wrote Roger Williams of the Narragansett Indians in 1643, "it seemes as hard to finde, as . . . the *Well-head* of some fresh *Streame,* which running many miles out of the *Countrey* to the salt *Ocean,* hath met with many mixing *Streames* by the way." That Williams, new to the shores of Narragansett Bay, likened the history of his Indian neighbors to the myriad inlets and outlets of an estuary was no coincidence. The Bay and its upland sources, its tidal ebb and flood, and even its seasonal winds largely defined the rhythms of Native American and settler life there. And like the greater estuary's indeterminate geography, the indigenous people who inhabited the western shore and islands of Narragansett Bay, at least to Williams's English sensibilities, lacked identifiable beginnings. "They say themselves," he wrote, "that they have *Sprung* and *growne* up in that very place, like the very *trees* of the *Wildernesse.*"[1]

As Rhode Island's founder and first ethnographer, Williams labored tirelessly to recover the history of the Narragansett people, but his observations suggest that the littoral, or coastal, environment in which they lived was something altogether unknowable. Trickling from the forest at one end and pouring into the ocean at the other, the moving waters of the estuary (from the Latin *aestuare*, meaning to fluctuate or boil), formed a meandering seam among the folds of which were tucked the mysteries of nature.[2] Teeming with life, the ocean's edge, one of the most fertile environments on earth, overwhelmed the senses. Clear, cold water coursed through vibrant green grasslands suffused with the smell of peat and filled with the trill of insects. For this deeply devout Puritan, the damp soil between America's wooded fringe and the vast ocean that brought him to it germinated, however inexplicably, the seeds of life. Like the clay from which God had molded mankind, the people of the Bay and all that surrounded them had been sculpted from a mixture of water and earth. So shrouded in mystery was this place that when Williams went in search of the "little island" from which he had been told the Narragansett name derived, he saw it nestled somewhere "on the sea and freshwater side" of southwestern Rhode Island but could neither reach it nor learn "why it was called Nahiganset."[3]

As Williams well understood, the communion of land and sea shaped much more than the sound of the surf. Indeed, estuaries had captivated some of the most imaginative early modern minds. Only a few years earlier, the Thames River had inspired what is widely considered the first English landscape poem. When during the winter of 1640 Sir John Denham climbed Cooper's Hill outside of London, he looked down upon "the most lov'd of all the Ocean Sons," as it flowed "Hasting to pay his tribute to the Sea / Like mortal life to meet Eternity." For Denham, a human Thames flowed into an eternal ocean, which, among the swirling waters of the estuary, formed the gateway of empire. This was a river that "Visits the World," returning with wealth and bestowing it "where it wants." "[T]o us," Denham explained, "no thing, no place is strange" because "his fair bosom is the World's exchange." So powerful was the Thames that Denham carefully noted it had made "both *Indies* ours." In its ability to merge human initiative and boundless nature, the estuary had ushered in a new era of globalization.[4]

As pools of communal identity and rivers of imperial ambition, estuaries seeped deeply into the human psyche. If land provided permanence, the sea was ephemeral. If interiors provided isolation, the ocean provided connection. Security on shore gave way to uncertainty at sea. Construction gave way to corrosion.[5] These coastal contrasts were so stark that some envisioned littoral spaces as "beginnings and endings," as places that "divide[d] the world between here and there, us and them, good and bad, familiar and strange."[6] But for others, the shore defied duality. As zones of cultural contact and exchange, these brackish borderlands were permeable and elastic, spaces of syncretism rather than separation.[7] Deeply complex, coastal frontiers often formed worlds of their own.[8] With a keen sense of balance, denizens of the shore deftly straddled the gap between punt and pier, living not simply "on" coasts but "with" them, carefully and perhaps intuitively attuned to the wind, weather, and tide.[9] So distinct was the littoral way of life that coastal communities often exhibited similarities regardless of their location.[10] Theirs was an amphibious existence, lived on the edge and in between.[11]

A deeper understanding of these littoral dynamics reveals the ways coastal cultures were often torn between two dominant epistemologies concerning the natural world. The first considered the ocean (and to some extent water in general) unchanging, eternal, and somehow exempt from human influence. The second believed that terrestrial land could be—and often must be—"improved." It was the push and pull of these two conceptions of nature that shaped coastal space. At one end of the estuary lay dry land easily measured by the surveyor's rod, and at the other, a trackless, eternal ocean that defied European understandings of ownership, jurisdiction, and even the passage of time. And somewhere in between, caught in the conceptual wrack that collects along the farthest reaches of the tide, lay European assumptions about the natural world. By reimagining an estuary in light of these ecological and cultural complexities, we can reconstruct littoral settlement and development, not as the opposition of English culture to pristine nature, but as spaces that were neither "wild" nor "civilized" in which people used nature and were shaped by it in return.

Yet, oceans and their estuarine arms have, until recently, been presented without attention to historical change.[12] As any sailor knows, the sea holds the ability to distort time and space, the principal currencies of the historian's trade. Something so fundamental as "distance," a function of speed, which is subject to so many outside variables on the surface of the sea, can only be measured with imprecision. Among the swirling currents and fluky winds so typical of coasts, time and space expand and contract, making littorals just as difficult to navigate for observers of the past as for contemporary captains. How then did coastal people bring coherence to such confusion? And what were the social and ecological ramifications of adding order to the unruliness of the ocean's edge?

A closer look at the ways humans settled and meddled with the Atlantic world's coastal margins recovers an important piece of the living sea's history. Early modern expansion exacted a heavy toll on the coastal ocean.[13] Fish, birds, and marine mammals were hit hard with simple technology.[14] Among inshore waters, these changes to the sea were even more pronounced. Around Chesapeake Bay, for instance, English efforts to wring a living from the land caused such dramatic soil erosion during the seventeenth and eighteenth centuries that deepwater ports became impassable mudflats.[15] Declining water quality, in turn, damaged fish stocks and human health.[16] By probing both the historical record and muddy sediments long trapped beneath the Bay, scholars have pieced together traces of this ecological drama.[17] Historians of early New England have done much the same, although their work has focused largely on the land.[18] But a more careful examination of the ways coasts were constructed imaginatively and then modified materially adds a narrative of renewal to what has often been told as a tale of tragedy or inevitable decline.[19]

Careful focus on estuaries and the people who inhabited them also adds fresh perspective to the conventional narratives of Atlantic exploration and continental settlement. Littorals, after all, set the stage for colonial expansion. Smith, Cartier, Verrazano, and countless others spent considerable time reconnoitering Atlantic estuaries. Navigation expanded as knowledge of coastal features circulated.[20] At once confused and coherent, protected and connected, estuaries, in contrast to environments farther inland, provided reasonable safety while remaining highly accessible.[21] Coastal margins provided sufficient stability to encourage investment but enough variability to create opportunity. As a result, trade funneled through littoral spaces, which became important points of interconnection where societies became firmly "entangled."[22] Estuarine harbors bridged vast Atlantic and continental networks, the intricacies of which historians are only beginning to understand.[23] If Atlantic history has profited from the study of "middle grounds," it will surely benefit from a closer look at "muddy grounds," where interactions between Europeans and Native Americans were often vexed by the complexities of coastal space.[24]

Among these marshy marchlands, littoral people set America in motion. Almost every voyage to the New World ended in an estuary. And most settlers clung to their shores. As late as 1776, Adam Smith noted that America's English plantations were largely confined to the coast and banks of navigable rivers, having "scarce anywhere extended themselves to any considerable distance from both."[25] Among the nicks and notches of the sea, cities formed as their inhabitants, in a "deliberate and enduring dialogue" with the natural world, drew resources from and made modifications to their inland and estuarine hinterlands.[26] Material changes to this private–public interface shaped where and how people lived, which affected the coastal environment in return.[27] So demanding was the process of building infrastructure in Boston, for instance, that it forged a new relationship between humans and nature, one that helped to define what it meant to be "urban" in America.[28] In New York the transformation of littoral space was so profound that efforts to digitally reconstruct its ancient appearance have captured the popular imagination.[29]

Harboring some of New England's earliest settlements during the seventeenth century and powering America's first industrial revolution at the end of the eighteenth, Narragansett Bay and the rivers that rolled into it set important precedents for coastal–human interactions. Among Indians and Europeans the eternal arms of the ocean often marked lines of social and political division. But at other times, the open, mutable nature of the estuary made it an important avenue of transportation and trade, a conduit for economic and cultural exchange. For slaves, the Bay's maze of tidal creeks and rivers often provided geographies of refuge. Pirates and privateers likewise sought cover among the estuary's innumerable coves. But mariners, their maps, and the "spatial forms and fantasies" they projected added order to the edge of the sea.[30] So too did scientific inquiry. When dam and canal builders began partitioning the upper reaches of the estuary, they transformed the shore like never before.

At the nexus of land and sea and the confluence of sweetwater and seawater, a host of political, legal, and cultural ambiguities were shaped by the tension between a desire to "improve" the land and a belief that the ocean was eternal. At times, the shores of Narragansett Bay were subdued. People living in the littoral caught fish, dug clams, and sent their animals out to graze its meadows. And their work changed the place. But at other times, the sea prevailed. Its ability to blur lines of ownership and notions of legality and to dissolve the spatial mechanisms of social control often thwarted the impulse toward improvement. By tinkering with and adapting to this world in flux, Rhode Islanders continuously renegotiated their relationship with nature. Much more than a geologic formation or the passive recipient of human action, Narragansett Bay was a cultural construct, created and recreated by the people who lived near and worked on its waters. If the people of the Bay believed there was a boundary between them and the natural world, it was as porous as the sands along their shores.

But that, too, changed over time. As the population of Rhode Island rose and land and resources diminished, Anglo-Americans changed the Bay to meet their needs. They reorganized coastal space both literally and figuratively by harvesting shellfish, drawing maps, building beacons and forts, and fighting wars, among countless other undertakings, thereby changing a porous, inchoate coastal margin into a more clearly defined edge, or "coastline."[31] What had once been a sprawling, uninterrupted watershed, sustained as much by the push of its rivers as the pull of its tides, was summarily split in two. This newly bifurcated Bay, one in which the freshwater portion was imagined as separate from the salt, bolstered the belief that nature itself was separate from people and could be mastered by them. In turn, the engineers of the industrial age modified the Bay irrecoverably. They built a new edge for modern America, which made its coast less resilient, or less capable of absorbing the blows of human initiative and natural variation. And this had far-reaching environmental and social repercussions.

Beginning in 1636 with the first European settlement of Narragansett Bay and ending in 1849 with the final dissolution of the Blackstone Canal Company, which completely reconfigured the upper reaches of the Bay's watershed, this book shows how people shaped and were shaped by the Bay as well as by the more distant shores across the Atlantic world. Our story begins with early European travelers who provided the first written visions of southern New England. Their encounters with Native Americans and the seventeenth-century wampum trade summarily transformed the estuary and the broader region. Home to some of the richest clam beds in southern New England, Narragansett Bay was an important source of shell beads, produced largely from several kinds of whelks and the hard-shell clam, or quahog, *Mercenaria mercenaria*. Cut, polished, and drilled, these shells were assembled by the thousands into intricate belts and traded among Indians and Europeans, often across long distances into the continental interior. The value these groups placed on wampum—a marine animal mined from the sea—illuminates the ways in which Native American and European cultures perceived nature more broadly. Closely tied to its ecological origins at the intersection of improvable land and an eternal sea, wampum's value rested on its ability to forge physical, conceptual, and economic continuities between the ocean and inland environments. Accordingly, it hastened the extirpation of beaver. As beavers were killed, their dams were destroyed, and water that had once been impounded on the landscape rushed into the estuaries. In short, wampum, much of which was produced on the shores of Narragansett Bay, drove a trade that made the Northeast a drier place and, at least in small ways, affected the estuary in return. Long before widespread European settlement, a few traders and the hunters they hired began to change the estuary and the entire region. What had been a soggy coastal margin began to harden into an edge.

As the trade in wampum waned during the second half of the seventeenth century, the people of Rhode Island saw new opportunities among the Bay's grassy fringes. Long known for their ties to seafaring, Rhode Islanders took to the sea not because the land was too poor to farm but because the fruits of the shore sent them in search of markets.[32] An important source of provisions for the Royal Navy and English merchant and fishing fleets, the rich coastal meadows of Narragansett Bay supported vast flocks of sheep and herds of cattle. But pastoral development led to Bay pollution. By the end of the seventeenth and beginning of the eighteenth centuries, Rhode Island's livestock had changed the Bay and nearby tidal lagoons. Animal waste led to algal blooms that, when combined with silt and sawdust, clogged waterways. In response to environmental issues, the colony, which had once been only a loose association of towns, began to pool resources toward large public works projects, many of which affected the Bay and its harbors either directly or indirectly. In some cases, livestock runoff chemically "improved" tidal ponds, while in others, the sea resisted human attempts to control it. Nevertheless, by modifying meadows, carving passable channels, and redirecting rivers, the people of the Bay added new definition to the ocean's edge.

If exchanges between land and sea changed the Bay biologically, they also shaped the way people perceived it. The watery nature of the estuary blurred geographic boundaries, forcing Rhode Islanders to continually fend off the territorial advances of outsiders. When in 1741 Rhode Island and Massachusetts went to court to decide the former's eastern boundary, which was legally defined in relation to Bay waters, the ability to "place" the Bay was of utmost importance. But bounding littoral space proved difficult. Court commissioners called scores of deponents to explain how they understood Narragansett Bay and its surrounding lands. Their testimonies revealed the extent to which littoral space was shaped, not geologically, but through expressions of political allegiance, the desire for economic opportunity, and interactions between the metropole (London) and colonial periphery. Narragansett Bay was a deeply human construct. And its borders, which had become more clearly defined than ever before, were established by way of testimony showing that the Bay was improvable space.

As the contest of coastal boundaries came to a close, the people of Narragansett Bay began to reevaluate coastal nature in light of more rational, scientific modes of thought. Rhode Island's coastal climate and daily weather drew considerable commentary. For some, atmospheric anomalies remained proof of God's enduring presence. But for others, the sky was a palette for human progress, a space shaped increasingly through careful observation and systematic notation. While Rhode Island's weather watchers organized the skies, others gazed deeply into Narragansett Bay's waters. The desire for natural knowledge about the Bay's biology led the Rhode Island General Assembly to enter protracted debate over the "nature" of Bay oysters during the first half of the eighteenth century. Attempts to classify tidal creatures, however, were often complicated by the complex ecology of the estuary itself. That oysters were planted and harvested like vegetables, pickled and eaten as animals, and often mined from the seabed as minerals (for producing lime) made them difficult to pin down. Placed within the wider context of eighteenth-century philosophical discussions about the "nature" of aquatic creatures such as mollusks and polyps, the debate on the Bay highlights some of the forces that shaped natural knowledge. In many ways, this attempt to classify the Bay's biology mirrored the wider efforts toward scientific and social improvement that came to define the age of enlightenment. As Rhode Islanders organized the Bay intellectually, they also increased their ability to modify it materially. As a result, the contours of the coast continued to change.

The most powerful forces of progress on Narragansett Bay during the eighteenth century were commercial expansion and war. The construction of coastal beacons, which projected light over long distances, extended control over littoral space. Stone forts equipped with long-range cannons also added order to the estuary and its most important harbors. Detailed maps of Narragansett Bay published by the English and reproduced by the French during the American War for Independence similarly served to organize littoral space. But in Newport and across Aquidneck Island, it was the guns, saws, and shovels of war that most dramatically reconfigured the coast. What had been the "Garden of New England" at the beginning of the eighteenth century was a wasteland by its end. The Bay had assumed new and more permanent shapes and roles to accord with shifting times.

But even more profound ecological changes would grip the upper reaches of the Bay watershed during the nineteenth century. As the birthplace of America's first industrial revolution, the Blackstone River played host to a series of pitched legal battles that, in the name of progress, saw the erosion of traditional ideas about access to natural resources—in most cases, river fish. The creation of the Blackstone Canal Company affirmed this trend toward the privatization of water. Upon opening in 1828, the forty-five-mile canal connecting Providence and Worcester sought to extend littoral space deep into the New England interior. Such an ambitious undertaking required a highly complex and incredibly expensive network of holding ponds and dams for managing the flow of water into the canal and over mill waterwheels. The Canal Company turned the Blackstone River watershed into an engineered system. But when the Canal Company failed, so too did its carefully managed network of ponds, dams, and diversions. So important had the corporation become to the movement of water that the Rhode Island state legislature stalled its dissolution for years. Ultimately, the canal became a sewer that flowed downhill and into the Bay. When the improvement of a watershed was privatized, its failure had dire consequences. If in earlier years Narragansett Bay had been shaped by the push of progress on land and the pull of a vast and enduring sea, industrialization gave material development the upper hand. Lined with brick buildings and granite walls, the Bay's new man-made coast formed a permanent, impenetrable edge, one less capable of absorbing the shocks of nature and society.

As Roger Williams, Sir John Denham, Adam Smith, and countless others could attest, the shore provided footing for the first two centuries of European settlement—though sometimes it was slippery and uneven. At once a barrier and bridge, Narragansett Bay was a product of both physical geography and human demands on that geography, a space shaped by its marginal location as well as by the religious beliefs, political allegiances, and economic aspirations of those who inhabited it. But when coastal communities severed their uplands from the forces of the sea, there were unintended consequences. Their attempts to bring "progress" to the strand made it more fragile.

As something long considered absolute and inevitable, "progress," therefore, needs to be reassessed. As sea levels continue to rise and "super storms" batter our coasts with increasing ferocity, an examination of past relationships between freshwater and saltwater, land and sea, and the spaces precariously perched between them may help us rethink the notion of progress in ways that account for instability and uncertainty. Environmental change in the littoral was anything but linear and far from predictable. And our efforts to fully stem the tide have yet to succeed. So perhaps leaving room for improvement along our shores is more sustainable than actually achieving it. Perhaps inviting the sea to flow freely into the streams of our collective imagination could even be restorative.

CHAPTER 1

Clams, Dams, and the Desiccation of New England

IN JULY 1636, the coastal trader John Gallop weighed anchor from a harbor in eastern Connecticut and steered his bark of twenty tons southwest toward Long Island. Sailing before what was likely a fickle, northerly wind across the placid waters of the Sound, Gallop, traveling with a man and two boys, was forced to abandon his intended destination when the wind shifted. Changing course, he sailed east into the open ocean toward Manisses, or Block Island, a pear-shaped splotch of land seventeen miles east of Montauk and twelve miles south of the Narragansett Country. About two miles north of the island they came upon a small boat, which Gallop recognized as that of fellow coaster, John Oldham. To his count, fourteen Indians were on deck, and several others paddled a nearby canoe toward shore. Upon hailing Oldham, Gallop and his crew received no reply.[1] A simple day trip to Long Island, Gallop soon discovered, would grow quite complicated in the coastal waters south of Narragansett Bay.

That such a bizarre scene was unfolding on a vessel owned by Oldham, whose lack of scruples had earned him wide notoriety, seemed not altogether impossible. A troublemaker through and through, Oldham had flirted with impropriety since the day he landed on American soil. Not long after arriving in Plymouth in 1623, Oldham, according to Plymouth Colony Governor William Bradford, "grew very perverse and showed a spirit of great malignancy." Later accused of religious subversion, Oldham responded with impertinence, hurling invective at his accusers and even drawing a knife on Captain Myles Standish.[2] Banished from Plymouth, Oldham fled to Massachusetts Bay, settling first in Nantasket, then Cape Ann, and finally Watertown, where he continued to indulge his penchant for mayhem. In July 1632, noted Massachusetts Bay Governor John Winthrop, he fired his musket loaded with pistol bullets and shot "three men, two into their bodies, and one into his hands."[3] Fortunately, he had been at such a distance that he inflicted only superficial wounds. A month later, however, Oldham burned down his own house at Watertown "by making a fire in it when it had no chimney."[4] So quick-tempered was Oldham that his contemporary Thomas Morton called him "a mad Jack in his mood."[5] Unflinching in the face of risk or confrontation and increasingly unwelcome ashore, Mad Jack soon became the most experienced and savvy coaster in New England.

Despite his unsavory reputation—or perhaps because of it—Massachusetts Bay sought Oldham's extensive knowledge of the New England coast when it asked him to retrieve a hefty ransom on the colony's behalf. Four years earlier, in 1632, Indians (likely Western Niantic) killed the English traders John Stone and Walter Norton, and the Pequots of eastern Connecticut were blamed. A Pequot delegation presented magistrates in Boston with two bushels of wampum and a bundle of sticks representing the number of beaver and otter with which they would compensate the English for the deaths.[6] They sought peace with the English and also requested help establishing concord with the Narragansetts, who bordered them to the east. The English, in turn, demanded the Indians responsible for killing Stone and Norton, a promise not to interfere with English settlement in Connecticut, and 400 fathoms of wampum and the pelts of forty beaver and thirty otter.[7] This final demand was oppressive. The pelts could be readily acquired, but 400 fathoms of wampum—2,400 feet, or nearly a half-mile comprising roughly 120,000 hand-carved shell beads, which promised to take ten skilled artisans almost a year to produce—was truly onerous.[8] And it was the irascible Oldham's job to collect it.

That Oldham was nowhere in sight while a band of Indians "armed with guns, pikes, and swords" unfurled his vessel's canvases signaled to Gallop that something was dreadfully wrong. By John Winthrop's account, Gallop responded swiftly by showering the Indians with "duck shot."[9] He then drove the stem of his bark into the aft quarter of Oldham's pinnace, nearly capsizing it. So frightened were the Indians that "six of them leaped overboard and were drown." Gallop then pulled away and his crew positioned their anchor so that when they rammed the pinnace a second time, the anchor pierced its bow, the two vessels thereby "sticking fast." At close range, Gallop's crew blasted the vessel with shot and then "raked her fore and aft," whereupon four or five more Indians leapt into the sea and drowned. Finally, Gallop stood off a third time, then sailed alongside and boarded. They bound one Indian and threw him in the bark's hold. They bound another and threw him into the ocean. They were unable to apprehend two Indians who had locked themselves with their swords in a small room belowdecks. While removing the sails and various provisions from the vessel, Gallop and his crew found, under a seine net, John Oldham's naked body, "his head cleft to the brains, and his hand and legs cut as if they [the Indians] had been cutting them off, and yet warm." They attempted to tow the battered boat toward the mainland (with the Indians still hiding inside), but facing nightfall and building winds, they set it adrift. Before casting off the line, they dragged Mad Jack's mangled corpse to the gunwales and "put him into the sea."[10]

For the first generation of settlers, danger lurked at the ocean's edge. Although estuaries were the natural sites of settlement, their entrances and passages were often difficult to discern. From the deck of a ship, important headlands blended into what appeared to be a seamless timberline, making it all too easy to lose one's way. Dotted with rocks and shoals and frequently harboring disease, coasts were known to kill with caprice.[11] But among early explorers and traders, it was the specter of Indian attack that instilled truly penetrating fear. So apprehensive was Frenchman Jacques Cartier when he first sailed into the Gulf of St. Lawrence in July 1534 that his first impulse was to fire on the local Micmacs, who most likely sought peaceful trade. After the first Spanish scouts rowed ashore in Tampa Bay in May 1539, they were quickly forced to retreat when the local Timucuan Indians attacked, killing two Spanish horses.[12] While surveying Cape Cod Bay for the first time, a group of Pilgrim scouts only narrowly escaped a barrage of Indian arrows on the beach early one morning.[13] And when trader Henry Spelman sailed up

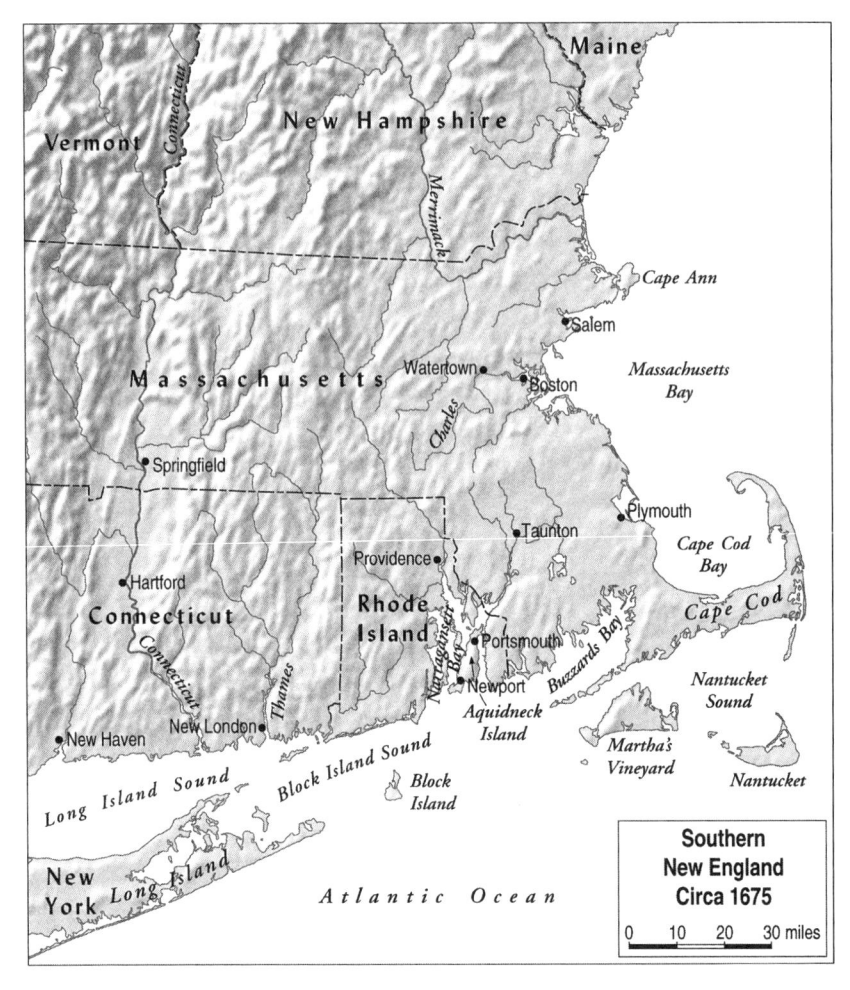

Southern New England, c. 1675, using modern political boundaries.

the Potomac in 1623, he was so quickly surrounded by Indian canoes that he and his men managed to fire only one shot before they were killed and one boy, Henry Fleet, was taken captive.[14] Similar accounts haunted European sailors up and down the coast of America. Invariably, anxiety was met with suspicion, and violence was met with retribution. Disorienting and uncontrolled, the coast played host to decades of struggle.

The assassination of the tribute collector, accordingly, led to calamity. First the English attacked Pequots on Block Island and then at Saybrook. The Pequots responded by attacking English towns in Connecticut. This chain of reprisals culminated in May 1637 when the English, with Narragansetts, Mohegans, and Niantics, surrounded and massacred the Pequots at Mystic, almost annihilating the tribe in the process. The bloody abandon with which the English slaughtered the Pequots left even their Indian allies appalled.[15] The ferocious manner in which they engaged their foes did much to establish English dominance across southern New England during the seventeenth century. But Oldham's murder and the chaos that followed also underscored the ways the English, and Massachusetts Bay in particular, had begun to exact tribute, often at usurious levels, which fomented rancor among indebted tribes and their allies. In short, shell beads had become a key source of conflict for the region.

Before Europeans adopted its use, wampum had played a more palliative role in Native American life. According to seventeenth-century observer Daniel Gookin, Indians used wampum to "pay tribute, redeeme captives, satisfy for murders and other Wrongs, [and] purchase peace with their potent neighbors, as occasion requires."[16] It was also used to confer social status. William Bradford noted that only "sachems and special persons . . . wore a little of it for ornament."[17] Wampum was essentially valued for its ability to generate political and social order. But as Europeans vied to become the arbiters of that order, they imbued these shell beads with their own value systems. Reflecting New England's growing radiance in the economic firmament of an expanding Atlantic world, William Wood observed of the Narragansetts, "The northern, eastern, and western Indians fetch all their coin from these southern mintmasters."[18] Wampum had become money, and this developing market mentalité had far-reaching social and ecological repercussions.[19]

As shell beads circulated inland they transformed the fur trade. More and more pelts were drawn to the coast.[20] By the end of the 1620s a veritable "wampum revolution" had invigorated the Dutch West India Company fur trade along the Hudson River.[21] Highly durable and easily portable, wampum soon spurred trade across southern New England and eventually across the entire Northeast.[22] Mounting demand for wampum, in turn, changed Native American patterns of subsistence, as more of them shifted their efforts toward wampum production.[23] Such efforts sent waves well beyond American shores. As wampum opened a wider economic exchange between the North American "periphery" and European "core," furs and other commodities purchased with shell beads were exchanged for European supplies and metropolitan credit.[24]

Although Native Americans and Europeans held very different understandings about the purposes of payment—the former believing it was the beginning of a relationship and the latter believing it was the end—their common conviction that wampum was worth something was rooted in their mutual need for a means of economic and social exchange. A persistent emphasis on wampum's economic role, however, has reinforced a narrative that Europeans sullied wampum by supplanting spiritual traditions with capitalism. The transformation of wampum into currency certainly led to bouts of discord and even outright violence between natives and settlers, but its ability to circulate unimpeded across cultures also reveals a meeting of minds: mounting contention over money presumes that both sides treasured it. That shared sense of value grew not only from economic necessity, but also from commonly held beliefs and the confidence it created.[25] At once a commodity and a gift, the torch of trade and solder of sociality, wampum developed value within highly variable cultural, political, and geographic contexts.[26] But one thing remained constant: as Europeans and Indians spun webs of meaning around shell beads, both groups placed mutual importance on the watery origins of the shells from which they were made.

The value relationship that developed in connection to wampum illuminates similarities between the ways Europeans and Native Americans understood land, sea, and nature in general. For both groups, wampum held value because it was mined from the edges of an "unknowable" sea, a space that embraced all the mystery of the divine. But wampum also nourished the European desire to subjugate the land. It was precisely this dual nature of shells and the mollusks that made them—at once "mysterious" and "intelligible," "hidden and manifest, placid and aggressive, flabby and vigorous," according to the philosopher Gaston Bachelard—that has long drawn people to cherish them.[27] Fueling the beaver trade that systematically extirpated beaver, wampum drained swamps and ponds and altered forest composition. Like axes, scythes, and ox-pulled plows, wampum's value, in this sense, lay in its ability to alter landscapes. For Iroquois and Algonquian tribes, mythic stories concerning wampum often celebrated its ability to effect terrestrial transformations. The value of wampum, therefore, lay partially in its watery origins and partially in its fantastic powers of "improvement." Ultimately, the value of wampum embraced mutually held beliefs about the wealth of water while assimilating, however reluctantly, European ideas of terrestrial change.

Born of the estuarine exchange between freshwater and saltwater, these whelks and quahogs in their cut and polished forms forged economic, spiritual, and material continuities between the ocean and the continental interior. As the wampum-for-fur trade dismantled the Northeast's vast network of beaver dams, the amount of time a single water molecule remained in the system dropped significantly. During storms, rivers ran faster, and during droughts, the earth grew dryer. This altered, among other factors, the distribution of plants, the biogeochemistry of rivers, and the rate of sedimentation, which in turn affected river fauna, including fish and fowl. In short, wampum initiated the process of transforming a broad, soggy coastal margin into an edge. These changes began and then radiated upstream from the place where wampum was made, which in time fundamentally reconfigured the Northeast's landscape. During the first half of the seventeenth century small ripples of environmental change blew across Narragansett Bay and soon developed into waves. For Mad Jack Oldham, the tribute collector bludgeoned to death with a hatchet to the head just a few miles south of the Bay's mouth, wampum evinced these connections between ideas, ecology, and culture in ways that were all too real.

Narragansett Bay in Time

The processes that transformed southern New England during the seventeenth century had been set in motion long before. Just over a century earlier, during the spring of 1524, the Italian explorer Giovanni da Verrazano sailed cautiously toward a hitherto-uncharted harbor, its mouth marked by a narrow strip of an island to the north and a low-lying bluff dominated by what he described as a "rock of freestone, formed by nature" to the south. Hired by the French Crown, Verrazano had been at sea for nearly three months, having left the island of Madeira in January and making landfall in "a new country, which had never before been seen by any one," near Cape Fear, North Carolina, in March.[28] Unable to find a suitable harbor to the south, Verrazano and his crew of fifty men turned north and continued along the coast, making several stops, until just east of Long Island Sound they reached a "very excellent harbor" in early May. Before entering the harbor's mouth, Verrazano was approached by about twenty small boats "full of people . . . uttering many cries of astonishment." Despite some hesitation, they eventually paddled alongside his vessel and several boarded, including "two kings more beautiful in form and stature than can possibly be described."[29]

Verrazano marveled at their appearance. He described their jewelry as like that of Egypt and Syria. He praised their ornamented deerskins and the "rich lynx skins upon their arms." Many wore their black hair, he observed, in braids and other intricate knots. They were, he avowed, "the finest looking tribe, and the handsomest in their costumes, that we have found in our voyage."[30] The warm welcome Verrazano had received from the local people combined with deteriorating weather encouraged him to enter the harbor at the mouth of Narragansett Bay that he called "Refugio" and the English later named Newport.

The master was impressed by his surroundings. Rich mudflats and lush marsh grass covered the harbor's south side. Sounding north, he and his men found deep water and a muddy bottom capable of holding an anchor fast in a gale. This refuge and its surroundings were so inviting that Verrazano and his men spent fifteen days exploring the area. Stretching twelve miles north from the coast, Narragansett Bay, he estimated, was "twenty leagues in circumference." Near the bay's southern end Verrazano observed "very pleasant hills" from which flowed "many streams of clear water . . . to the sea." Farther north, he described five small islands "of great fertility and beauty, covered with large and lofty trees." So protected was this bay that, "Among these islands," he wrote, "any fleet, however large, might ride safely, without fear of tempests or other dangers."[31]

The land surrounding Narragansett Bay was similarly striking. Traveling into the interior upwards of eighteen miles, Verrazano "found the country as pleasant as is possible to conceive, adapted to cultivation of every kind, whether of corn, wine or oil." He and his men saw vast plains "entirely free from trees or other hindrances" for "twenty five or thirty leagues in extent," suggesting the region had been heavily managed by its native inhabitants. The forests, he observed, were filled with oaks and cypresses (among others he did not recognize) and "might all be traversed by an army ever so numerous." That a large army could pass through this type of wooded terrain further suggests that Native Americans had cleared the undergrowth, likely by burning, and that the trees were enormous, creating a dense canopy whose shade prevented new growth from clogging the forest floor. Thomas Morton, a seventeenth-century observer of southern New England (but mostly in Massachusetts) explained that the Indian custom of "firing" the land had made it parklike and "very beautifull and commodious."[32] One of his contemporaries, Edward Winslow, noted that in the forests surrounding Narragansett Bay there was enough space between trees that "a man may well ride a horse amongst them."[33] In the gaps between these impressive stands, Verrazano saw numerous plum, apple, and filbert trees and animals, including stags, deer, and lynxes "in great numbers." These, he noted, the indigenous people hunted with snares as well as bows and arrows, which were "wrought with great beauty," the arrow heads carefully carved from "emery, jasper, hard marble, and other sharp stones." Also made of stone were their axe heads, which the Indians used to fell trees used for making canoes. These were constructed of single logs hollowed out to "contain ten or twelve persons" who, using short, wide oars, rowed "by force of the arms alone, with perfect security, and as nimbly as they choose."[34]

Narragansett Bay and its surroundings, as Verrazano's observations suggest, supported a large and mobile human population. Canoes certainly provided important means of transportation on and around the Bay. The Indians there frequently moved entire villages as well. They lived, Verrazano noted, in round dwellings "about ten or twelve paces in circumference" that were capable of housing twenty-five to thirty people. Made of split logs and thatched with hay "nicely put on," their wigwams were used only temporarily, for they moved seasonally to take advantage of local resources. In addition to hunting and fishing, Verrazano noted, they subsisted by growing beans, which they "carefully cultivated." That the impermanence of their dwellings and, as Verrazano believed, their lack of building skill, caused them to live in such simple shelters

The watery maze of southern New England as imagined by Giovanni da
Verrazano in 1524. *La Nuova Francia* (Venice: Giunti, 1606) by Giovanni
Battista Ramusio (1485–1557) and based on the cartography of Giovanni da
Verrazano (1485–1528). Courtesy of the John Carter Brown Library at Brown
University.

NTANA
TERRA DE LABORADOR
ISOLA DE DEMONI.
TERRA NVOVA.
LEVANTE
Bonne uiste.
Bacalaos.
C. Breton
C. de ras
Isola de Bretoni.
Isola della rena.
TRO

was a cause of lament for the explorer. Upon the beaches of Narragansett Bay, he observed, were scattered the raw materials to build "stately edifices." In its entirety, the shore, he attested, "abounds in shining stones, crystals, and alabaster."[35]

For Verrazano, his "Refugio" and its surroundings were a veritable Eden. The mainland and islands abounded with animals and towering trees. The rivers, abundant and running clear and cold, teemed with fish and fowl. And a shelter, comfortable enough to house two dozen people, could be built by simply gathering materials from the landscape. So ripe for the taking were the fruits of Narragansett Bay that the Native Americans who cultivated beans, corn, and squash in its adjacent soils, who cleared and maintained broad upland meadows, and who in canoes frequently plied its coastal marshes and islands, and its salt creeks and rivers, lived, at least in Verrazano's estimation, in blissful ease.

That Narragansett Bay was the backdrop to what Verrazano suggested was his favorite New World stopover speaks to the richness of its surroundings. Roughly twenty-eight miles from the ocean to its northern end and eleven miles at its widest point, the saline portion of Narragansett Bay covered 147 square miles with an average depth of about twenty-seven feet.[36] An estuary, or, by definition, a semi-enclosed body of water fed at once by the ocean and freshwater sources, Narragansett Bay comprised numerous islands, the largest of which were Aquidneck (later called Rhode Island) and Conanicut dominating the Bay's southern end, and four smaller islands in the Bay's center that Roger Williams later named Prudence, Patience, Hope, and Hog. In its western, northern, and eastern reaches, the Bay was met by nine river basins, the largest of which were later called the Taunton, Blackstone, and Pawtuxet. Also important was the Wood-Pawcatuck River watershed that fed the marshes and salt ponds of the Narragansett and Pequot Country, what would become southwestern Rhode Island and eastern Connecticut. If one added this vast network of rivers to the tidewater, the Narragansett Bay watershed covered more than 2,000 square miles.[37]

Moving water defined this coastal space. As tidal water from the ocean flooded Narragansett Bay from the south, roughly 260 million cubic feet of water per day rolled down its rivers into the upper estuary.[38] The flow

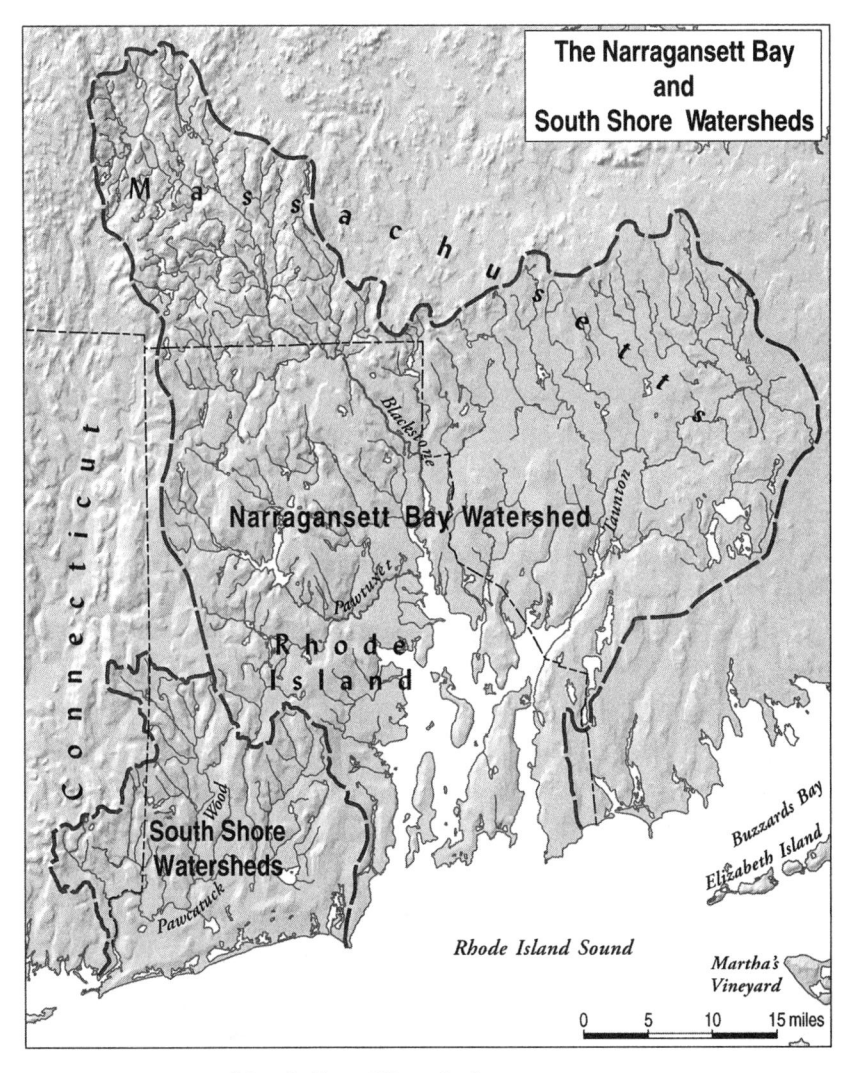

Narragansett Bay and South Shore Watersheds.

of water was dramatic, particularly during the winter months when rainfall was high and strong winds accelerated the Bay's circulation. During these periods of high flow, the bay's flushing time, or the amount of time it took to exchange all of its water, was only ten days. During the summer, when freshwater input was low, the flushing time could reach thirty-five days.[39] Nevertheless, these high rates of exchange made the waters of the Bay clear, clean, and nutrient rich. And as a result, the Bay was teeming with life.

With ready nutrients, plants burst from the Bay's shores, which in many ways mirrored those of other New England coastal areas. Wide swaths of eelgrass swayed among the sandy shallows which, as William Wood noted in 1634 of neighboring Massachusetts, provided habitat for a "great store of salt water eels." Using baskets baited with lobster, early settlers could in a night catch a bushel of eels, which were often cleaned and salted for winter. The rocks crawled with lobsters, "some," Wood noted, "being twenty pound in weight." So abundant and easily acquired were they that "their plenty makes them little esteemed and seldom eaten." Only the Indians partook "when they [could] get no bass."[40] Thomas Morton noted that he had seen Indians take 500 or 1,000 lobsters to "eate, and save dried for store."[41]

There were enormous quantities of fish in Narragansett Bay. Wood, writing about nearby waters, noted that sturgeon up to eighteen feet long plied coastal rivers, and salmon were in "great plenty." Fishermen frequently caught halibut "two yards long and one wide and a foot thick" as well as "Thornback and skates," which were fed to the dogs. The most highly prized coastal fish was striped bass, which Wood described as "one of the best fishes in the country." There were so many bass, "some be[ing] three and . . . four foot long," that, using a hook and line baited with lobster, Wood explained, "a man may catch a dozen or twenty of these in three hours." Bass typically followed the bait. And in the spring, Wood observed, they chased runs of spawning alewife so thick that the rivers turned black with fish. These alewife, Wood explained, ran "in such multitudes as is almost incredible, pressing up in such shallow waters as will scarce permit them to swim."[42] Thomas Morton marveled that so many bass pass through the salt creeks that one could walk across their backs "drishod."[43] On the outgoing tides, the English blocked the creeks with seines to trap bass, catching "sometimes two and three thousand at a set." When lobsters migrated inshore, the bass could be found in the rocks, and when giant schools of mackerel pushed into the bays, again, the bass followed close behind.[44] During the summer and fall, when the sun was high and the water warm and algae and zooplankton growth in the Bay peaked, menhaden traveling in frothing schools spanning dozens of acres—schools so big that their smell drifted downwind for miles—were driven into the bay by bass, bluefish, and squeteague and harassed from above by screeching osprey, terns, and gulls, which were, in turn, molested by bald eagles. These oily, foot-long members of the shad family played a key role in the Bay's food chain but also served as invaluable seasonal custodians. Each one of these filter-feeding fish sifted upwards of eight gallons of water per minute.[45] Arriving by the tens of millions, these menhaden scoured the waters of Narragansett Bay, keeping it brilliantly clean.

But the true kidneys of the estuary were its shellfish. Prodigious beds of blue mussels held fast by tough bissell threads covered the intertidal zones, particularly in rocky areas, such as the southern tip of Conanicut and the shores of Warwick Neck. Wood noted that "Muscles be in great plenty." So numerous were they that mussels were "left only for the hogs."[46] Across the bottom of the bay, particularly at the mouths of the tidal rivers, vast reefs of oysters, some covering hundreds of acres, had formed over the millennia, with young "seed" oysters propagating on the shells of others. The banks were so large that during spring tides, Wood noted, they were exposed to the open air. The individual animals were enormous as well. Wood observed that they typically took the shape of a shoehorn and were upwards of a foot long. "The fish without the shell," he noted, "is so big that it must admit of a division before you can well get it into your mouth." The Bay's vast mudflats were packed with soft-shell clams. Their numbers were so great, he explained, that "a man running over these clam banks will . . . be made all wet by their spouting of water." Sharing these same beds but also extending into deeper water were hard-shell clams or quahogs, "some as big as a penny white loaf." Like mussels and their smaller, soft-shelled counterparts, the sheer abundance of quahogs left them largely ignored by settlers. They, Wood noted, "were great dainties amongst the natives and would be in good esteem amongst the English were it not for better fish."[47] The siphoning action of so many millions of bivalves carpeting the Bay's floor created clean water capable of supporting an extraordinarily productive ecosystem.

In terms of the broad sweep of time, the ecologically rich environment Wood and Morton observed was a relatively recent development. Following the end of the last ice age, roughly 10,000 years ago, the massive glaciers covering New England began to recede, causing sea levels to rise. What had been an enormous tundra-covered peninsula spanning more than 20,000 square miles and forming the southern flank of the Gulf of Maine was slowly submerged, becoming Georges Bank, the incredibly productive fishing grounds east of Cape Cod. As sea levels stabilized around 2,500 years ago, a prominent river valley to its west flooded with seawater, and the area that now constitutes Narragansett Bay became "coastal."[48]

The high biological productivity of the estuary that formed provided a nutrient-rich environment for shellfish growth. Archeological studies of various sites along the coast of Narragansett Bay and particularly Greenwich Cove, on the Bay's western shore, showed that as the bay developed, Native Americans living in the area began to diversify their diets by including marine mollusks.[49] At about the same time, southern New England also experienced one of the largest population expansions in prehistory.[50] It is difficult to know whether the exploitation of marine mollusks began in response to population growth or whether dietary diversification, which incorporated marine mollusks, actually caused it.[51] But the archeological record does show that southern New England's coastal Native Americans integrated shellfish into their lives in important ways as soon as natural environmental changes had made it available.[52]

Such powerful draws were the clams, quahogs, and mussels of Narragansett Bay that they altered the ways Native Americans interacted with their environment and each other. Abundant and easily acquired, clams were harvested by coastal Indians year-round.[53] William Wood noted of Indian women, who were the principal harvesters of shellfish, that "In winter they are their husband's caterers, trudging to the clam banks for their belly timber." Only three or four days after having given birth, one "bare-footed mother . . . ," carrying her child wrapped in beaver skin, "paddle[d] in the icy clam banks."[54] Although, as Verrazano observed, Indian bands moved periodically, they probably moved between coastal sites, rather than simply moving inland with the coming of winter.[55] This coastal sedentism, which began somewhere between 4,700 and 2,500 years ago, led Native Americans to diversify their use of plants and animals, placing new environmental pressures on the estuary.[56] The archeological record shows a reduction in the size of shells during the period just prior to European arrival, which when combined with an increasingly diversified diet, suggests increased stress on coastal resources.[57] In other words, Native Americans continued to be mobile, but they increasingly concentrated their ecological footprints on discrete coastal locations. In small ways, these changing patterns of subsistence began the long process of consolidating the coast. As Indians traveled less, their communities grew in size. And over time, they reached out to their neighbors to acquire things unavailable to them locally. As a result, trade networks formed that continued into the historic period.[58]

The Narragansett Indians were key players in these burgeoning trade networks. Archeologists working in southern New England have uncovered, among numerous other items, copper beads from the Great Lakes region and flint from Ohio and New York.[59] One important prehistoric trade item among Rhode Island's Native Americans was stone pipes, cut from steatite, or soapstone, which was abundant in the area. Narragansett pipes were found in archeological sites across southern New England and as far west as Indiana.[60] Roger Williams noted that the Narragansetts "sometimes ... make such great *pipes,* both of *wood* and *stone,* that they are two foot long, with men or beasts carved, so big or massie [*sic*], that a man may be hurt mortally by one of them."[61] William Wood observed that the Narragansetts were, among Native American tribes, "the most industrious, being the storehouse of all such kind of wild merchandise as is amongst them."[62]

In the years following European settlement the most valuable of the Narragansetts' "wild merchandise" were tiny beads hand hewn from the shells of marine mollusks. "They that live upon the Sea side," wrote Roger Williams, "generally make of it, and as many make as will," referring to the carefully crafted white and purple beads known to his neighbors, the Narragansett, as *wampompeage,* to the Dutch as *sewan* or *zeewan,* and the English as *peage* or simply *wampum.* "The *Indians,*" Williams explained, "bring downe all their sorts of Furs, which they take in the Countrey, both to the *Indians* and to the *English* for this *Indian* Money: this Money the *English, French* and *Dutch,* trade to the *Indians,* six hundred miles in several parts (North and South from *New-England*) for their Furres, and whatsoever they stand in need of from them: as Corne,

Venison, &c."[63] Wampum, noted Williams, had, during the first third of the seventeenth century, catalyzed a dynamic network of long-distance trade between Native Americans and the various European powers that had settled in New England, New York, southern Canada, and beyond.

These shell beads, which in earlier centuries had been prized among Native Americans for their spiritual legacy of the sea's powers, had during the early seventeenth century developed into a medium of exchange, and those tribes with ready access to the shells from which the beads were made grew powerful. "It [wampum] is made principally by the Narragansett black [Block] islanders and Long Island Indians," wrote Daniel Gookin in 1674. "Upon the sandy flats and shores of those coasts the wilk shells are found."[64] William Wood observed that the Narragansett "are the most curious minters of their wampompeag and mowhacheis."[65] His use of the word "minters" suggests that the Narragansetts were producing large amounts of wampum by the 1630s, that they largely controlled its production, and that it circulated widely as currency. Strong in numbers, the Narragansetts held considerable influence among neighboring tribes, thereby establishing a major stake in wampum production.[66] Although fewer in number and occupying a geographically smaller area west of the Narragansetts, the Pequot tribe, living along the Pawcatuck and Mystic Rivers, maintained close trading partnership with the Mohegans and were likewise important wampum producers.[67] William Bradford observed, "Only it [wampum] was made and kepte amonge the Narrigansets, and Pequents [Pequots], which grew rich and potent by it."[68] With ready access to an increasingly valuable marine resource, both tribes grew powerful, trading finished wampum for furs and European goods.

The actual production of wampum required considerable skill. Along the sandy shores of Narragansett Bay and Long Island Sound, Native Americans gathered the northern whelks, *Busycon canaliculatum* and *Busycon carica*, to produce white wampum from, Roger Williams observed, the "stem or stocke of the *Periwincle,* which they call Meteaûhock, when all the shell is broken off." Williams explained that these white beads, worth six to an English penny, were often strung into bracelets. Worth twice as much as the white, dark-colored beads—or as Williams explained, "black, inclining to blew"—were produced from

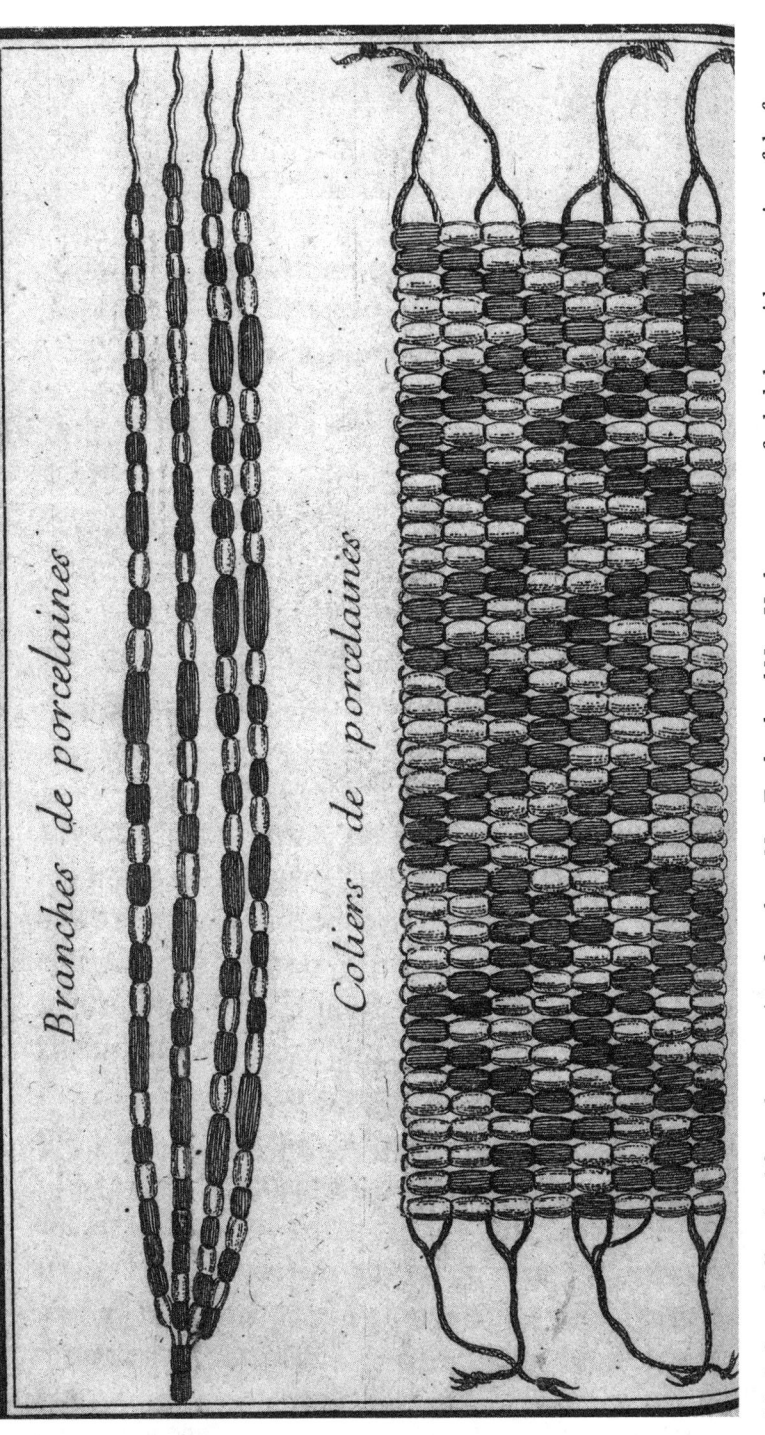

Made from shells mined from the estuaries of southern New England and New York, wampum fueled the rapid expansion of the fur trade, which transformed the ecology of the region. Claude-Charles LeRoy, M. de Bacqueville de La Potherie, *Histoire de l'Amerique Septentrionale* ... (Paris: Jean-Luc Nion and François Didot, 1722). Courtesy of the John Carter Brown Library at Brown University.

the hard-shell clam, or quahog, *Mercenaria mercenaria*.[69] From the deep purple rim of the quahog's inner shell known to clam biologists as the pallial sinus, a segment of shell was cut and smoothed into tiny cylinders a quarter-inch long and an eighth-inch in diameter. These were then carefully drilled into beads that the Narragansett called suckáuhock, meaning black money, or, as William Wood attested, mowhacheis.[70] Using animal sinews or bark threads, Indian women assembled them by the thousands into strings, intricate belts, and other decorative garments. Some, Williams noted, strung wampum into necklaces and bracelets. Others wove their wampum beads into girdles several inches wide that were worn around their waists, shoulders, and breasts. "Yea the Princes," he noted, "make rich Caps and Aprons (or small breeches) of these Beads thus curiously strung into many forms and figures: their blacke and white finely mixt together."[71]

Wampum was undoubtedly a symbol of social prominence, but its value was rooted in its mythic meaning and otherworldly origins. For the Iroquois, much of wampum's spiritual significance resided in its "orenda," a supernatural force inherent in shiny things that seemed to come from outside the natural world.[72] Among many Algonquian tribes, water, shells, wet rocks, and polished metals, among other reflective materials, held supernatural qualities because, it was believed, they provided a window into the soul.[73] One of the most important places that the Indians of southern New England came into direct contact with the spiritual world was along the ocean's edge.[74] Their god Cautántouwit lived somewhere to the southwest, past the salt ponds of the Narragansett Country and beyond the ocean and summer breezes. "[T]o the *Southwest*," wrote Williams, lay the souls of their forefathers, and there "they goe themselves when they dye."[75] According to Narragansett tradition, the dead were buried with earthly things they needed in the afterlife; for their journey over the seas, these often included wampum drills and cutting tools as well as beads ceremonially placed on the body, which, William Wood noted, were used "to purchase more immense prerogatives in their Paradise."[76] These spiritually powerful "under(water) *grandfathers*," moreover, were widely considered "sources of precious gifts, medicines, and charms" in the world of the living.[77] Finally, the estuary from which wampum was plucked had been created by and was home to important deities. In addition to their "Sea-God," called Paumpágussit, Williams noted, the Narragansetts "have many strange Relations of one *Wétucks,* a man that wrought great *Miracles* amongst them, and *walking upon the waters* . . . with some kind of broken Resemblance to the *Sonne of God.*"[78] Of central importance to almost all southern New England coastal bands, Wétucks, Ezra Stiles noted during the eighteenth century, was also called Maushump by Long Island Indians and Maushop among the Wampanoags.[79]

The Indians of Cape Cod and the surrounding islands developed a particularly vibrant estuary-centered creation story. According to one Wampanoag myth, the giant Maushop, the first to visit the coastal plains, carved the region's vast network of salt lagoons as he wandered through the dunes and among the marshes while dragging his heavy foot.[80] Accordingly, he created the estuaries of southern New England and then, as other accounts attest, settled on Martha's Vineyard, where he feasted on fish and whale meat. Maushop was particularly fond of enjoying his pipe, the smoke and ashes from which brought fog to Cape Cod and formed the island of Nantucket. Sand from his moccasins formed the Elizabeth Islands. When the English came, it is said that he disappeared below the waves, but before doing so, Maushop hurled his wife, Saconet, toward the mainland, where she landed and remained to collect tribute at the promontory marking the southeast corner of Narragansett Bay.[81] That an important deity like Wétucks or Maushop was known to have created some of southeastern New England's most prominent coastal features and weather phenomena, and that he retreated below the waves with the arrival of Europeans, suggests the sea was a spiritual sanctuary. That Maushop's wife was sent to collect tribute, likely wampum, at her seaside station at the mouth of Narragansett Bay suggests that estuarine shell beads were an important medium between the human world on land and spiritual world at sea.

Its watery origins imbued wampum with value for Europeans as well. For Williams, however, wampum, like precious metals, marked a departure from the path of righteousness. "The Sonnes of men having lost their Maker, the true and onely Treasure," he wrote, "dig downe to the bowels of the earth for gold and silver; yea, to the botome of the Sea, for shells of fishes, to make up a Treasure, which can never truly inrich nor satisfie."[82] Only God, Williams attested, held true value, but having strayed from him, man burrowed deep into the earth for ores and below the sea for shells. Like gold and silver, wampum was plucked from the deep unknown and held significance because of it. If Williams was suspicious of wampum's cosmological value, he nevertheless saw beneath the waves of the Bay a touch of the divine. Of one particularly frightening incident, he wrote:

> *Alone 'mongst* Indians *in Canoes,*
> *Sometime o're-turn'd, I have been*
> *Halfe inch from death, in Ocean deepe,*
> *Gods wonders I have Seene.*[83]

The ocean, he conceded, held divine mysteries. And when his canoe overturned, Williams, echoing the Psalms, saw "the deeds of the Lord, his wondrous works in the deep." Describing his trans-Atlantic crossing, William Wood noted that ships "seldom doth . . . sink or overturn because it is kept by that careful hand of Providence by which it is rocked." For Williams and Wood alike, the author of the oceans who had delivered Noah from the Deluge and saved St. Paul from shipwreck, still operated among the waves.[84]

The Lord spoke through the sea's creatures as well, including shellfish. "How many thousands of Millions," Williams asked, "of those under water, sea-Inhabitants . . . preach to the sonnes of men on shore [?]" Williams referred to marine animals as "Christs little ones," which after being "Devour'd," will "rise as Hee."[85] Blessed in the brackish waters of Narragansett Bay, these fruits of the estuary delivered to man the word of God and the body of Christ. For both Europeans and Native Americans, the sea was a source of the divine and wampum benefitted from its glow. That its production required highly specialized skill and hours of labor and that the resultant beads were beautiful, durable, and easy to handle would have also led to a mutual appreciation of it. But along the coast of southern New England, where wampum was physically produced and then socially constructed, it was widely acknowledged that there was something special about its origins below the surface of the sea.

As wampum circulated deeper into the continental interior its watery origins continued to underpin its value. Inland mythic traditions surrounding wampum, however, began to emphasize a supernatural presence among rivers and lakes.[86] In one myth, Hiawatha, while traveling to establish the Iroquois League, saw a flock of ducks take flight from a lake. Upon their departure, they left the lakebed dry and scattered with shells. Using them, Hiawatha assembled the belts that forged the bonds of the confederacy.[87] With wampum's arrival the birds disappeared, the lake dried up, and a powerful political union was formed. This suggests that among some Native Americans, wampum's value was rooted in its ability to transform not only social interactions but the physical landscape as well. As it was traded farther inland, wampum's value reflected the power of improvement.

For other Native American groups, wampum's ability to transform the land honored important spiritual obligations. One Cree myth justified the slaughter of beaver by claiming the Great Spirit had, for some unnamed offense, banished them from dry land into the rivers where they became numerous and expanded their watery domain. The Great Spirit countenanced the beaver's destruction, and man, literally defending his turf, obliged.[88] In both the Iroquois and Cree myths, the transformations are environmental, involving the reclamation of dry land. These stories suggest that some Native American groups were aware of the ways the beaver trade affected water resources. For the Iroquois, wampum desiccated lakes. For the Cree, the exchange of shell beads for furs validated a spiritual duty to drain the swamps and bogs created by a rodent that had fallen from grace.

Although there are still other myths attributed to the genesis of wampum, most show wampum moving from the spiritual world to that of human construction. Many of these stories involve some form of preternatural bird that sheds wampum or leaves it as a gift in its wake, perhaps referencing its distant or supernatural origins or its great mobility. That the birds were either scared off or killed, leaving wampum behind for human use, perhaps acknowledges that wampum was inextricably tied to changes in animal populations. That water was often central to wampum mythology suggests either that contemporary Indians saw changes to the rivers and surrounding landscape or that subsequent generations who retold those stories had witnessed those types of transformations. Over time, the value attributed to wampum in coastal areas based on shared belief in an eternal sea gave way to one that reflected wampum as a tool of improvement. Dug from the shores of Narragansett Bay and fashioned by the Indians who lived there, wampum, once ferried upstream by canoes and dragged inland on sleds, was reevaluated by new groups of Indians and the European traders with whom they interacted. If wampum still held a touch of the divine, it came to reflect the power of human needs and aspirations. In consequence, wampum was tied to and in some cases the cause of material changes that began on the coast and stretched deep into the continental interior.

Draining the Land

Less than a year after the Dutch West India Company was established in 1621, one of its own, a trader named Jacques Elekens, kidnapped and held hostage a Sequin sachem. The lands of New Netherland were still largely unsettled and the Dutch commercial war with Spain in America provoked outright violence from the West India Company and its traders.[89] Not limiting physical aggression to his European competitors, Elekens embraced the mercenary nature of his mission with grim enthusiasm. Demanding a hefty ransom, Elekens threatened to cut off the sachem's head if his people did not deliver.

Although Elekens's countrymen had landed in the New World more than a decade earlier, New Netherland was still a crude outpost along the margins of the Atlantic world. Henry Hudson first landed in the area in 1609, exploring the island of Mannahatta and the impressive river that would later bear his name. In 1614 the New Netherland Company was granted a trade monopoly patent, but it wasn't until 1623 or 1624 that colonists began to people the area between the Delaware River and Narragansett Bay in earnest. To lay claim to the land, the Dutch spread out, a few settling near Saybrook Point, others at Burlington Island, a few at *Nooten Eylandt* near Manhattan, and several along the North River at Fort Orange, or Albany.[90] As early as 1625 the Dutch West India Company established a trading station at Quotenis, later named Dutch Island, at the southern end of Narragansett Bay, as well as two others on the mainland.[91]

The promise of colonial settlement in the New World captured the imagination of the Dutch at home. Following these developments closely, the historian, physician, and publisher Nicolaes van Wassenaer launched the first edition of the *Historisch Verhael* in 1622, an annual periodical dedicated to compiling the notable events of the year. Impressed by the extraordinary happenings abroad, Van Wassenaer dedicated the entire second issue to descriptions of New Netherland.[92] His coverage continued until 1630, providing vivid reports as relayed to him from his correspondents, many of whom traveled deep into New Netherland's interior in search of furs.

For Van Wassenaer, the number of correspondents multiplied after Elekens seized the sachem, for that act played a pivotal role in energizing the fur industry. In exchange for the sachem, Elekens was paid a ransom of 140 fathoms of zeewan, or wampum, which Van Wassenaer characterized as "small beads they manufacture themselves, and which they prize as jewels."[93] With hundreds of feet of valuable shell beads in their possession, Elekens and Dutch authorities parlayed their ill-gotten winnings into the fur trade along the Hudson River, which Van Wassenaer explained, was the most important trade route to points north.[94] Prized by Native American hunters and so easily stored and transported, wampum turned a largely localized, small-scale trade in furs into a region-wide mad dash for pelts. But the removal of so many beavers and the subsequent disintegration of their dams fundamentally changed the way water rolled down hill.

Early Dutch West India Company correspondents described a watery world north of New Amsterdam. Fifty leagues above Manhattan, Van Wassenaer carefully noted, was "very swampy, [with] great quantities of water running to the river, overflowing the adjoining country." So wet was the terrain that Dutch settlers at Fort Nassau "frequently lay under water," and as a result the site was abandoned for nearby Fort Orange (later Albany).[95] One prominent New Netherland landowner, Adriaen van der Donck, a careful observer of nature who spent considerable time exploring the countryside during the 1640s, noted, "The rivers have their origin in sprouts which flow from valleys, and in springs which connected form beautiful streams." He also saw "numerous small streams and sprouts throughout the country, serving as arteries or veins to the body, running in almost

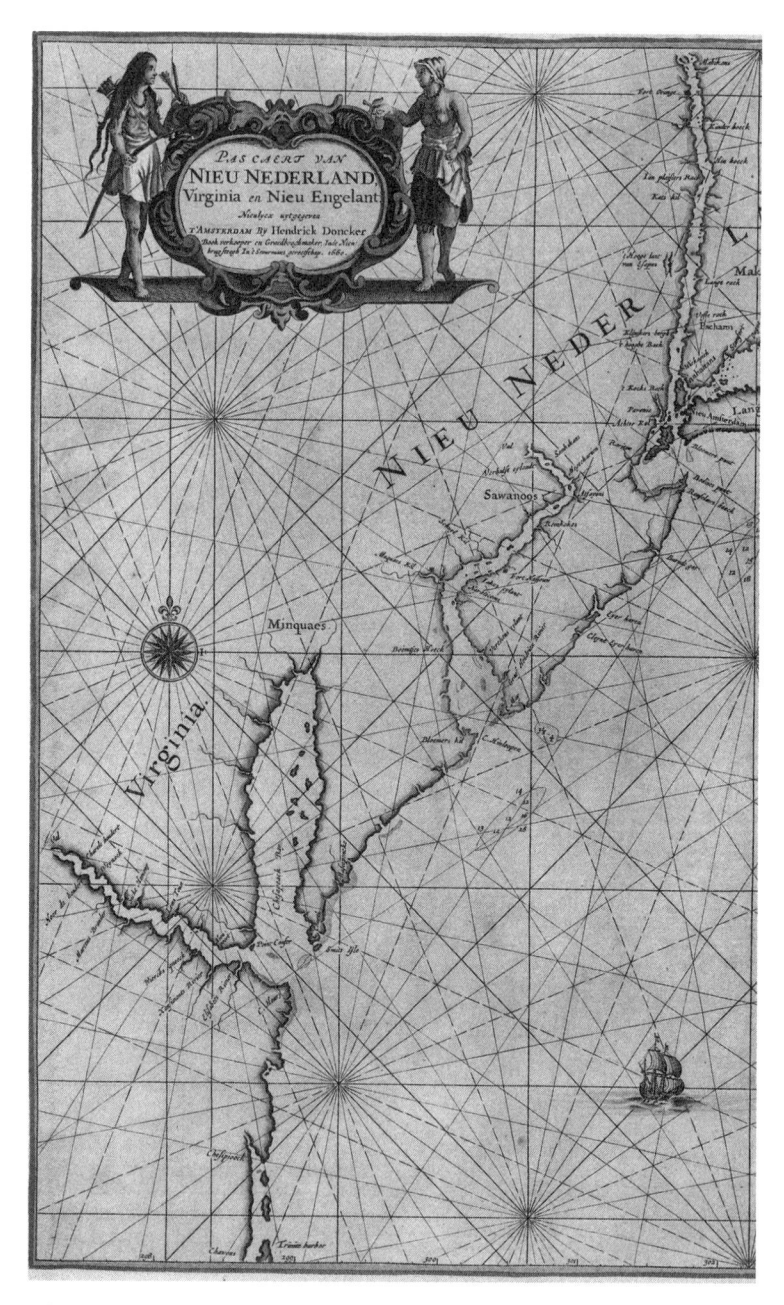

The seventeenth-century lands of New Netherland and New England. This may be the first mention of the town of Providence on a printed map. *Pascaert van Nieu Nederland, Virginie en Nieu Engelant* (Amsterdam: Hendrick Doncker, 1660). Courtesy of the John Carter Brown Library at Brown University.

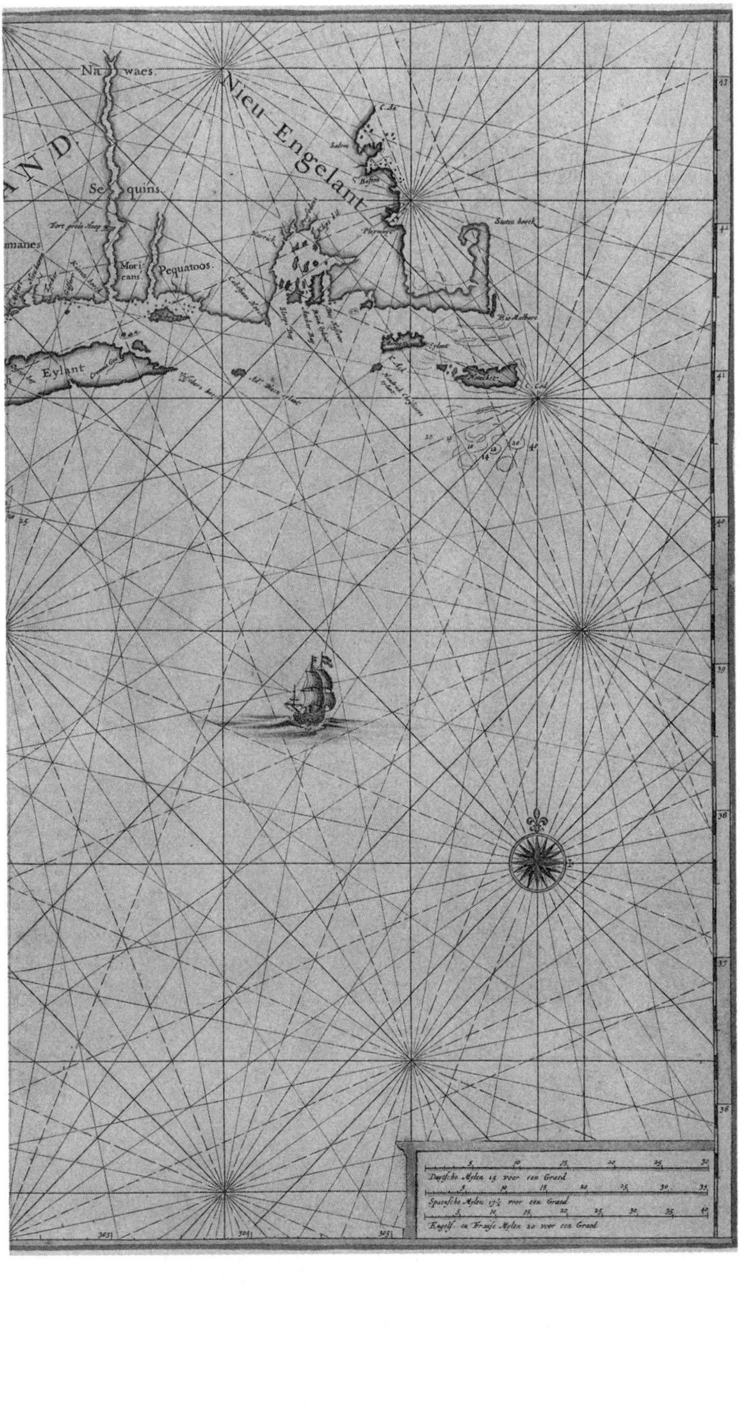
Nieu Engelant
Na waes.
Se quins.
Mori cans
Pequatoos.
Eylant

every direction, and affording an abundance of pure living water." Still other streams, he wrote, "rise in bushy woods, through which the summer sun never shines, which are much trodden by wild beasts, and wherein the decayed leaves and rotting vegetation falls."[96]

This flooded landscape shaped a rich riparian ecology. Van Wassenaer described vast wetlands that included "all sorts of fowls, such as cranes, bitterns, swans, geese, ducks, [and] widgeons." And this soggy landscape extended deep into the forests, which echoed with their calls. "Birds fill also the woods," he wrote, "so that men can scarcely go through them for the whistling, the noise, and the chattering." Teeming with other riparian fauna, the forest floors, he explained, crawled with small tortoises and the "most wonderful . . . dreadful frogs, in size about a span, which croak with a ringing noise in the evening." Dotted with bogs, ponds, and vernal pools, the land to the north was so wet that Van Wassenaer remarked it was "surprising that storks have not been found there, since it is a marshy country."[97]

The Dutch in New Netherland expanded their wampum-for-fur trade into the wet interior, while keeping a watchful eye on the English neighbors to the east, who remained largely ignorant of this incredibly lucrative exchange. Perhaps realizing that the English would eventually catch on, in 1627, Isaack de Rasieres, the Secretary of New Netherland, traveled to Plymouth Colony to talk with William Bradford. According to Bradford, Rasieres sold him wampum worth fifty English pounds so that Bradford might trade it for furs with the Indians in the English-controlled Kennebec region.[98] According to Rasieres, he sold them the wampum "because the seeking after sewan by them is prejudicial to us, inasmuch as they would, by so doing, discover the trade in furs; which if they were to find out, it would be a great trouble for us to maintain."[99] Rasieres's generosity was motivated by his interest in keeping the English out of New Netherland's trading hinterland, the eastern end of which was Narragansett Bay. But Rasieres's attempt to distract the English ultimately drew them squarely into the trade. "[T]hat which turned most to their [Plymouth's] profite," wrote Bradford, "in time, was an entrance into the trade of Wampampeake."[100]

Narragansett Bay played host to this imperial power struggle, for where wampum was readily available, pelt traders set up shop. While the Dutch West India Company maintained its trading stations on the Bay, Plymouth established trading posts in 1632 at Sowamset on eastern Narragansett Bay and Aptucxet, at the northern tip of Buzzards Bay. In 1636 Roger Williams established a permanent settlement in northern Narragansett Bay at Providence. Soon after, Richard Smith, in the words of Roger Williams, "broke the ice at his great charge and hazard, and put up in the thickets of the barbarians, the first English house amongst them."[101] Smith built a trading depot at Wickford on the western shore of Narragansett Bay along the coastal overland route between Boston, Connecticut, and New Amsterdam, and was for several years accompanied by Roger Williams, who did the same.[102] By 1638 increasing numbers of settlers had moved into the Connecticut River Valley, and the first wave of settlers near Boston had begun trading furs. As coastal beaver were decimated, traders looked farther inland. In 1636 William Pynchon settled Springfield at the junction of the Connecticut River and the overland Connecticut Path in south-central Massachusetts. There, he established trading relations with Agawam and Woronoco Indians, commencing a lucrative trade in pelts that he continued into the early 1670s.[103]

As wampum circulated deep into the continental interior, hunters pulled increasing numbers of beaver from the Northeast ecosystem. Whereas only ten years earlier the coasts had served as the fur-trading frontier, by 1630 traders extended their reach up the Concord and Connecticut Rivers. By 1640 they had pushed as far as Springfield, Massachusetts, and had penetrated the interior along the Blackstone and Merrimac Rivers.[104] Although the full extent of the beaver hunt is unknown, scattered records indicate that vast numbers were killed and that beaver populations began to decline. Between 1652 and 1657 Thomas Pynchon alone shipped 8,992 beaver skins, weighing 13,139 pounds, to England. But in the following sixteen years between 1658 and 1674, he exported only 6,480 beaver pelts weighing roughly 9,000 pounds, suggesting that due to declining numbers they were much more difficult to acquire and the beavers that remained were smaller in size.[105] Combining Pynchon's efforts with those of countless others, this mass extraction of beavers caused dramatic changes to the ways water moved across the land.

a. a. Höhlen der Biber.
b. b. Teiche der Biber.
c. c. c. ein Damm, zu Erhal-
tung des Wassers.
d. d. d. Wasser, so über den Dam̄
hinein fällt.
e. e. e. e. wie selbige zur Arbeit
gehn.
f. wie sie die Bäume
fällen.
g. g. wie sie die
Bäume schwim̄-
mend ziehen.
h. der Bach.

i. ein Wilder, wie er den Biber
mit einem Schuß tödtet.
k. ein Wilder, so den Biber mit
dem Wurffspieß tödtet.
l. ein Wilder, so den Biber mit
dem Degen tödtet.
m. Art, wie die Biber mit
dem Netze gefangen
werden.
n. wie selbige in einer
Falle gefangen
werden.
o. o. o. Hunde, so die
Biber erwürgen.

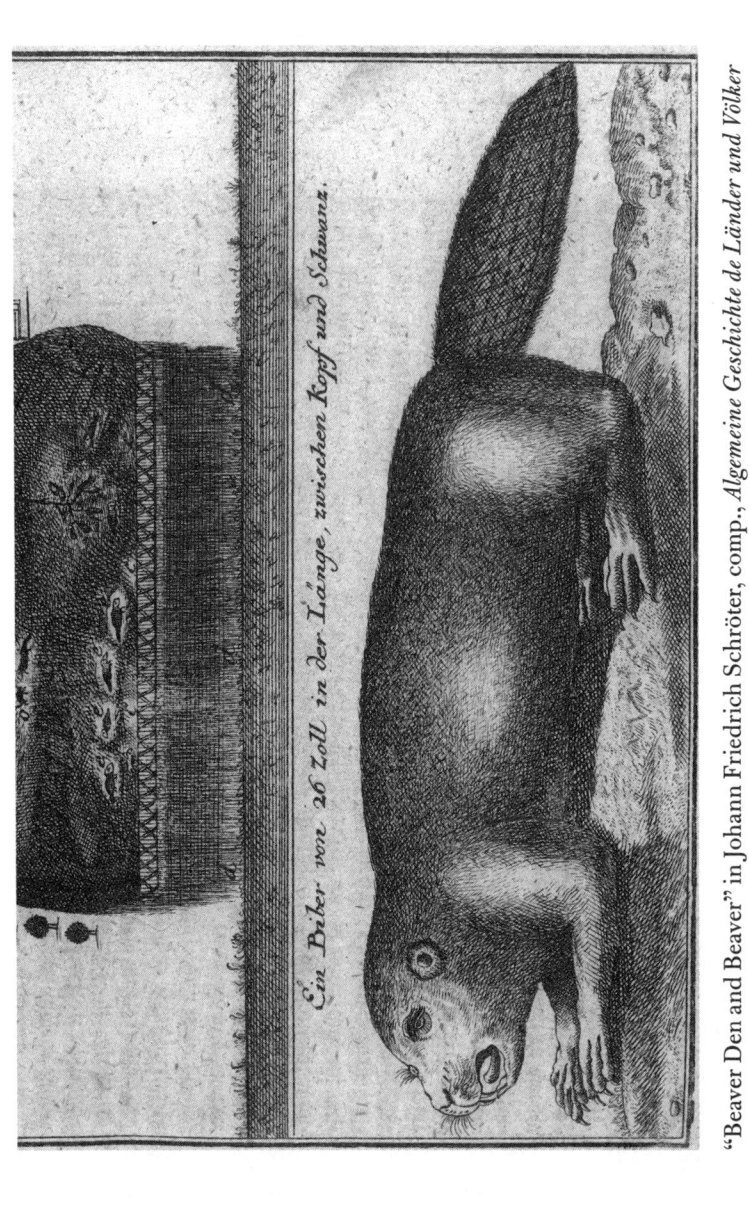

"Beaver Den and Beaver" in Johann Friedrich Schröter, comp., *Algemeine Geschichte de Länder und Völker von America. Zweiter Theil. Nebst einer Vorrede Siegmund Jacob Baumgartens* (Halle: Johann Justinus Gebauer, 1752). Courtesy of the John Carter Brown Library at Brown University.

Prior to European entry into the fur trade, the beaver, *Castor canadensis,* had shaped the Northeast landscape. So extensive was their engineering that Roger Williams characterized beavers as "beasts of wonder" that could "draw of great pieces of trees with his teeth, with which and sticks and earth I have often seen, faire streames and rivers dammm'd and stopt up by them."[106] Echoing Williams's seventeenth-century observations, modern scientists have shown that beavers fundamentally altered river morphology. Beavers built dams, which created ponds, in the center of which they constructed lodges of mud and wood. Those beaver ponds, when maintained, often lasted for decades. As a result, the dam- and lodge-building activities of *C. canadensis* established vast wetlands, which retained sediment and organic matter. These broad swaths of standing water shaped the ways vegetation decomposed and nutrients cycled through forest systems. Beaver dams also affected the chemical composition and amount of water transported downstream. In turn, the activities of beavers shaped the species composition and diversity of plants and animals of the forests in which they lived.[107] Accordingly, water-loving plants and the insects and animals that ate and made their homes among them were abundant. Ultimately, the beaver is what ecologists call a keystone species, one that was ecologically integral to the healthy function of forest and riparian ecosystems.[108] But as the hunt for beavers escalated, the forests and rivers—the entire ecosystem—began to change.

The broad geographic range and sheer numbers of beaver in North America made their effects on the waters profound. Before European arrival in North America, beavers inhabited all of Canada and the territory that now includes most of the continental United States, barring the border regions with Mexico. Although the exact numbers of beaver are unknown, in 1929 Ernest Thompson Seton, compiling figures from numerous beaver surveys, estimated that before European contact there were between 60 and 400 million beavers colonizing the rivers of North America, a calculation that numerous scientists have subsequently used for their own estimates.[109] Before Europeans began targeting beavers, almost all lower-order streams—small rivers, brooks, and creeks—and almost every lake or pond in New York, New England, and southern Canada was occupied by them.[110] If these estimates are accurate, one can assume that during the seventeenth century beavers were almost as plentiful as the gray squirrel is today.[111]

The abundance of beavers was incredible. Recounting the fur trade's history, eighteenth-century explorer of southern Canada David Thompson explained:

> Every River where the current was moderate and sufficiently deep, the banks at the water edge were occupied by their houses. To every small Lake, and all the Ponds they built Dams, and enlarged and deepened them to the height of the dams. Even to grounds occasionally overflowed, by heavy rains, they also made dams and made them permanent Ponds, and as they heightened the dams [they] increased the extent and added to the depth of water; Thus all the low lands were in possession of the Beaver, and all the hollows of the higher grounds.[112]

The ubiquitous beaver, Thompson noted, shaped a watery landscape in which every river, stream, and creek was dammed, creating broad ponds and marshes that covered lowlands for miles. In a letter to the Congregational minister Jeremy Belknap of New Hampshire, Joseph Peirce explained that God "has a farther design in this little animal . . . which stops the water from pursuing its natural course, and makes it spread over a tract of land from five to five hundred acres in extent." Marveling at a beaver dam's ability to transform the landscape, Peirce explained, "[A]ll trees, bushes and shrubs are killed. In a course of time, the leaves, bark, rotten wood and other manure, which is washed down, by the rains, from the adjacent high lands, to a great extent, spread over this pond, and subside to the bottom, making it smooth and level."[113] As Peirce observed, beavers and their dams built a watery world.

Weighing up to fifty-five pounds, beavers, like humans, engineered the landscape extensively. Using powerful incisors that grow throughout their lives and sharpen continuously, beavers typically target medium-sized trees that will fall toward and slide into small riverbeds. Once a tree is felled and divided into more manageable pieces, it is either floated to the dam site via canals constructed by the beaver or dragged overland. Trees are integral to beaver dams, which can range from two to twenty

Beavers hard at work near Niagara Falls. Beavers and their dams created a swampy continental interior. But when beavers were killed and their dams destroyed, water that had been impounded on the landscape flooded downhill toward the sea. In consequence, the Northeast became a dryer place. Image inset in *A New and Exact Map of the Dominions of the King of Great Britain on ye Continent of North America* . . . (London: Herman Moll, 1726). Courtesy of the John Carter Brown Library at Brown University.

feet high and can stretch over a hundred feet in length. The logs are piled at the dam site and the spaces between them clogged with mud and other debris. When the dam is complete a pond develops.[114]

Quick to multiply, beaver families populated North America in staggering numbers and built millions of ponds in the process. A typical beaver family comprised between four and eight members.[115] And most beaver families constructed between two and five ponds. As such, the minimum number of pre-European beaver ponds was between 15 and 100 million, with a maximum range of between 37.5 and 250 million ponds.[116] Observing these numbers—whether at the more conservative or more aggressive end of the estimates—one can conclude that before Europeans began hunting beavers in earnest, the Northeast region of America was decidedly wet.

But as beavers were removed from the landscape, patterns of water flow and impoundment changed. In 1624 Van Wassenaer noted of Indian traders who had traveled "far from the interior" that they "declare there is considerable water everywhere and that the upper country is marshy," which suggests that near the coast where beavers had already been removed, conditions were drier.[117] By the 1640s Adriaen van der Donck observed that "beavers keep in deep swamps, at the waters and morasses, where no settlements are."[118] By mid-century, beavers had been reduced to the far reaches of New Netherland, and the watery environment that beavers created was far removed from Dutch settlements. On the coast, remarked Van der Donck, the terrain was pleasantly dry. "It is a great convenience and ease to the citizens of New Netherlands," he wrote, "that the country is not subject to great floods and inundations, for near the sea, or where the water ebbs and rises, there are no extraordinary floods."[119] Along the coast where beavers were first removed, Van der Donck's observations suggest, rivers flowed obediently within their banks. In fact, a lot of beavers had been removed: between 1624 and 1626, the initial years of Dutch settlement, they shipped 16,553 of them.[120] And those numbers soared. Between 1641 and 1650 Van der Donck estimated that "about eighty thousand beavers have been killed annually, during my residence."[121] So sure of these numbers was Van der Donck that he was careful to note he had "frequently eaten beaver flesh, and have raised and kept their young." In addition, he avowed, "I have also handled and exchanged many thousands of skins."[122]

As tens of thousands of beavers were killed, their dams were destroyed, which drained the land. The second half of Peirce's letter to Belknap explained the effects of beaver hunting. "[T]he water is drained off, and the whole tract, which was the bottom of a pond, is covered with wild grass, which grows as high as a man's shoulders, and very thick." He continued:

> These meadows doubtless serve to feed great numbers of moose and deer, and are of still greater use to new settlers, who find a mowing

field already cleared to their hands; and though the hay is not equally as good as English, yet it not only keeps their cattle alive, but in tolerable order; and without these natural meadows, many settlements could not possibly have been made, at the time they were made.[123]

Although Peirce's observations were made a century after the fur trade's heyday, the stages of dam-to-meadow succession in newly settled areas likely followed the same general patterns. After beavers were hunted, their dams deteriorated, and what had been vast ponds became dry meadows, which attracted settlement. Adrien van der Donck observed similar patterns during the mid-seventeenth century. "Near the rivers and water sides there are large extensive plains . . . ," he noted, "which are very convenient for plantations, villages and towns." And these meadows, he noted, appeared deep within the forests as well. "We also find meadow grounds far inland," he explained, "which are all fresh and make good hayland."[124] As early as 1634, the Englishman Thomas Morton observed that among the "great conflux of waters as are there gathered" in the northern reaches of the Iroquois country, there were "many fruitfull and pleasant pastures all about it."[125] That these meadows were formed near waterways and particularly in patches within the heavily wooded interior suggests that a generation of hunting beavers had begun to transform the land. As Peirce observed, once those meadows were settled, they were frequently mowed, the natural grasses sapping nutrients and moisture from the soil. Grass growth slowed and settlers responded by planting European hays, which further changed the soil's nutrient composition. With nowhere else to go, water that had once inched lazily across the country coursed swiftly into narrow riverbeds.

Surging rivers and streams caused sedimentation, and in some places on a massive scale. When beaver dams were removed, stored sediment and nutrients were released into faster-moving fluvial systems, which changed downstream water quality. When one dam breaks, faster-moving water can damage those downstream.[126] And sediment loss can be so extensive when a beaver dam breaks that it can kill riverbed plants and smother fish eggs.[127] In the Mohawk country, the Reverend Megapolensis observed in 1644, "The soil is very good, but the worst of it is, that by the melting of the snow, or heavy rains, the river readily overflows and covers that low land."[128] That rivers consistently overflowed their banks suggests that as the marshy buffers surrounding beaver ponds were removed, rivers had begun to run faster and more violently during periods of high runoff. It is possible, too, that the rich lowland soil Megapolensis observed was sediment that had been moved there from farther upstream. Similarly, Van der Donck observed of these lowlands that "Sometimes the water may wash out a little in places, but the land is manured by the sediment left by the water." It is impossible to know whether dam destruction was the precise cause of that sedimentation, but for these Dutch observers the movement of mud was noticeable and even remarkable. When combined with their accounts of capricious rivers, their observations suggest the system was reeling from the effects of widespread dam removal. "Those floods do not stand long," Van der Donck wrote of the wild fluctuations in water level; "as they rise quick, they also again fall off in two or three days."[129] Although sediment transport volume is highly dependent on sediment size and hill slope, all indications are that sedimentation increased—and in some cases quite dramatically—when beavers were removed from the landscape.[130]

Ever the ecological keystone, beavers shaped not only the movement of water, but also the surrounding geology and patterns of forest succession. Any disruption to this carefully balanced system—namely, killing beavers—lowered the water table, changed the floral and faunal composition of the surrounding forest, and altered the course of water and the speed at which it moved. If the introduction of wampum to the fur trade sparked a "wampum revolution," then one must also acknowledge the ecological revolution that followed.[131]

Changes to inland rivers impacted the coastal estuaries into which their waters flowed, and in a relatively small watershed like that of Narragansett Bay, where hunting pressure on beaver was particularly strong, the effects on sedimentation were dramatic. Drawing on earlier work that calculated a 673-square-kilometer watershed retained 3.2 million cubic meters of sediment behind beaver dams, estuarine ecologists calculated that in Narragansett Bay, which was seven times larger, beaver dams would have retained 22 million cubic meters of sediment.[132] If those dams were removed, a full ten centimeters, or nearly four inches, of sediment could have theoretically covered the entire bottom of Narragansett Bay.[133] At the very least, such a volume of mud and sand would have surely fouled Bay waters. But it is important to note that beavers were not removed all at once, and that sediment would have become trapped in beaver ponds downstream. Nevertheless, with less time to steep in numerous beaver ponds, river water flowing into Narragansett Bay at higher rates deposited dissolved nitrogen, phosphorus, and a cocktail of organic compounds that spurred the growth of chlorophyll-producing algae, the foundation of the food chain.[134] This nutrient-rich water released by the destruction of beaver dams, experts believe, caused the "primary production" of chlorophyll to more than double on Narragansett Bay.[135] It is possible that in some sheltered inlets this flood of nutrients caused harmful algal blooms that consumed oxygen and choked Bay creatures. But it is also likely that the introduction of organic matter caused dramatic increases in microscopic plant and animal life in the estuary.[136]

Accordingly, beaver dam removal could have made Narragansett Bay more productive. Although estuarine ecologists admit that the variables affecting coastal plant and animal growth are legion, they nevertheless believe that the introduction of nutrients can increase marine fauna.[137] Ecologists have demonstrated that there is a "strong correlation" between elevated chlorophyll and "the reported landings of finfish and shellfish."[138] In short, the doubling of nutrients flowing into the estuary likely increased fish productivity. It is conceivable that the prodigious fish runs and sprawling clam banks observed by settlers like William Wood, Roger Williams, and Thomas Morton, among others, were to some extent accentuated by human action. In this case, for a brief time humans made things better for themselves before they made them worse.[139]

Although the amounts of freshwater that flowed into Narragansett Bay before beaver dams were removed are unknown, it is highly likely that more freshwater flooded the Bay after those beavers were killed and their dams deteriorated. As a result, more nutrients entered the Bay and the flushing time decreased, changing the chemical composition of Bay waters, particularly those in the estuary's brackish arms where shellfish proliferated. Conceivably, the shells mined from Narragansett Bay and later transformed into beads changed the very environment from which they had come. The imposition of Native American and European cultures onto a simple estuarine resource—the shells of whelks and quahogs—ultimately, following a tortuous route, reshaped the environment of that resource itself. The value attributed to wampum's timeless, trackless, and even divine estuarine origins came to reflect the powers of "improvement" that fueled the relentless search for furs farther and farther into the continental interior. As beavers were "gleaned away," as William Hubbard, a contemporary, observed, and water drained off the land, the eternal sea felt the impact.[140] Even before human and animal populations around Narragansett Bay began to grow in earnest and large-scale forest clearing ensued, a handful of fur traders and the Indians with whom they developed shared notions of value initiated the systematic destruction of the beaver. This changed not only Narragansett Bay but the entire region as well. Southern New England's soggy coastal margin had begun to harden. Within a generation, the Northeast became a dryer place and one less likely to contain European settlers on its shores.

CHAPTER 2

Shoveling Dung Against the Tide

"THE PRODUCE OF this Colony," wrote the Anglican minister James MacSparran of Narragansett in 1753, "is principally Butter and Cheese, fat Cattle, Wool and fine Horses, that are exported to all parts of the *English America*." Although MacSparran praised the horses for their "fleetness and swift Pacing," his description of Rhode Island was also disparaging. More than three hundred vessels "from 60 tons and upwards," MacSparran observed with reproach, called the waters of Narragansett Bay home, and some of them shipped this "produce" of Rhode Island's plantations to points throughout the colonies. "[B]ut," he groused, "as they [the ships] are rather Carriers for other Colonies than furnished here with Cargoes, you will go near to conclude that we are lazy and greedy of Gain, since, instead of cultivating the Lands, we improve too many Hands in trade. This indeed is the case." Although Rhode Island was well known for its cows, mares, and mutton, this emphasis on livestock exporting and other forms of inter-colony trade, the good reverend believed, was shady business. Rejecting the blisters of an honest-day's work, all "hands," he insisted, had been greased by the drippings of commerce, and as a result had grown ethically soft.[1]

MacSparran was correct in his assertion that Rhode Island farmers had largely emphasized livestock over crops. He was also correct when he observed that by the middle of the eighteenth century Rhode Island ships played a central role in Atlantic world networks of trade. But in his haste to show that Rhode Islanders, who had taken "Liberty of

Conscience . . . to an irreligious Extreme," had succumbed to sloth and avarice by a collective failure to plow and plant, MacSparran suggested that Rhode Island's graziers and shipping magnates had simply traded honorable work in the fields for ill-gotten gain at sea. For MacSparran, a missionary working among some of Rhode Island's largest plantations to sustain an Anglican parish surrounded by "Quakers, Anabaptists, . . . [and] Independents, with still a larger Number . . . devoid of all Religion," the mercenary nature of Rhode Island's livestock plantations and their close ties to offshore trade was just one of many examples of the colony's corrupt character.[2]

Although mostly free from MacSparran's heavy-handed moralism, modern historians of early Rhode Island have likewise highlighted connections between the shores and ships of Narragansett Bay. Chasing the backstory of the often-repeated caveat "except in Rhode Island," they usually emphasize the ways Rhode Island's economic (as well as social and religious) development differed from that of other English colonies in the North.[3] Most notably, Carl Bridenbaugh challenged the assumption that across New England poor soils forced coastal people to turn to the sea. Blessed with rich coastal farmlands, Rhode Island, he showed, developed instead a maritime infrastructure with an eye on exploiting the colony's terrestrial resources. "[It] was the prospect of marketing a lucrative agricultural surplus . . . ," he wrote, "that forced local merchants to build wharves and warehouses, . . . ketches, barques, and sloops for the youth of the colony to sail to faraway ports."[4] The development of animal husbandry on the islands of and coastal mainland surrounding Narragansett Bay, in other words, shaped the colony's maritime character. Rhode Island's identity—one symbolized by a ship's anchor by the mid-seventeenth century—was rooted not simply in its ties to the ocean, but in its ability to facilitate exchanges between land and sea.[5] Although many Rhode Islanders drew their livelihood from the shell and finfish of Narragansett Bay and the many tidal lagoons adjacent to it, the colony had during the seventeenth and eighteenth centuries developed an economy that relied primarily on the grasslands of its coastal fringe. In essence, Rhode Islanders established their place as major maritime players because of their collective ability to work in and draw from an estuarine hinterland.

But by emphasizing the connections between Rhode Island's plantations and distant ports, perhaps the colony's chroniclers drove its sheep, pigs, cows, and horses too quickly from the stockyards to the quays without considering the ways ecological changes on shore affected the sea closer to home. If eighteenth-century observers like MacSparran emphasized the moral continuities between livestock plantations and Atlantic world markets and later historians emphasized commercial continuities, a closer look at the ecological relations between Rhode Island's shores, Narragansett Bay, and the coastal ocean reveals that human ideas, institutions, and economies were inextricably tied to the natural environment. In all their infinite complexities, the spiritual and material worlds were firmly entangled.

MacSparran's disparaging account of Rhode Island commerce suggests that eighteenth-century coastal people, or at least the ornery, aging ministers who preached to them, saw important conceptual differences between land and sea, between workable soil that nurtured the righteous and a defiant ocean that harbored the avaricious. For MacSparran, confining livestock to their deck pens and driving them toward the sea's endless horizon was anathema. They were the fruits of the bucolic—of dirt, mud, and grass—not salt spray and the twisting, pounding, infinite tumult of the waves. Cattle represented time-honored tradition and methodical stewardship. Thrust upon an unrelenting ocean, these cows, pigs, sheep, and horses and their profane attendants were, for MacSparran, forced to endure the same physical and spiritual displacement to which he himself had been subjected while living along the Narragansett Country's godforsaken arms of the sea. Purposefully worked, dry land held hope for redemption through constructive works. Unchanging and beyond control, the sea, by contrast, was arbitrary and wicked.

The Rhode Islanders who lived and worked on the Bay developed attitudes and values about nature based in part on their proximity to the littoral. For farmers who mowed salt hay by daylight and speared eels by torchlight, every facet of quotidian life straddled—sometimes deftly and at other times with noticeable hesitation—the psychological and material boundaries between land and sea. Mirroring the complex ecology of the shores on which they lived, a space defined by seasonal winds, weather, and the ebb and flood of the tide, these littoral people sometimes sought to master nature and at other times submitted to its demands.

That shifting or even ambivalent relationship with nature had environmental repercussions for the coastal zone. In the interstices between land and sea, between a geography of control and one that was literally and conceptually unfathomable, the boundary between man and nature could became porous. At times, an enduring ocean thwarted the mastery of man and made him adapt to its will. Every twelve hours, the rhythmic renewal of the tide reminded the littoral people living there that they, too, were subject to the powerful forces of nature, which could take away as much as they could give. But as the sea slipped toward shore, its waters growing brown with tannins and green with algae, the shallow, sheltered ponds into which it flowed grew increasingly malleable. Alongshore, the ocean lost some of its opacity. The estuary became a placid harbor of refuge and a cradle of incredible fecundity. And the impulse to "improve" it was strong. The people of Narragansett Bay tinkered with the shore, sometimes to dramatic effect. As the years passed, the estuary, like an aging face weathered by work and stress, slowly and imperceptibly began to show the deepening creases and fading pallor of environmental change.

Settling the Shores

The bitter winter of 1635 had descended on Salem, a small, coastal Puritan settlement fifteen miles north of Boston, and schism, like the icy winds and knee-deep snows that had blanketed the town in early December, had forced the religious community there, at least a sizeable portion of it, into isolation. With the intense heat of the hearth in his face and bitter cold at his back, Roger Williams, the vociferous minister who, according to the Puritan powers in Boston, had infected Salem with separatist sentiments, was, not unlike the clashing climes of his chambers, emotionally and spiritually divided.[6] For months, Williams had been torn between betraying his conscience for the comfort of community and remaining true to his principles but submitting to banishment. To make matters worse, he had fallen gravely ill in August and had scarcely recovered.[7] After closing a private letter from John Winthrop, whose advice he later characterized as "a hint and voice from God," Williams turned his back to the fire and began to pack his things.[8]

Since making the decision to immigrate to New England, Williams had been no stranger to conflict. Born most likely in 1603 in Wales, raised in London, and educated as a minister at Cambridge, Williams was a Puritan who, like so many other English Protestants, believed the Church of England had not distanced itself far enough from Catholicism. In 1629 Williams married Mary Barnard, and a year later the couple boarded the *Lyon* with twenty other passengers intent on leaving England and the increasingly anti-Puritan policies of Canterbury for one of the newly established Puritan communities abroad. Departing the port of Bristol, the young clergyman and his new wife arrived to a warm welcome on February 5, 1631 at Nantasket, Massachusetts.[9] But the Puritan authorities in Boston soon grew critical of Williams after he turned down a job as a minister because he felt they had not distanced themselves enough from Anglican ways. Williams, it seemed, sought not only to purify the Church of England but also Puritanism itself. And this threatened the authority of the church fathers in Boston. He tried to find a community in Salem and then tried again with the separatist Pilgrims in Plymouth, with limited success. Smarting from Williams's defiance, Boston leveraged political power to alienate the radical parson wherever he went. Although Plymouth governor William Bradford admitted Williams was "a man godly and zealous, having many precious parts . . . ," he was nevertheless, Bradford conceded, "very unsettled in judgment." Eventually, Bradford explained, Williams began to "fall into some strange opinions, and from opinion to practice."[10] In 1633 Williams and his wife returned to Salem where, no doubt frustrated, he began to express his thoughts in writing.

Heated rhetoric poured from his pen. Williams was particularly critical of the ways the English flagrantly seized Native American property. "[W]e have not," he wrote in a published response to a letter from John Cotton, "our Land by Pattent from the King, but that the Natives are the true owners of it."[11] So egregious was the seizure of land that Williams called it "a sin of unjust usurpation upon others' possessions."[12] Williams further embittered Boston officials when he raised the issue of women wearing veils during prayer. Debate ensued there, and although the matter was ultimately dismissed (contrary to Williams's opinion), officials saw in the polemical minister's willingness to broach the subject the seeds of separatist radicalism. In 1634 Williams was again drawn into contentious debate when Salem freeman John Endecott used scissors to cull what he saw as popish symbolism from the English ensign. "He," recorded John Winthrop, "judging the cross, etc., to be a sin, did content himself to have reformed it."[13] Although Williams did not participate, his adversaries concluded that Endecott's brash behavior was yet another manifestation of Williams's perverse ministry.[14] In 1635 Williams opposed a law requiring oaths of allegiance to the Massachusetts Bay magistrates and then wrote two particularly inflammatory letters "complaining of the magistrates for injustice, extreme oppression, etc."[15] The tipping point, noted an exasperated John Winthrop, was when Williams urged his parishioners to "renounce communion with all the churches in the bay, as full of antichristian pollution."[16]

In response, officials in Boston had no choice but to take drastic measures. For a short time, the General Court contemplated his execution.[17] But cooler heads prevailed, and upon hauling Williams into court, on September 3, 1635 the governor and magistrates banished him instead. Williams, the court concluded, "hath broached and divulged dyvers newe and dangerous opinions, against the authoritie of magistrates." In addition, they noted, he had written "letters of defamacion . . . and maineteineth the same without retraccion."[18] And this did not sit well with those in power. For his contrarian ways, the General Court ordered that "Mr. Williams shall departe out of this jurisdiction within six weekes" or its members would "send him to some place out of this jurisdiccion" themselves.[19] Due to Williams's illness, the General Court granted him stay in Salem until spring if he did not "draw others to his opinions."[20] Despite the birth of his second child in October, Williams persisted and was called to court again. Citing his illness, Williams did not appear. In response, the General Court dispatched Captain Underhill to apprehend Salem's incendiary teacher and place him on a ship bound for England.

Warned of the impending arrest and left with few options, Williams fled. "I was," he later wrote, "unkindly and unchristianly . . . driven from my house and land and wife and children."[21] Guided by a compass decorated with a winged hourglass supporting a human skull, Williams, laboring through deep snow and freezing temperatures, was no doubt reminded of his own mortality with each check of direction.[22] He was,

Williams recalled, "sorely tossed, for . . . fourteen weeks, in a bitter winter season, not knowing what bread or bed did mean." The private letter Williams had received from Winthrop had advised him "to steer . . . to Narragansett Bay" because of the "freeness of the place from any English claims or patents."[23] As governor of Massachusetts Bay from 1630 to 1634 and Deputy Governor in 1636, Winthrop was clearly part of the Puritan establishment that had run Williams out of town. In fact, he had voted for Williams's banishment and cast any separatist settlement on Narragansett Bay as a threat, for if Williams and his followers succeeded, "the infection," he wrote, "would easily spread into these churches."[24] But privately he supported it. Williams was certainly an indefatigable critic of the Puritan establishment in Boston and perhaps even a threat to its religious authority in New England, but Narragansett Bay presented untapped economic potential. Although Plymouth held jurisdiction over most of the Bay's eastern shore, an alliance with Williams, who was eminently capable of acquiring land to the west and followers to fill it, promised Winthrop substantial dividends.

After surviving a harrowing journey of more than three months, Williams arrived at the head of Narragansett Bay. During his exodus, he had sheltered with the Wampanoags. Nevertheless, upon reaching his new home, Williams felt he had vanquished the angels of the wild. "Unto these parts . . . ," he later wrote, "I may say Peniel, that is, I have seen the face of God."[25] Having learned the Indian language during his time in Plymouth and his stay with the Wampanoags, Williams arranged with Ousemequin, or Massasoit, for land at Seekonk. But Edward Winslow, the Governor of Plymouth, informed Williams that he was still within Plymouth colony's jurisdiction. Brokering a deal with the Narragansett sachem Canonicus and his nephew Miantonomi for £30, Williams, in May or early June, packed his things once again and with five others traveled west by canoe around Fox Point and up the next largest river. There, by a spring flowing from a ridge, they established a settlement that "having in a sense of God's mercifull providence unto me in my distresse . . . ," Williams later wrote in the land's confirmatory deed, he named "Providence."[26]

In short order, Williams and his small band of followers began transforming the wilderness into a town. Providence was primarily situated on

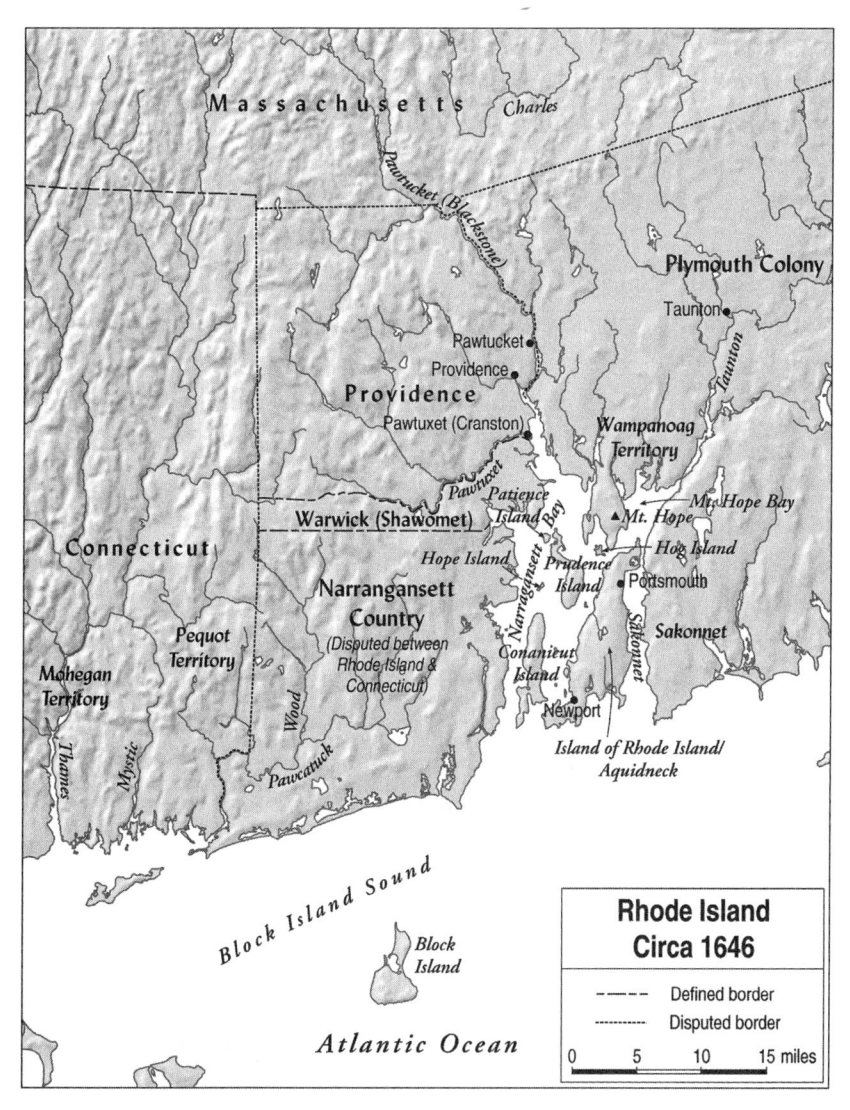

Rhode Island during the first decade of English settlement.

the "lands and meadowes upon the two fresh rivers, called Mooshausic and Wanasqutucket," but the tract extended northwest along the Blackstone River to Pawtucket and south along the Bay to the Pawtuxet River, which was surrounded on both sides by rich salt meadows.[27] In Providence Williams and his "loving friends" carved out Town Street along which they began building houses, the roofs of which were thatched with grasses from the Bay's broad marshes.[28] So determined were Williams and his fellow townspeople to develop the area that, by the end of 1636, the town council mandated fines for inhabitants "in case they do not improve their ground at present granted to them, viz.: by preparing to fence, to plant, to build, etc." In response, the settlers at Providence subdued the land aggressively, and within a year new laws were necessary to rein in the wanton clearing of forests, for settlers had begun felling trees faster than they could use or remove them. "[A]ny timber . . . lying on the ground above one yeare after the felling," the council mandated, "shall be at the Towne's disposing."[29]

Although the tract of land surrounding Providence was generous, Williams saw great economic potential in the vast salt meadows ringing the rest of Narragansett Bay and moved quickly to secure property on its islands. In partnership with Winthrop, Williams approached Canonicus about purchasing the five-square-mile island called Chibachuwese, which, shaped "spectacle-wise" in the middle of Narragansett Bay, had been previously (and tenuously) owned by the late John Oldham. Desirous of Oldham's trade goods and hoping to induce him to make the Bay his base of operations, the Narragansetts had offered him the island. Mad Jack never settled there, but Williams, quick to capitalize on an entrepreneurial opportunity, maneuvered to fill the gap that had opened in Oldham's absence. "[T]urning their [the Narragansetts's] affections towards myself," wrote Williams to Winthrop in October 1637, "they desired me to remove thither and dwell nearer to them." Although Williams did not want to settle there, he sought a partnership with Winthrop, who had "motioned . . . desire" to fill some of the Bay's islands with swine. Knowing Winthrop's interest in the proposition, Williams himself expressed "more desire to obtain it." Canonicus was only willing to sell half of the island "because of the store of fish," likely rich shellfish beds from which the Narragansetts mined the raw materials for wampum. Nevertheless, Williams felt confident that if he approached Canonicus on the island in person, he "shall obtain the whole."[30] True to his word, Williams acquired the entire island and renamed it Prudence, later purchasing two more nearby that he named Patience and Hope.

As counterpoint to the dogmatic intransigence of Massachusetts Bay, Williams christened the islands composing the geographic center of Narragansett Bay with names that reflected the philosophical and economic dexterity with which he set out to develop his new colony. In short order, Williams, in partnership with Winthrop, began populating the islands with livestock. Natural corrals, the islands made the Bay particularly well-suited for raising animals. In contrast to the more exposed shores of the Chesapeake, the islands of Narragansett Bay prevented animals from straying while keeping them safe from wolves, which rarely swam such great distances.[31] So secure were the islands that they required only periodic oversight. "I have a lusty canoe," wrote Williams to Winthrop in January 1637, "and shall have occasion to run down often to your Island (near twenty miles from us) both with mine own and (I desire also freely) your worship's swine."[32] Even in the "lustiest" canoe, twenty-mile paddles down the Bay would be infrequent.[33]

But Williams could not maintain the serenity he envisioned for his island pastures when in 1638 he brokered the purchase of Aquidneck, or "Rhode-Island," for the followers of Anne Hutchinson, who had also fled religious persecution in Massachusetts Bay. Hutchinson had been excommunicated and banished from the Bay colony for promulgating an understanding of predestination and grace that theoretically made earthbound laws unnecessary, thereby threatening the church hierarchy. The lawless conclusions to which Hutchinson's ideas led prompted her critics to brand her an "Antinomian." Under the leadership of William Coddington, William Aspinwall, and John Coggeshall, Hutchinson and more than eighty families fled to Narragansett Bay and settled on Aquidneck Island's northern end, where they built a town at Pocasset, which they later renamed Portsmouth.

A mix of farmers, tradesmen, seaman, and merchants, the community at Portsmouth was largely agricultural, but with mounting religious strife among the group and growing disparity between economic interests, the settlement fractured. Banished from Plymouth, Samuel Gorton, a particularly zealous Puritan who came to Portsmouth and began preaching ideas similar to but even more extreme than those of Hutchinson, rankled community leaders. And in 1641 they ran Gorton out of town. Crossing the Bay, he purchased land at Shawomet, where he established his own community just south of the Providence patent on Warwick Cove.[34] The desire for religious uniformity and greater economic opportunity on Aquidneck prompted the merchants, led by Coddington, to move south, establishing Newport around the same deep, secure harbor that Verrazano had visited a little over a century earlier. Around a small, spring-fed brook, they built their town, each settler receiving a small house lot along the harbor, a piece of meadowland, and a tract of farmland.[35] The wealthiest of the Newport settlers rewarded themselves with sprawling estates that flanked the harbor to the north, south, and east.

Although most histories of Rhode Island have emphasized the colony's role as a religious refuge, far fewer have examined how a lack of central authority on Narragansett Bay—whether religious or civil—opened the door to economic opportunity during the seventeenth century. The omission is understandable. During the 1630s, Roger Williams's letters, which provide detailed insight into his early years on Narragansett Bay, are overwhelmingly concerned with spiritual matters and the specter of Indian attack, with little attention paid to economic development. But amidst the religious squabbling and Williams's tireless efforts at diplomacy, a quiet scramble for land surrounding Narragansett Bay ensued.

At about the same time that Portsmouth, Newport, and Warwick were settled, Richard Smith with his wife and children arrived in Cohannock, or Taunton, a small village resting at the northeastern corner of Narragansett Bay. Following the Nonconformist clergyman Francis Doughty, who had sparked the antipathy of religious authorities near Smith's home of Thornbury in Gloucestershire, Smith and his family abandoned their possessions and, most likely departing Bristol, set sail for Plymouth Colony. At Taunton, Smith took an oath of allegiance on December 3, 1638 and by 1640 was made a freeman in a growing town that, observed Edward Winslow, had "very good [ground] on both sides, it being for the most part cleared."[36] The forests were filled with timber and upon the river, Winslow explained, "a shipp may goe many myles up it."[37]

Despite his adopted town's natural advantages and having established himself, according to Roger Williams, a "leading man in Taunton," Smith quickly pursued opportunity farther afield.[38] Seeking his fortune in the vast salt meadows and rich uplands that formed the western shores of Narragansett Bay, Smith became, in late 1637 or early 1638, the first European to purchase property there.[39] Called Cocumscussuc by the Narragansett Indians and later named Wickford, Smith's tract included a sheltered harbor fringed with salt marshes and mudflats and frontage on the main overland route through the region, known at various times on sundry deeds as "The Pequot Path," "the road to Pequot," or simply "the country road."[40] His longtime business associate and spiritual confidant, Roger Williams, testified in 1679 that "[F]or his conscience sake ... [Smith] left Taunton and came to the Nahigonsik countrey, where (by the mercy of God and) the favour of the Nahigonsik Sachims, he broke the ice (at his great charges and hazards), and put up in the thickest of the barbarians, the first English house among them."[41]

Initially, however, Smith did not reside there. When Doughty fell out with Pilgrim leaders during the early 1640s, he was obliged to leave New Plymouth. Smith and his family, "for his conscience sake," followed. They spent a brief time in Portsmouth, Rhode Island, and then traveled west into the Dutch territory, whereupon Smith purchased a 13,000-acre tract with Doughty and several others on Long Island in what is now Maspeth, Queens. Smith also purchased property on Stone Street in New Amsterdam and a lot abutting the East River. His daughter soon married a Dutchman named Gysbert op Dyck in 1643.[42] But during that same year Maspeth was attacked by Indians, and after a falling out with Doughty, Smith had pulled out of that investment altogether by 1645. While his family remained in New Amsterdam, Smith left most likely in 1645 or 1646 and moved to the Narragansett Country, thereby opening some of the most fertile lands in New England to a much larger European presence.[43]

Smith built a trading post along the principal travel route along the west side of Narragansett Bay. Within several years, Roger Williams also built a trading house nearby.[44] But in 1651, he sold it to Smith, along with two big guns and a small island for goats.[45] Smith soon expanded his landholdings. On March 8, 1656 he entered a sixty-year lease with the sachem Coginiquant for the land extending south from Cocumscussuc Harbor to the Annaquatucket River and east to the western shore of Narragansett Bay. On June 8, 1659 Coginiquant leased him a slightly larger but more clearly defined parcel, including meadows at Sawgoge and Paquinapaquoge and a spit of land east of Smith's house on the northern side of Cocumscussuc Cove, all for a thousand years. On October 12 of the following year, two other sachems, Scultob and Quequaganuet, confirmed the agreement.[46]

With the help of Williams's skills in negotiating with Indians, Smith had established himself as the first permanent European resident in the Narragansett Country, but his time alone there was short lived, and soon other investors poured into the area. In 1657 four men from Portsmouth, Rhode Island—Samuel Wilbore, Thomas Mumford, John Porter, and Samuel Wilson—and a fifth, John Hull, a goldsmith and the Mintmaster for Massachusetts Bay, combined to purchase a tract of land along the Pettaquamscutt River, a thin, four-mile-long estuary spilling into the West Passage of Narragansett Bay across from the southern tip of Conanicut Island.[47] The Pettaquamscutt saw prodigious annual runs of alewife and bass and was lined with valuable salt meadows. "[The] country on the west of the bay of Narragansett," observed John Winthrop of the area surrounding the Pettaquamscutt, was "all champaign for many miles, but very stony and full of Indians."[48] That western Rhode Island was blanketed with sprawling, grassy levels, or "champaigns," and was home to a large Native American population was widely known. The Narragansetts, although increasingly molested by disease, had been largely spared from the devastation wrought upon the Wampanoags on the Bay's eastern shores. Aware of the Narragansetts's peaceful reputation and willingness to sell land, by the 1650s English colonists began to purchase tracts. But because the consent of so many Indians was required to secure the land, title to the Pettaquamscutt lands wasn't finalized until 1660.[49]

The rocks dotting the Narragansett Country's grasslands did not altogether deny the plow. Corn grew throughout the area, and within a decade of the Pettaquamscutt Purchase southwestern Rhode Island played host to some of the most productive livestock farms in New England. But initially the Pettaquamscutt purchasers were most likely after mining rights. The deed specified that the sellers—the Narragansett chief sachems Quassaquanch, Kachanaquant, and Quequaquenuet—sold not only the land for "£16 and other reasons" but also "grant[ed] them all the black lead in this title."[50] Although lead claims in the area produced few results, the vast tracts of coastal marshland drew considerable interest.

Productive farms and the (hollow) promise of valuable mines were highly attractive to investors, but the zeal with which they entered and subsequently partitioned Narragansett lands destabilized the region.

Williams contended that one important cause of the Pequot War was the insatiable thirst for land among the English. Men with a "depraved appetite after the great vanities, dreams and shadows of this vanishing life," he wrote, placed themselves in "as great necessity and danger for want of great portions of land, as poor, hungry, thirsty seamen have, after a sick and stormy, a long and starving passage."[51] Blinded by avarice, these desperate land speculators, Williams believed, had raised hackles across the region. Williams's use of a seafaring simile to characterize this ravenous pursuit of property was no coincidence. European explorers often employed marine metaphor to establish a "semiotic *tabula rasa*," or an empty, oceanlike space that could be constructed and controlled according to their own desires.[52] It is conceivable that Williams saw such large tracts of land inhabited by so few Europeans as vacant terrain, ripe for improvement, even if he disapproved of the cupidity with which it was obtained. That it was bordered by the Bay and flooded by so many arms of the sea made the comparison all the more fitting. In his willingness to couch the push of progress in oceanic language, to employ the profound nature of the sea to explain development on shore, Williams reveals some of the ways in which the sea surrounded and suffused the settlement of Narragansett Bay's littoral space. But if emptying coastal lands through symbolic sleight of hand only improved the land theoretically, the powerful investors who swooped into the Narragansett Country took control of it in more substantive ways.

The unwavering determination with which these speculators targeted coastal lands led, as Williams observed, to contentious dealings with the Indians and even among the English. In 1659, under the leadership of Humphrey Atherton, a second company purchased land in the Narragansett Country and sought to absorb those properties into Connecticut and Massachusetts.[53] So aggressive were their maneuvers that Williams characterized the ensuing ill will between Rhode Island and its neighbors "like a prodigy or monster." He chastened Connecticut and Massachusetts for their greed, contending that they already had ample rich land. Rhode Island's English neighbors were so well endowed that he compared their resources to "platters and tables full of dainties." He felt their incursions into the Narragansett Country were akin to "snatch[ing] away their poor neighbors' bit [of] crust," which was "a dry, hard one, too because of the natives' continual troubles, trials, and vexations."[54] The great value of Rhode Island's littoral, which lay in its lush grazing lands and ready access to Atlantic world markets, raised political tensions for everyone involved. It made for covetous English neighbors and exacerbated tensions with the Indians who lived within its boundaries. Nevertheless, by the end of the seventeenth century the English had spread to every corner of Narragansett Bay and had begun exploiting that "bit [of] crust" with spectacular results.

Stockyards of the Sea

When Thomas Hazard emigrated from England to Portsmouth, Rhode Island, and later helped found the town of Newport in 1638, his son, Robert, was four years old. As an adult in the 1670s Robert moved to the Narragansett Country, acquired his first parcel of land, and even served as surveyor of Kingstown village.[55] Robert's son, Thomas, became a great landholder in the Narragansett Country. In April 1698 he purchased for £700 roughly 1,000 acres in Scituate, along the Saugatucket River, and on Point Judith Neck, from Samuel Sewell, the son-in-law of John Hull, who was a key player in the Pettaquamscutt Purchase.[56] In 1710 Thomas Hazard bought more land from Sewell along the west shore of the Great Pond and Pettaquamscutt Cove.[57] By the beginning of the eighteenth century, Robert's great-grandson, Robert, was said to have amassed 150 cows and twelve African female slaves working as dairywomen. On Boston neck, a thin spit of land between the Pettaquamscutt River and Narragansett Bay, where the grass was said to have grown waist high, he kept upward of 4,000 sheep.[58]

In a little more than fifty years, Rhode Island had become the most important producer of livestock in New England. In 1661 Royal Commissioners tallied upwards of 100,000 sheep grazing the coasts of New England.[59] And the highest concentrations were in Rhode Island. In 1665 the King's commissioner George Carr explained of Narragansett Bay, "The best English grass and most sheep are in this Province, the ground being very fruitful, ewes bring ordinarily two lambs [per year]," which was nearly double the rate of England at the time.[60] Many of Rhode Island's flocks, however, were lost during King Philip's War in 1675 and

1676, when fire raged from Westerly, in Rhode Island's southwest corner, to Providence, destroying most of the established farms in its path. Edward Randolph estimated that across the region 1,200 houses were burned and 8,000 head of cattle were destroyed.[61] But Rhode Islanders were quick to transplant animals from elsewhere as soon as the smoke had dissipated. Conanicut and Block Island, which had been purchased for the express purpose of raising sheep and cattle, helped fuel the recovery on the mainland. But it was on Aquidneck Island, particularly at Newport, where the numbers of sheep were prodigious. As early as 1662 flocks had grown so large on Aquidneck Island that Portsmouth, responding to "the greate Negligence in many persons, for not takinge there Sheepe Rames from the Ewes," made it lawful to kill any rams on the town common between August 10 and November 10.[62] It was said that William Brenton pastured upwards of 11,000 sheep on his Newport farm.[63] In 1673 he bequeathed 1,500 of them to his heirs.[64] In April 1675 William Harris characterized Aquidneck Island as the "Garden of New-England" and observed, "at a Town called Newport . . . , wch thrives very well," there are "more sheep than in any place in New-England."[65] Passing on knowledge from experience, William Coddington of Newport explained to Jonathan Winthrop, Jr., who had begun raising sheep on Fisher's Island, that the sheep "do ordanierly duble in a yeare, and more of the Lambes have Lambes when they are a yeare ould." One reason for their proliferation, he explained, was the lack of wolves on Aquidneck, there being only one or two.[66] Assessing New English settlements for the French crown in 1692, La Mothe Cadillac explained, "They say that the settlers [of Rhode Island] own two hundred thousand sheep or lambs."[67] So important were sheep by century's end that in 1696 Newport made one the centerpiece of its town seal.[68]

Sheep had brought economic prosperity to Rhode Island, but grazing so many of them along its shores had a profound environmental impact. Sheep eat grass down to the roots, making it necessary to move them. Narragansett Bay farmers were often forced to expand pastures by cutting forests and in some cases filling marshes.[69] Even the finishing stages of wool production impacted the environment. Likely one of the first wool manufacturers in the Narragansett Country, Colonel George Hazard gave to Thomas Culverwell in 1719 "a Little part of my farme belonging to my now Dwelling house . . . for ye Promoting of ye Wooling Manufactury which may be for my benefit and the Publick Good." Culverwell accepted a half-acre lot abutting the Saugatucket River near Rose Hill, whereupon Henry Gardner commenced construction of a dam "to be made for ye fulling of Cloth, and to ye Promoting of a fulling Mill." Hazard carefully noted that Culverwell was also to receive the land that "shall be Drowned by making of ye said Dam."[70] Sheep would not only denude Narragansett Country meadows, but the act of processing their wool would also change, at least in small ways initially, the movement of water across the landscape.

In addition to sheep, vast herds of horses and cattle roamed Rhode Island's shores. William Harris noted in 1675 that the "country is healthy and well replenished with people and cattle and so many horses that men know not what to do with them."[71] So numerous were horses on Aquidneck Island that a Portsmouth law made it legal to kill any horse over a year old left "unfettered or unshackled" on the town common.[72] The horse population grew so great in the Narragansett Country that on June 24, 1686, Rhode Island courts mandated that "thirty or any less number of wild or unmarked horses of two years old or upward shall be taken up, and . . . sold."[73] In Plymouth Colony, which extended to the eastern shores of Narragansett Bay, John Josselyn noted that horses had multiplied rapidly after mid-century.[74] There were so many horses and other animals roaming the countryside that horse traders took pains to prevent interbreeding. A landowner on Point Judith Neck, John Hull, explained to a business partner that if they "did fence [the Neck] with a good stone wall at the north End . . . noe kind of horses nor Cattle might gett thereon . . . [so] that noe mungreel breed might come amonge them."[75] This measure, he believed, would protect the integrity of their export business, which specialized in coach, saddle, and draft horses.

The proliferation of southern New England livestock was so spectacular that Samuel Maverick, writing in 1660, was awestruck. "In the yeare 1626 or thereabouts," Maverick explained, "there was not a Neat Beast Horse or sheepe in the Countrey and a very few Goats or hoggs, and now it is a wonder to see the great herds of Catle belonging to every Towne." In addition to the "brave Flocks of Sheepe [and] The great number of Horses," he also marveled at the numbers of "those many sent to

Barbados and other Carribe Islands, and . . . how many thousand Neate Beasts and Hoggs are yearly killed . . . for Provision in the Countrey and sent abroad to supply Newfoundland, Barbados, Jamaica, @ [*sic*] other places, As also to victuall in whole or in part most shipes which comes there."[76] As Maverick explained, the shores of southern New England raised the livestock that fueled the movement of ships and their crews around the Atlantic world. Devoid of trees and fertilized naturally by the sea, Bay meadows produced fodder with far less human effort than was required farther inland. Livestock raised on Bay islands was at once protected from predators and connected to the wider world via Newport. Either salted and barreled or transported live, Rhode Island's animals fed the slave plantations of the West Indies and fueled the expansion of New England fisheries.

So integral to England's Atlantic world network was Narragansett Bay that officials maneuvered to keep it tightly within the imperial fold. Imploring Whitehall to absorb Rhode Island and Connecticut into the Dominion of New England, Joseph Dudley explained in 1686 that "they are the Principall parts of the Countrey whose Corne and Cattle are raised for the supply of the Great Trade of fishing and Other shipping belonging to this his Majestyes Territory." Without those valuable livestock plantations, Dudley fretted, "wee shall not be able to support our Trade with bread." He further argued that if they were not annexed metropolitan officials should "at least Command a free and uninterrupted trade without Duty for Cattle and Corne . . . without which we shall be greatly distressed." Narragansett Bay and its coastal farms had become the engine of England's Atlantic empire by dint of its geography and environmental attributes, and it was worth bending the rules to keep the machine humming.[77]

By the end of the seventeenth century Rhode Island's role in the Atlantic provisioning trade was beginning to rival that of the empire's other important exporter, Ireland. In its 1663 Staple Act, the English Parliament identified the growing importance of Irish provisions to sugar producers in the English West Indies. As sugar prices soared during the second half of the seventeenth century, Caribbean planters carpeted nearly every inch of their islands with cane, and with little room to produce food locally, they looked to Ireland to feed themselves and their slaves.[78] Its ships laden with barreled beef, ham, and herring, among many other provisions, cities like Galway, Dublin, Cork, and Belfast became key players in the expansion of the English plantation complex. But London, fearful that direct, unregulated trade between Ireland and the West Indies would reduce customs revenues, began to enforce controls. Although the Staple Act had supported direct trade from Ireland to America, it limited the return trip, stipulating that American goods first pass through English ports. Modified in 1671 and reinforced in 1685 and again in 1696, these mercantilist mandates increasingly trimmed the Irish trade. If Ireland was the most important provider of provisions for the British Caribbean before 1690, mainland North America had partially eclipsed it by the turn of the century.[79] And increasingly, Rhode Island's champaign shores were meeting the West Indian demand.

By the beginning of the eighteenth century the business of raising animals in Rhode Island was firmly established. Contemporary observer William Douglass noted that in addition to butter, cheese, and lumber, Rhode Island was known to "export for the West India islands, horses [and] live stock of several kinds."[80] So many animals were moving around the Bay that congestion became a problem. In 1748 John Gardner was granted authority to run a new ferry service from Narragansett because trade had increased to the point that the boats had become "crowded with men, women, children, horses, hogs, sheep and cattle to the intolerable inconvenience, annoyance and delay of men and business."[81] The coastal farms of Narragansett Bay were so productive that, as Harris and Maverick had witnessed, they consistently produced a surplus that fueled a brisk offshore trade.

But many Rhode Island farmers, especially during the first decades of settlement, kept their animals closer to home. In 1638 Portsmouth allocated to each man "one acree of medow for a Beast, one acree for: 5: sheep, & one acree & a half for a horse," which suggests that even the most modest households maintained livestock.[82] And within a generation, most freemen surpassed these early allotments. When Adam Mott, a farmer in Portsmouth, died on August 12, 1661, he left to his family four oxen, five cows, a bull, two calves, three horses, thirty ewes, two rams, and six pigs.[83] The number of cattle increased on Aquidneck Island to the point where it became necessary to establish a committee of cattle overseers. In

1656 Portsmouth appointed Richard Bulger, Thomas Cornell, Jr., John Trip, and William Hall to "survaie and view all Cattell that shalbe henceforth transported of the Island." The law also levied fines to anyone who transported cattle without being surveyed.[84] By the 1660s the allocation of earmarks on cattle dominated the Portsmouth town records; there were so many beasts roaming common pastures and nearby forests that authorities needed a way to keep track of them.[85] The growing confusion over cattle was understandable, for almost every household had a cow. They appeared in nine of every ten farm inventories in New England, and in Massachusetts there were roughly two cows for every five people by the middle of the eighteenth century. On the largest estates in southern Rhode Island herds often comprised between fifty and one hundred cows.[86]

Goats and pigs also dotted Narragansett Bay's coastal landscape, where they typically ran wild according to English custom. Although initially some goats were kept for milk, their time was short-lived. Left to scour Bay islands and mainland plantations, goats were notorious for destroying hedges, trees, and just about any other crop that lay in their path. They were, however, largely eradicated by about 1650 after towns like Warwick made it lawful to kill them should they wander into the commons.[87] Capable of fending off predators and foraging for acorns, roots, and shellfish, swine were left to roam the mainland forests, islands, and mudflats of Narragansett Bay. William Wood noted that the vast beds of clams were "a great commodity for the feeding of swine both in winter and summer." So drawn to the intertidal zone were they that, Wood explained, "once used to those places, they will repair to them as duly every ebb as if they were driven to them by keepers."[88] Coastal pigs even took to the sea to avoid enclosures. In New Haven, "younge cattle & hoggs" hemmed in by a gate, swam around to "doe damadge."[89] Although it is impossible to know exactly how many pigs roamed the shores of Narragansett Bay and the rivers running into it, estimates for Massachusetts Bay in 1735 approach eighteen swine to each one hundred people.[90] At Dartmouth, on nearby Buzzard's Bay, 772 households had 383 swine at mid-century.[91] The numbers were likely similar on Narragansett Bay, although they could have been higher on the largest plantations that packed salt pork for the West Indies and the Atlantic fishing fleet.

The sheer number of animals scouring the countryside took a toll on the Bay and its feeders, particularly in the Narragansett Country, where water dominated the landscape. In the southern portion, there were upwards of fifty ponds, one of which, Worden's, or Pesquamscut, was the largest in the colony. The forested uplands, swamps, and meadows were laced together by a matrix of streams, springs, and rivers, the largest of which include the Pawcatuck, the Pettaquamscutt, Usquepaug, and Shickasheen.[92] To the south, coastal barrier beaches dominated the Atlantic shoreline. Covered with white sand and sloping upward to grass-covered dunes, these beaches varied from year to year, depending on weather conditions and the corresponding movement of sands. Behind the dunes lay "fastlands," comprising pockets of grass, shrubs, and intermittent forests stands. And behind these lay tidal wetlands ringing a network of brackish bays.[93] The Narragansett Country was a watery world of winding tidal rivers, salt ponds, and meadows. And the vast herds of animals that drank, ate, waded, and wallowed there were inextricably tied to it.

As the lands of South County (as the Narragansett Country came be to known) were settled and the farms and their inhabitants expanded, the rivers, ponds, and estuaries changed in response. When in 1795 "Nailer" Tom Hazard, a blacksmith and farmer in Narragansett and a descendent of the great Hazard planters of the area, noted in his diary that "Philip Shearmn workt for me digin Dung," he identified a chore that had been central to coastal farming communities for generations.[94] Manure, along with seaweed and fish, had been used by the English to fertilize fields since first European settlement. But as the numbers of animals and people increased along Rhode Island's coastal margin, the estuary was fertilized as well. In many cases, cattle strayed into the intertidal zone, or the space between high and low tide, which introduced nutrients to the estuary and coastal ocean directly. In a 1785 court case over property rights in Newport, Josiah Arnold affirmed that during his grandfather's days in the early part of the century "cows and horses that were put to pasture upon his land, went down to feed upon the beach and marsh."[95] Multiplied by many times, the effects of these cattle in Rhode Island's coastal stock farms transformed the biogeochemical makeup of the watershed, which in some cases affected estuarine ecology, forcing the people who lived around the bay to contend with the problem.

Although an exact population count for seventeenth-century Rhode Island remains unkown, records do exist for the eighteenth century. By 1708 Rhode Island had a total population of 7,181 people, 30.7 percent of whom lived in Newport, 20.1 percent of whom lived in Providence, while the remaining 49.2 percent resided in other towns around the colony. By 1730, the total population was 17,935. That number doubled to 31,778 in 1748–1749 and to 40,536 in 1755. By 1774, 59,607 people lived in Rhode Island. As the population increased, the number of people who lived outside of Rhode Island's two major cities soared. By 1774 only 15.3 percent of Rhode Islanders lived in Newport, 7.3 in Providence, while 77.3 percent lived in other towns.[96]

As the population increased, so too did the amount of waste flowing into the Bay. Human waste typically produces between 2.2 and 6.2 kilograms of nitrogen per year, with a value of 4.4 kilograms of nitrogen per capita per year being the most frequently cited figure.[97] It is difficult to determine just how much of this nitrogen would have made it into watercourses and ultimately into the Bay, but it is possible to make calculated estimates. A cesspool system without a leach field situated less than 200 meters from a stream—probably exhibiting similar (albeit wetter) characteristics to the privies of the seventeenth and eighteenth centuries—will contribute roughly 62 percent of its nitrogen into a given watershed.[98] About 4.4 percent of that nitrogen would likely be lost in streams before it reached the Bay.[99] Scientists studying Narragansett Bay have estimated that under these conditions between 0.7 and 3.3 kg of nitrogen per person per year would be introduced into Bay waters. Therefore, with a population of roughly 20,000 people in 1735, between 14,000 and 66,000 kilograms of nitrogen would have made its way into the Bay. The closer people lived to the estuary, the more nitrogen was introduced.[100] With a relatively small human population, this nonpoint source introduction of waste was not in and of itself enough to cause meaningful change. But when combined with animal waste, the effects increased dramatically.

Crowded with cattle of all kinds, the shores of the Bay felt the impact. Agricultural census data for Rhode Island did not begin until 1865 nor for Massachusetts until 1885.[101] As a result, most scientific studies that examine the introduction of nitrogen from agricultural sources for Narragansett Bay have focused their efforts on the late nineteenth and early twentieth centuries. But by mining anecdotal evidence, it is possible to make some estimates for the seventeenth and eighteenth centuries. Scientists examining agricultural pollution employ a measurement called the "animal unit." A horse or cow equals one animal unit. A pig is 0.25, and a sheep is 0.1. Mid-eighteenth-century estimates for Massachusetts Bay assume two cows for every five people, four cows to each horse, eighteen pigs for every one hundred people, and a three-to-one ratio of oxen to horses.[102] In addition, if one considers La Mothe Cadillac's observation that in 1692 Rhode Island had roughly 200,000 sheep, it is possible to construct rough estimates for the livestock population.[103] Based on a human population of 20,000 people for 1735, it is likely that Rhode Island and its waterways played host to 8,000 cows, 2,000 horses, 6,000 oxen, and 3,600 pigs. All told, the total reaches just under 37,000 animal units, each of which produced on average 50 kilograms of nitrogen per year.[104] Typically, between 16 and 32 percent of animal waste nitrogen is introduced into any given watershed.[105] As a result, livestock would have introduced between 296,000 and 592,000 kilograms of nitrogen per year into Narragansett Bay. In turn, in 1735 humans and their animals introduced between 310,000 and 658,000 kilograms of nitrogen. It is likely that the numbers would have been on the higher end of this range because most people lived by and most farms were situated on the shores of the Bay or along the rivers flowing into it.[106] In addition, the historical record shows that Rhode Island had higher concentrations of animals than its neighbors, so that, again, higher levels of nitrogen were likely.

These numbers are noteworthy when one compares them to those of the late nineteenth century, when Narragansett Bay began experiencing the first of its large-scale fish die-offs.[107] These occurred when the introduction of excess nutrients into Bay waters caused dense blooms of algae. When the algae died and decomposed, they consumed the oxygen in the water, suffocating the animals in it. In 1885 there were about 30,000 animal units living in direct proximity to Narragansett Bay. At about the same time, Rhode Island farmers were fertilizing their fields with guano that introduced roughly another 180,000 kg of nitrogen per year, of which between 10 and 40 percent leached into Bay waters.[108] In turn, animals and fertilizers introduced between 258,000 and 552,000 kg of nitrogen into the Bay—numbers quite similar to those of the mid-eighteenth century. These nutrient loads—whether in 1735 or 1885—were not enough to transform the Bay in its entirety. (The devastating anoxic events of the late nineteenth century were most likely the result of industrial pollution combined with untreated human waste from a population that had climbed to over 300,000.)[109] But they were enough to alter tidal lagoons and the most sheltered estuarine creeks by clogging them with algae and subsequently starving them of oxygen.

By the mid-nineteenth century, English chemists had acknowledged that manure polluted water. "They [the chemists] say," noted an 1886 Rhode Island health commission report that addressed similar issues on the waters of Narragansett Bay, "water from cultivated lands, and from under drains of cultivated lands, is always more or less polluted from the manure, even after it has been stopped in ponds or reservoirs."[110] In 1902, an inspection of the tidal Kickemuit River, a portion of which had been dammed to provide water for the town of Bristol, revealed, "cows had passed through the brook, and deposits of cow excrement had been left on the edge of the stream." In addition, surveyors noted, "Two pigs ... had established a wallow-hole on the side of and in the stream."[111] These were animal behaviors and pollution trends that had existed from time immemorial. But when animal concentrations were high, as they were in early eighteenth-century Rhode Island, their impact on sheltered coastal waters would have been substantial. The noticeable effects of animal pollution on the Great South Bay and Moriches Bay tidal lagoons on Long Island during the 1950s prompted biologist John Ryther to conduct the first study of marine algal blooms, which had destroyed the once-prosperous oyster industry there.[112] Ryther's study linked the blooms to the area's large duck industry, which comprised forty farms lining Moriches Bay and the streams flowing into it. The combination of shallow waters, low flushing rate, high temperatures, and the steady influx of duck waste led to such high levels of algae and bacteria that it destroyed the lagoon's bottom-dwelling animals.

With the bulk of seventeenth- through early nineteenth-century agriculture located in lower Narragansett Bay, any excess nitrogen would have become concentrated there. It is likely that nutrient loads climbed so high that algal blooms occurred during the summer months. On June 8, 1729 the professional surveyor James Helme filed "a Draft of

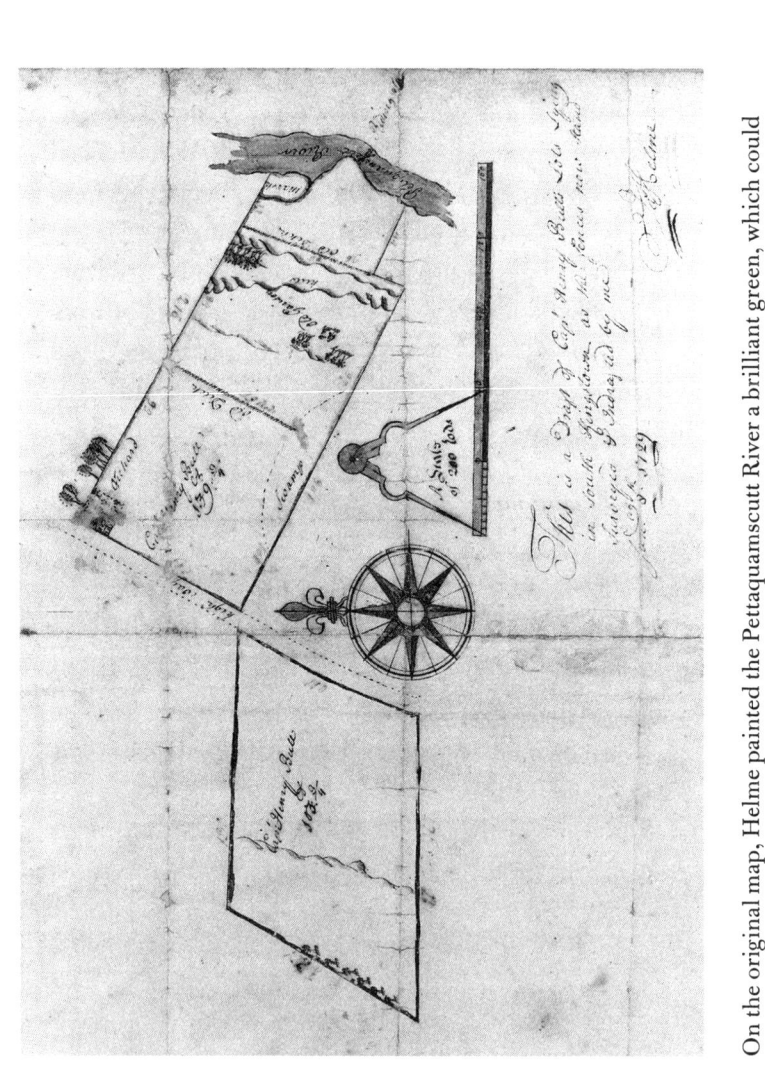

On the original map, Helme painted the Pettaquamscutt River a brilliant green, which could indicate that during the early eighteenth century high levels of animal waste combined with other forms of pollution were beginning to change some of Rhode Island's more enclosed coastal lagoons. James Helme, "Draft of Capt. Henry Bull's Lots," S. Kingston 1729, Map 0150, Cartography Collection, RHi X3 2471. Courtesy of the Rhode Island Historical Society.

Capt. Henry Bull's Lots L[y]ing in South Kingstown as the fences now stand."[113] Adjacent to the upper end of the Pettaquamscutt tidal pond, Bull's lots, according to Helme's rendering, were a mixture of hills, swampland, and intermittent marshes. Sparing no detail, Helme's map showed a broad highway running southwest to northeast. Orchards had been planted, and an "old garrison" had probably been abandoned for a proper house on the main highway. He showed a ditch ninety-nine rods long running across the property. Another, seventy-six rods long, ran to the river, and a third ran thirty-three rods into the marsh along the river's western shore. Their proximity to the marshes and the river suggest they had been dug to drain the adjacent meadows for livestock grazing. Helme painted the Pettaquamscutt River itself, flanked by Bull's land to the west and a plot marked "Ridington" to the east, a deep, vibrant green. This could have been merely an artistic decision. But it is more likely that Helme, a careful surveyor who took great pains to account for every detail of the landscape, would have painted the map's most dominant feature with purpose. His choice of bright green suggests that the river had become highly enriched, perhaps naturally, but also perhaps due to the introduction of animal nutrients. Marine scientists studying Narragansett Bay have claimed that by the second half of the nineteenth century, the high volume of manure and lack of buffer zones around tidal lagoons would have impacted them considerably.[114] But some simple quantitative estimates and even a brightly painted survey map suggest that these changes could have occurred more than a century earlier.

Changes in the Bay

Fueling the dramatic growth of animal populations was the introduction of English hayseed. Although early settlers had been attracted to the shores of Narragansett Bay because of its abundant salt meadows, they soon discovered that the salt hay and even native upland grass did not have the same nutrient content as traditional English hays. In most cases, they carefully cultivated English grasses to replace native species.[115] It is possible that cows simply could not absorb the nutrients in salt hay as efficiently, passing it instead of converting it into growth.

Having witnessed this phenomenon, the Reverend Mellen noted in 1794 that "The Manure made of cattle fed on salt hay is much more fertilizing than that made from fresh."[116] With the goal of fattening animals as fast as possible by stuffing them with English hay, Rhode Islanders converted tidal marshes into upland meadows by damming and diking.[117] Inland swamps were likewise drained and their timber removed. The historical record shows that English grass, for which farmers and merchants were willing to pay handsomely, was ubiquitous in Rhode Island, where royal commissioners and other observers agreed that the best grass grew.[118] But English grass sapped the soil of its nutrients and in many cases transferred those nutrients to the watershed. After receiving a batch of English hayseed from Robert Williams, Roger Williams's brother, John Winthrop, Jr., asked for planting directions. Roger Williams provided a list of instructions, explaining that three bushels of seed would cover one acre of land. Among his directions, Williams warned Winthrop to "Sow it not in an Orchard neere fruit trees for it will steale and rob the Trees etc."[119] English grass was so efficient at drawing nutrients from the soil that it hampered fruit production in nearby trees. This fortified grass digested by livestock provided a means through which important landbound nutrients were, via cattle, transferred to the rivers, streams, and salt ponds besides which they grazed. He also instructed Winthrop to let the "grasse stand until it seede and the wind disperse it . . . up and down," which suggests the English method of dispersal quickly transformed grasslands indiscriminately across the entire region.

The clearing of trees and draining of swamps surely facilitated this nutrient exchange as well. As the land was cleared, the residence time of water—or the amount of time water is retained on the landscape—decreased. And forests were razed at a blistering pace. The average household used between twenty and forty cords of wood per year for cooking and heating. Trees were also cut for ship masts, construction lumber, charcoal, and for use in lime production and brick making.[120] As timber was removed, more water flowed into the estuary and more nutrients flowed with it.[121] And in some places, particularly those where fields were plowed for growing corn, wheat, barley, and even, for a short time, tobacco, sediments washed downstream as well.

The combined effects of excess algae and plant growth and shifting mud and sand transformed the estuarine environment, often to dramatic effect. In 1735 the inhabitants of Westerly petitioned the Rhode Island General Assembly for financial aid to remedy a clogged harbor entrance. They complained that they were "destitute of a harbor there, by reason of a breach (that formerly used to be open in the largest salt pond in Westerly, aforesaid) being shut or filled up."[122] The petition explained that the opening had been shallow but that over time it had become blocked, rendering their harbor unnavigable. It is likely that a combination of algae and plant growth, shifting sand, and accumulating debris caused the blockage. Surrounded by farmland, stone and brick quarries, timberlands, and sawmills, the salt pond opened to the ocean with only a small entrance.[123] High nutrient levels would have spurred algal blooms and the rapid growth of marsh grasses, including *Spartina alternaflora*, which is particularly responsive to nitrogen increases. As algae and various macrophytes proliferated over the years, they contributed to the sedimentation of the salt pond. Coursing downstream, farmland silts and sawmill dust also settled to the bottom as the water slowed, became trapped in the thick mats of marsh grass, or both, which clogged the pond entrance.

In response, Westerly residents realized they had to fix their ailing harbor. Their solution, a massive undertaking for which they needed financial support, was to redirect a branch of the Pawcatuck River into the pond, which would open the silted breach and deepen the anchorage, providing them with a "very commodious harbor" that was "navigable as well for small sloops as boats; and that it would be likewise very convenient for the catching and making of cod-fish, which would be of great service to this colony."[124] Westerly's farms, quarries, and sawmills had changed the estuary. But the people who depended on its salt pond, harbor, and fish sought to fix the mess they had created by completely transforming it. By diverting a river into a placid lagoon, Westerly's residents reconfigured not only the salinity gradient but also the size of the harbor and the species that would have existed there. But of equal importance were the ways this response reflected the changing relationship between a littoral people and the sea. The exact causes of the clog were, like the sea, unknowable, but the people of Rhode Island nevertheless concluded that the harbor could, through much labor and ingenuity, be "improved." The shifting sediments of the estuary were wholly unpredictable—in a sense, very much "wild"—but the harbor's subsequent transformation by the people of Westerly suggests that they believed it was capable of being "civilized" as well. Wielding mattocks and hoes, they, in turn, resolved to rebuild the ocean's edge.

Rhode Island's livestock plantations affected other estuarine harbors as well. In 1661 Dr. John Alcock and several investors from Roxbury bought Block Island off Rhode Island's southern coast and began raising cattle, which multiplied quickly there.[125] As early as 1665 the Rhode Island Governor, deputy Governor, and John Clarke were ordered to Block Island to ascertain whether the 1,000-acre salt pond in the north-central part of the island could be transformed into a harbor. In 1670 the General Assembly ordered "that the . . . inhabitants are authorized to use all fitting indevers to accomplish the same, and doe very much commend their worthy intentions therein."[126] Surely made possible by the harbor's construction, the town of New Shoreham was established on the island in 1672. But over the course of the next century, the harbor became clogged. Naturally shifting sands were likely the primary culprit, but a formal request to widen what had become a trickling tidal creek into a legitimate harbor entrance illuminates the ways human actions had transformed not only the tidal lagoon but the coastal sea as well. When Edmond Sheffield and Joseph Spencer petitioned the General Assembly in 1762, they explained that when the salt pond and the sea were connected, "the fishing ground for cod was well known, and bass was there to be caught in great plenty." They were adamant, however, that "since the creek has been stopped, the fishing ground for cod is uncertain, they being scattered about in many places; and the bass have chiefly left the island." They testified that if "this communication can be made, the fishery will again become sure and certain."[127] Sheffield and Spencer's petition reveals that harbor improvement to facilitate fishing and livestock exportation during the seventeenth century had established a rich fishery. But when the estuary became clogged—whether from shifting sands or the introduction of mud, animal waste, and other suspended solids from the cow pastures—that fishery declined. Terrestrial and marine environments were inextricably linked. And humans held the power to shape them.

When large-scale environmental problems arose—particularly those concerning waterways—colonists increasingly turned to the Rhode Island General Assembly for remedies. From the colony's beginnings, Narragansett Bay was a common geographic point of reference and shared means of transportation for the people who lived there. In 1643 the four original towns of the Bay—Providence, Portsmouth, Newport, and Warwick—entered into a federal compact in part to protect themselves from neighboring Massachusetts Bay and Connecticut. As such, they imbued the Bay with the power of central authority. Physically bound together by brackish arms of the sea, Rhode Islanders, who neither fished nor shipped ocean-going cargo at levels even close to those of Boston, adopted the anchor as the most prominent device in their colonial seal.[128] But the emphasis on protection overshadows the ways these littoral people also sought progress along their shores and therefore sought one another's support to improve them. Although modifying the coastal zone had become a common part of Rhode Island's littoral culture, it was also complex and expensive. By unifying, Rhode Islanders certainly staved off their aggressive neighbors, but they also established a common pool of resources that transformed the coastal environment in which they lived. In the case of Westerly's harbor, the townspeople understood that their breach could only be repaired if they secured considerable funds. Having lost some of its infinite qualities, the sea—or at least the coastal part of it—could, through federal mobilization of money and manpower, be reshaped to meet the collective needs of the community.

Although Rhode Island's central authority was an important arbiter of environmental change, individual towns nevertheless maintained control over their own waterfronts. On May 28, 1707, the General Assembly passed a law that granted each town "full power and authority to settle such coves, creeks, rivers, waters, banks, bordering upon their respective townships . . . by building houses warehouses, wharfs, laying out lots, or any other improvements."[129] By allowing individual towns to "improve" as they pleased, the law caused, in some cases, environmental problems. In October 1719 the General Assembly passed an act urging town councils "to take care to preserve and improve the fishing of the several rivers in their respective jurisdictions, and to prevent obstructions from being made, to hinder the same."[130] Assembly members recognized that when towns were left to their own devices, wanton development and reckless fishing harmed not only those towns in question but also the colony as a whole. In 1755 Major Ebenezer Brenton and others of South Kingston presented a petition to the General Assembly to "prevent people from fishing with seines in the breach and channel" of Point Judith Pond. Fished "at all seasons," the seines, Brenton claimed, had led to a situation in which "the course of the fish is daily obstructed." So acutely were fish stocks affected, complained Brenton, that "if not speedily prevented [the fish] will be totally turned some other way, to the great damage not only of the said town, but of the greater part of this colony." To remedy the situation, the General Assembly agreed to increase the fine, which in June 1736 was levied at 40 shillings—an amount so low that fishermen paid it as a cost of doing business—to £50.[131] A local environmental concern, it was recognized, could impact the entire colony and so required General Assembly action.

By the middle of the eighteenth century new environmental pressures had begun to change Narragansett Bay, albeit in fits and starts. The human population had increased significantly and was continuing to rise. The animal population grazing the Bay's shores was one of the largest in New England. Human waste and livestock manure fertilized not only coastal farms but also the Bay itself. Some of the most sheltered inlets and salt ponds became so suffused with algae that the water, if the survey of Henry Bull's lot was any indication, turned green. In response, a loose association of littoral towns began to aggregate their efforts.

In no way immune to improvement, the Bay bowed to human demands. In some cases, environmental changes threatened to close harbors. The people who depended on them responded, perhaps brashly, with environmentally transformative solutions. Although in the case of Westerly and Block Island these fixes improved harbor accessibility, the ecological ramifications were profound. The Anglican minister MacSparran expressed deep-seated concern that in Rhode Island's intense focus on livestock production for international trade, it had failed to improve its shores. His adopted colony's fields and even its people, he felt, had sadly gone to seed. What he failed to recognize, however, was that as the autumn trees shed their leaves and rivers rose, as melting winter snow coursed into stream beds, and as spring storms flooded coastal pastures, the people and animals of Rhode Island had actually done a great deal to change the place. By reshaping their harbors and mowing their meadows and organizing themselves politically, legally, and economically, they had added a new level of order to the edge of the sea on which they depended.

Pockets of progress notwithstanding, the complexities of the coast sometimes stymied their efforts. Harbors became clogged time and time again. Conservation laws remained difficult to enforce. And political agreement was hard to come by. But most of all, the Bay's deeply ambiguous geography came to challenge Rhode Island's sovereignty. This was something that by the middle of the eighteenth century colonial officials had resolved to change.

CHAPTER 3

The Geographic Quicksilver of Narragansett Bay

As the smoke lifted following King Philip's War, in 1678 or 1679 a curious man with a curious name settled at the head of Pettaquamscutt Pond at the northern tip of the eponymous tidal river, the original seat of the Narragansett sachem Miantonomi. According to Yale President Ezra Stiles, who at the end of eighteenth century interviewed a neighbor, the man known as Theophilus Whale "built himself a little under-ground hut in a high bank, or side hill, at the north end or head of the pond."[1] He subsisted, Stiles noted, by fishing, weaving, and writing.[2] Whale spoke Hebrew, Latin, and Greek and revealed little of his past to anyone, even his children. His neighbor, Willet, remembered that his own father, with a group of notable Boston gentlemen, periodically visited Whale in the evenings, whereupon they usually talked behind closed doors and sometimes gave Whale a little money. During Queen Anne's War, from 1702 to 1713, a captain of an English warship who shared the surname Whale called upon his reclusive countryman and invited him to dine aboard his vessel. But fearing a "snare laid for him," the Whale of Pettaquamscutt declined.[3]

For the people of Narragansett, this was proof that the enigmatic hermit was in fact Edward Whalley, one of the three surviving judges who had condemned Charles I to death, and who, upon the Restoration, had fled to America to hide amidst the swamps, marshes, and meadows of Narragansett Bay. Whale's peculiar erudition, his close associations with men of prominence, and even the similarity between his name, Whale, and that of the judge, Whalley, led the people of Rhode Island, according to Stiles, to "uniformly believe" that Theophilus Whale was the regicide. Although Stiles ultimately concluded that Whale was most likely not the judge, the story—or even mystery—underscores the extent to which the bay borderland was, at least according to popular belief, a geography capable of camouflaging identity or even harboring a fugitive. It was a landscape, or rather waterscape, where a man could hide his true self.

That the supposed outlaw Whale sought safe haven among the watery folds of Narragansett Bay reflects the ways the estuary's complex ecology and indeterminate geography often resisted definition. Living at the edge of a seemingly endless maze of salt creeks, Whale eluded his own past at the edge of settled society and the sea. Flood, flux, and shifting sand—these were the characteristics of an estuary in motion. And although stone walls, stump rows, and pickets ran through coastal forests and into the black grass meadows that sprouted along the farthest reaches of the tide, the Bay defied most attempts to impose nomenclature and erect jurisdiction among its waves.[4] That Whale's good neighbors waxed wary of his protean identity reflects a broader anxiety over the porous and mutable bounds of the Bay by which he had settled and they had lived all their lives. Anything but fixed, the littoral was physically and conceptually slippery, geographic quicksilver prone to skidding away from any surveyor's protractor that tried to pin it down.

As a result, Rhode Island, in the belly of which lay one of the largest estuaries in settled North America, had been mired in territorial disputes since Roger Williams had pitched his tent on the wrong side of the Seekonk River. By its watery, indeterminate nature, in its location at the nexus of the known and unknown, of the definite and indefinite, Narragansett Bay was in every sense a borderland, a place that blurred identity, defied ownership, and marked the "contested boundaries between colonial domains."[5]

Colonial competition was particularly bitter. During the seventeenth century, the English inhabitants of Rhode Island sparred with Connecticut over their shared boundary along the Pawcatuck River. So acrimonious grew the dispute that in 1670 some inhabitants of the Bay borderlands resorted to murder, which precipitated a chain of violent outbreaks.[6] Although relations between the two colonies were fraught

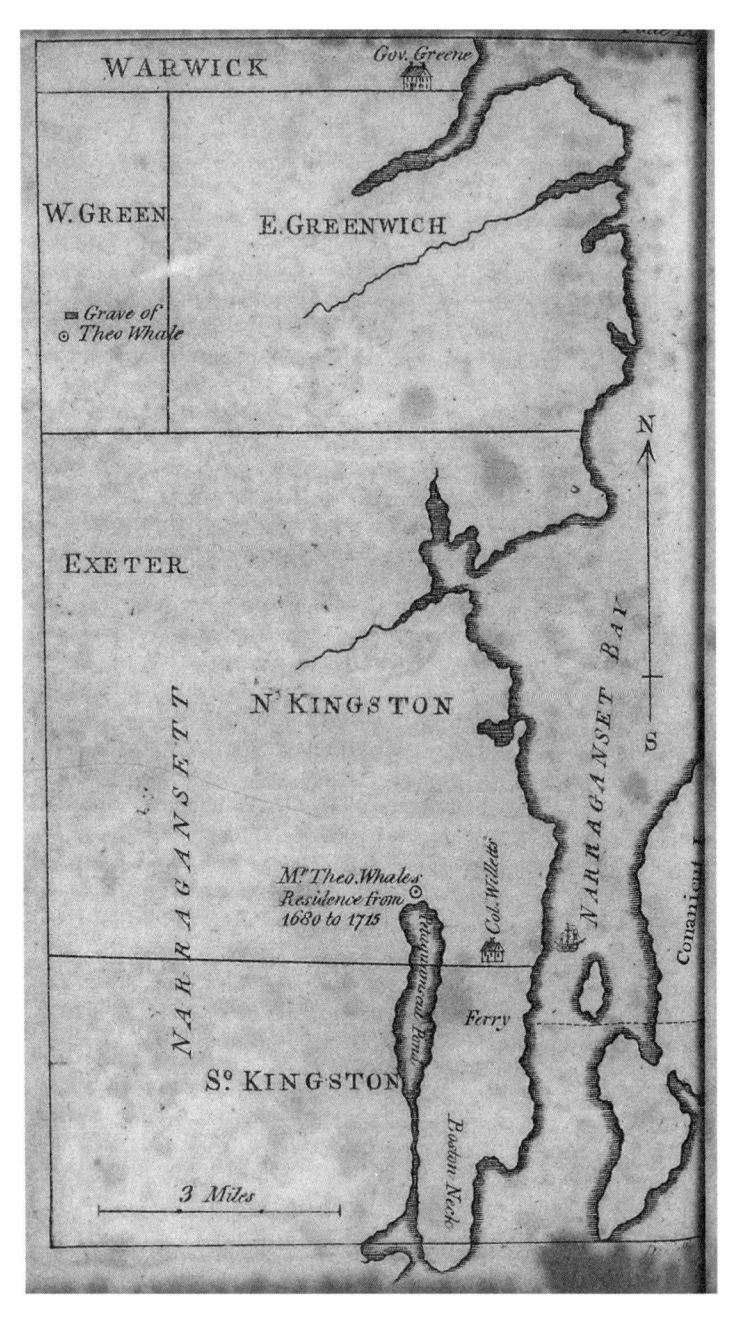

The home of the mysterious Theophilus Whale is marked at the head of the Pettaquamscutt River in the lower middle of the map. From Ezra Stiles, *History of Three of the Judges of King Charles I* . . . (Hartford, CT: Elisha Babcock, 1794), 344. Courtesy of the John Carter Brown Library at Brown University.

with contention, they began working toward a solution and made a tentative agreement in 1703. When that failed, however, the Board of Trade in London became so exasperated that it recommended their charters be withdrawn and their lands annexed by New Hampshire.[7] Naturally, this spurred renewed efforts to settle the line, but it was not until 1742 that the colonies reached a firm agreement. Maneuvering for lands in the former Pequot territory, on Block Island, and in the Narragansett Country, Massachusetts was never far removed from the debate.[8] Although the Bay Colony's claims to lands in Rhode Island's western precincts (and Block Island) were ultimately denied, the two colonies would come to blows during the middle of the eighteenth century over their shared border to the east.

If good fences make good neighbors, a maze of marshes and meandering tidal rivers makes for litigation. As Rhode Island's border deliberations with Connecticut finally wound down, the former began new court proceedings in 1741 with Massachusetts to decide who owned the eastern shores of Narragansett Bay. Citing its 1663 royal charter, Rhode Island claimed the lands "extending . . . three English miles to the east and north-east of the most eastern and north-eastern parts of . . . Narragansett Bay," which, Rhode Island contended, Massachusetts had acquired unjustly when it absorbed Plymouth Colony in 1691. Massachusetts rejected its neighbor's claim, arguing that the contested space did not actually touch Narragansett Bay. The question of who held jurisdiction, therefore, depended on Narragansett Bay's location.

But where was Narragansett Bay? As the ensuing court battle proved, the boundaries of this coastal space were anything but clear. If the people of Rhode Island had begun to organize the ocean's edge materially, they had yet to organize it geographically. More than a century after Europeans had settled the region, the people who lived there still understood its soggy landscape in very different ways. Informed by political allegiance, historical understanding, and tensions between imperial mandate and Native American and European vernacular knowledge, among other factors, the geography of Narragansett Bay was a deeply human construct. To define their shared border, both colonies called scores of deponents from all walks of life, including, among others, merchants, sailors, coastal farmers, and Indians, who explained to the court how they understood the boundaries of land and sea, saltwater and freshwater, and in many cases how those ideas had changed over time. The proceedings, held mostly in the high heat of summer, were long and messy.

The testimonies reveal that the people of Narragansett Bay developed cultural assumptions, patterns of settlement, and notions of jurisdiction that mirrored the physically and conceptually murky environments in which they formed. As fluid as a flooding tide, the borders had never been and would never be set in the proverbial stone. Rhode Island's eastern boundary was ultimately chosen and agreed upon because English arbitrators were willing to entertain the idea that the arms of the sea were spaces that could be modified by human effort. Well known to the people who moved among them—fishermen, sea captains, and Wampanoag Indians—the salt rivers and creeks of eastern Narragansett Bay were spaces capable of supporting work and therefore spaces capable of upholding jurisdiction. Nevertheless, these decisions were made with great ambivalence. After all, the boundary between Rhode Island and Massachusetts was defined in relation to the edge of the Bay, and to run that line required one to define the ever-shifting edge of the sea. To a littoral community that willfully accepted the myriad identities of Theophilus Whale, that saw the world not as black or white or wet or dry, but in many muddy shades of gray, the process of drawing boundaries was fraught with challenges. But by sounding the depths of historical memory, by confronting past anxieties over littoral division, they erected a more durable boundary between colonies and a new perceived edge to the Bay.

Defining the Bay

On the first Tuesday of April in the year 1741, representatives from New Jersey, New York, and Nova Scotia were ordered to convene in Providence, "as being the most conveniently situated," for "settling adjusting and determining the Boundaries of . . . [the] Colony of Rhode Island in America eastwards."[9] Although the two colonies had made earlier attempts to settle their shared border, their efforts had met with mixed results. Massachusetts claimed that (through its annexation of Plymouth) "they have always enjoyed" a boundary that began where it bisected the mile-wide salt river that separated Rhode Island (Aquidneck) and Little Compton "where it runs into the Main Ocean." The line, according to officials at Massachusetts Bay, ran north "up the middle of the said River to the . . . Seaconk River, and from thence up the . . . Patucket River," also called the Blackstone, north to a "heap of stones on the East Bank." There, forming Rhode Island's northern border, it ran due west along a "Line of mark'd Trees and Monuments of Stones through the Wilderness to the Colony of Connecticut."[10] But citing its 1664 Royal Charter, Rhode Island disagreed, claiming its "just and lawful right to the Jurisdiction of all the lands lying within and bordering on the Narragansett Bay from three English miles East North East." Rhode Island officials argued their claims "extended from a place called Assonet," which they believed was in the "most Eastern and North Eastern part" of Narragansett Bay and followed the contours of the Bay's eastern shore three miles inland in a southerly direction "to the Ocean." From Assonet, they also drew a line west to Providence. The boundary, Rhode Island claimed, then followed the "Easterly Side or Bank" of the Seekonk River to Pawtucket Falls, where it then ran due north to its northern border with Massachusetts.[11]

The dizzying descriptions aside, disagreement over the boundary arose because neither party could agree on the location of an arm of the sea. This impasse was rooted not in geographic ignorance—representatives from both colonies clearly understood the extent of and relationships between the region's numerous waterways—but in an inability to establish an ocean vocabulary. When it came to imposing all-important definitions on coastal space, a fixed lexicon had, by the middle of the eighteenth century, yet to be established. Of global hydrography, geographer Martin W. Lewis has written that continents and oceans have been largely considered "nonproblematic features of the natural world, features that have been discovered rather than delimited by convention." But the extent to which continents and oceans are broken into their individual units, he explained, are "as much intellectual constructs as they are given features of the natural world."[12] Even as late as the mid-eighteenth century, geographers used place names interchangeably. For instance, Emanuel Bowen's 1744 *World Atlas* alternated between "Southern Ocean" and "Atlantic Ocean," and numerous other cartographers alternated between "Pacific Ocean" and "South Sea" to refer to the ocean basin between the Americas and Asia. Not until the nineteenth century did geographers begin to codify spatial terminology, particularly that pertaining to the ocean.[13]

Coastal nomenclature was equally murky. The definition of a "bay" proved to be a stumbling block for deponents on both sides of the line. And the numerous bays nested within greater Narragansett Bay made deliberations downright confusing. That competing definitions emerged, however, was understandable, for among medieval and early modern English writers the word "bay" had many connotations. The Oxford English Dictionary attributes the word to the late Latin *baia*, circa A.D. 640, and later the French *baie* meaning simply "an opening."[14] But use of the word doesn't appear in writing among English authors until the Benedictine monk Ranulph Higden published his *Polychronicon* in Latin in 1385, describing a "grete mouthe and baye" that forms the Aegean.[15] During the fifteenth century, following the siege of Calais, a political poem outlining English policy concerning "the see enviroun" extolled the promise of colonizing Ireland when it explained "they have . . . grete and godely bayes Sure, wyde, and depe."[16] For Higden, a bay was a broad, deep, piece of the ocean bounded by islands and promontories. The unknown author who penned "The Libel of English Policy" defined a bay as an interface with land. Here, the importance of a bay rested in its ability to extend imperial dominion and facilitate commercial extraction. Even the Great Bard of Stratford vacillated between definitions over the years. In *The Merchant of Venice* (1600), Shakespeare depicted bays as shelters from the storm:

> The scarfed bark puts from her native bay,
> Hugg'd and embraced by the strumpet wind!
> How like the prodigal doth she return,
> With overweather'd ribs and ragged sails,
> Lean, rent, and beggar'd by the strumpet wind![17]

Here, a bay represented home, safety, and even purity from which the bark departed for a world of temptation. Outside of the bay, however, the lure of the sea stripped even the most prepared—"scarfed"—sailor of innocence. But in *As You Like It,* written about the same time but not published until 1623, Shakespeare's Rosalind explained her love for Orlando was so deep that "it cannot be sounded: my affection hath an unknown bottom, like the bay of Portugal."[18] Here, the bay is something unfathomable, an abyss incapable of measurement. If Shakespeare defined a bay only loosely in the early seventeenth century, Daniel Defoe was decidedly more descriptive a century later. After having "coasted the shore," Robinson Crusoe "came to a very good bay about a mile over, which narrowed till it came to a small rivulet, where I found a convenient harbour, and where she lay, as if she had been in a little dock, made on purpose for her: here I put in, and having stowed my boat very safe, went on shore to see where I was."[19] For Crusoe, the bay served as a safe haven. But it also had specific geographic features. At its mouth, his bay was a mile across and narrowed as it extended inland until a freshwater stream tumbled into the upper reaches of what was undoubtedly an estuary. There, at the head of Crusoe's bay, was a harbor perfectly suited to his and his ship's needs. In this sense, Defoe agreed with his predecessors that bays reflected human needs and desires. Sustenance and security, both physical and psychological—these were the threads that braided the bays of Hidgden's history, colonial policy, adventurers and lovers, and a lonely shipwrecked sailor together.

But when competing strands of self-interest were introduced, when the demands of private property required a geographically finite description of a bay, little consensus could be had. That Massachusetts Bay, the colony, was named after a broad arc of shoreline that in no way resembled the more penetrating Narragansett Bay likely added to the confusion.[20] Nevertheless, the authorities from both colonies charged with laying a line between them had little patience for the poetics of space. Rather, they sought definition among the islands, salt rivers, and harbors. A bay was no metaphor for unrequited love. Rather, it was a place where people lived, worked and, as everyone involved was well aware, paid taxes. For Massachusetts, Narragansett Bay lay west of Aquidneck Island; the Sakonnet River that flowed along Aquidneck's eastern shore was simply that—a river. For Rhode Island, which sought a large chunk of Plymouth's former territory, Narragansett Bay included the saltwater Sakonnet River and lands to the east of it. So confusing were the claims that the court needed professional help, and specifically, a map. On April 30, 1741, the court appointed three surveyors, Cadwallader Colden of New York, James Helme of South Kingstown, Rhode Island, and William Chandler of Connecticut, and on May 2 commissioned them to describe and draft maps of the Rhode Island coast and islands "to illustrate the Bounds in controversy."[21]

But the dispute over the boundary had been longstanding. In 1663, even before Rhode Island had received its charter, and again in 1664, commissions convened to determine the boundary between Rhode Island and Plymouth.[22] But they made little progress. In 1666 Rhode Island complained to the Crown that their neighbors "could not content themselves but incroached upon this small corner not only dispossessing, molesting, captivating and fining Your Majesty's liege people here living, but also claiming all the country by strange pretences of free purchases and gifts, by forced Mortgages from the Indians therein."[23] Through shady deals and slippery language, Rhode Island's neighbors, colony officials complained, had usurped their lawful property. Rhode Island felt so threatened by its English neighbors that it felt compelled to pen a defense. In a letter to the Earl of Clarendon dated 1666, Rhode Island officials explained that they had "intrencheth not on Plimouth" because the "Narragansett River," in relation to which Plymouth's patent identified its boundaries, did not exist. They also explained, "Rhode Island lyeth as inclosed and in a manner embayed within the Land" and that it was "therefore good reason that the main land inclosing and so near adjoining to the Island should pertain to it." Their description of the space suggests that even within the first generation of European settlement, Rhode Islanders saw Narragansett Bay as an integral part of their colony's cultural and geographic identity. It also suggests that they conceived of a bay as one continuous body of water comprising myriad rivers and smaller bays. But ultimately, they deferred to the King, who, in his "express words" as codified in the charter, granted Rhode Island the lands extending "three miles to the East of the most Easterly and North Easterly part of the said Bay." They explained to Clarendon that because Aquidneck Island was so small, its inhabitants were forced to pasture animals on the nearby mainland. Some Rhode Islanders had even built farms there. And finally, they explained, "the land hath other ways never been improved by Plymouth but it hath lain waste near forty years since they first began that Plantation."[24] In sum, Plymouth had failed to cultivate the Bay's shores, whereas Rhode Island farmers had with great industry and expense made their mark there, which pointed toward Rhode Island ownership. In other words, he who improved the estuary should own the estuary. But in fact the watery boundaries between the colonies could, literally, never be cut and dried.

Nearly a century after the protracted quarrel between Rhode Island and Plymouth began, the boundary was still blurry, and the people who lived near it were still angry. In response, the King's commission demanded the line be drawn once and for all. Ordered to "make return thereof at the first sitting of this Court," Colden, Helme, and Chandler, the Boundary Commission's surveyors, set to work immediately in May 1741. Per the court's request, they began on the Pawcatuck River at Rhode Island's western boundary. According to local custom, they coupled their survey with a narrative description of the process, a relatively new American literary form called "geodesy" that had come into widespread practice after 1690 when Parliament had revoked the New England charters, thereby requiring everyone to survey their lands.[25] Per their own prerogative, they described what they saw in verse. Clearly designed to win public appeal, the 177-line poem was printed as a broadside, though likely in limited numbers, as only two known copies still exist.[26] It is unknown whether the poem grew out of a more formal report or whether this was their sole narrative submission to the boundary commission. By order of the court, they did, however, also "compleat a plan of all that part of the Continent which the Inhabitants of the Province of the Massachusets Bay apprehend doth belong to the Colony of Rhode Island."[27] Drawn to the scale of 150 chains to an inch and filed on June 24, 1741, the map was quite detailed, including the location, for example, of a small cluster of rocks at the tip of Rumstick Point in Barrington that might have been omitted had the their work been less exact.[28] But beyond the poem and the plan, no other account of their survey exists, and the expediency with which they conducted their assessment—the survey took no more than six weeks if they began the second week in May and submitted their map during the third week of June—suggests the map was their first priority and the poem

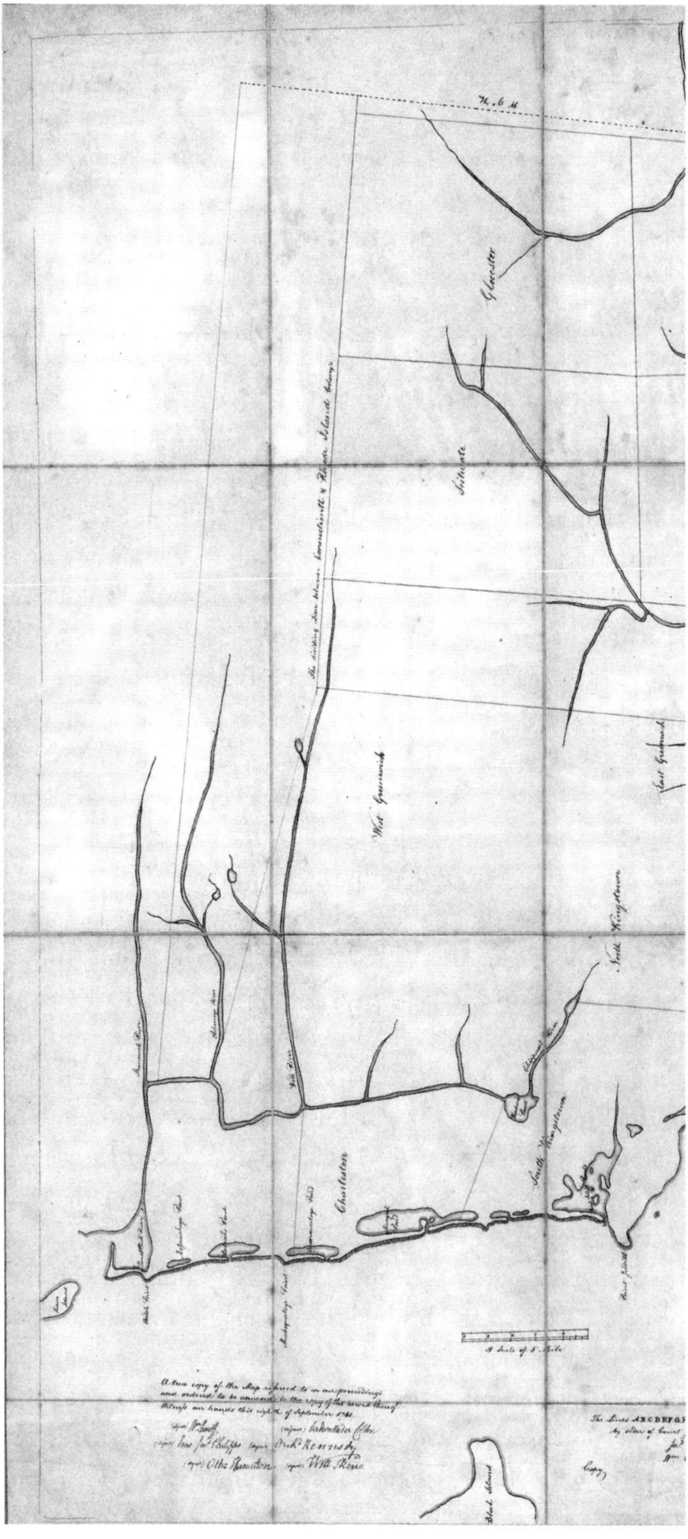
West Greenwich
North Kingstown
South Kingstown
Charlestown
Black Island
A Scale of 5 Mile
A true copy of the Map referred to in our proceedings
and ordered to be annexed to the copy of the record thereof
Witness our hands this eighth of September 1741.
Copy

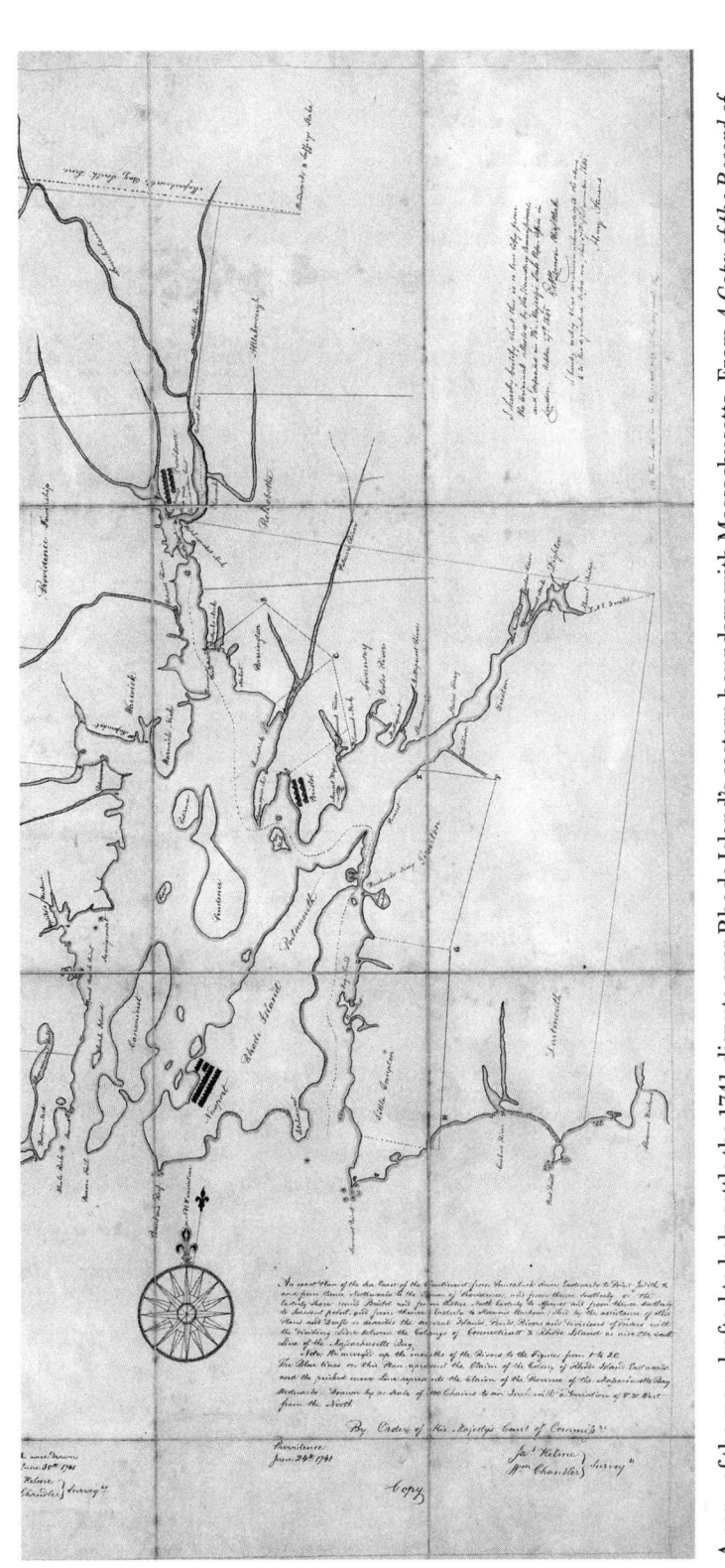

A copy of the map drafted to help settle the 1741 dispute over Rhode Island's eastern boundary with Massachusetts. From *A Copy of the Record of the Proceedings of the Commissioners for settling adjusting and determining the Boundary of the colony of Rhode Island and Providence Plantations Eastward toward the Province of the Massachusetts Bay*, Henry N. Stevens copy (1845) produced from James Helme original. Courtesy of the John Carter Brown Library at Brown University.

simply a way to justify its expense.[29] Although they did not weigh in on where the lines of jurisdiction should lay, these impressions nevertheless paint a detailed picture of Narragansett Bay in its entirety during the late spring of 1741. In some stanzas, they even hinted of the ways quantitatively minded surveyors were forced to grapple with the metaphysical complexities of the littoral.

Chandler and his crew favored narrative description over philosophic essay, and as a result their *Journal* followed a simple chronology. In what was likely a shallop or small pinnace—light, maneuverable inshore vessels—Colden, Helme, and Chandler left the village of Pawcatuck during what was most likely the second week in May 1741, sailing south past salt meadows and cleared upland fields. Chandler's poem noted:

> From *Pawcatuck* we steer'd our Course away,
> And to *Watch Hill* we went without delay,
> Which gave a Prospect of the Neighbouring Shore
> And distant Isles, where foaming Billows roar.[30]

Sailing east toward Watch Hill, their account explains, they could see across the shallow entrance to Little Narragansett Bay toward Stonington, Connecticut, and beyond to Fisher's Island, Montauk at the end of Long Island, and Block Island to the east. The shores of southwestern Rhode Island, they explained, were dominated by sandy shoals and beaches pummeled by heavy surf. So powerful were the "Raging waves" rolling into the southern end of the Narragansett Country that they forced the surveyors to sail through strong rip currents funneling from "the Breaches in our way / Made by the Surges of the raging Sea."[31]

Beyond the surf and over the dunes, however, the surveyors witnessed a placid estuarine world that dominated so much of Rhode Island. The team recalled:

> Where in the Land Calm Ponds we here espy'd
> Which rise and fall exactly with the Tide.
> Within these ponds are Fish of Various Kind,
> Which much delight and please[.][32]

Behind the barrier beaches and across the coastal fastlands was a tranquil world of tidal salt ponds. The rookeries of the sea, these ponds teemed with fish of all sizes. They would have seen black clouds of minnows scudding across the shallows, and in May and early June would have seen vast hatches of clamworms, red and white legged polychaetes, floating through the water column, particularly at night. In frothing schools at the water's surface, juvenile striped bass feasted on the worms and small baitfish. Huge runs of flounder and fluke did the same across the bottom. Buzzing with insects and crawling with crustaceans, broad Spartina grass meadows played host to the frenetic darting of countless sparrows and wrens. Egrets, herons, willets, ibis, rails, and oystercatchers would have poked among the weeds and rushes of the shallows and sand banks. So filled with fowl were these salt marshes that the surveyors noted the "Industrious Archer" would surely reap rich rewards.

Although their survey was meant simply to describe the Rhode Island shore, the style of their observations nevertheless illuminated the ways the edge of land and sea was anything but clearly defined among these tidal lands. While exploring the salt ponds of the Narragansett Country on horseback, Chandler noted, parenthetically:

(Here in a Pond, our Caution to oppose
A Horse did launch and wet his Owners Cloaths,
The frighted Jade soon tack'd himself about
Which made us laugh as soon as he came out.)

In some areas, the passage explains, the surveyors used horses to explore this world of saltpans and ponds, marshes and meadows. To travel by boat, Chandler suggested, was in some areas too cumbersome. So too was overland travel on the spongelike peat, which denied their horses secure footing. While wading through the estuary, one horse, the author explained, stumbled and threw its rider into the mud and muck. Navigating the hybrid environment of the littoral, the horse, according to Chandler, became a sea-going vessel that "tack'd himself about." Wide-eyed and wet, the horse had tripped off the edge of the earth and, panic-stricken, blundered into the liminal space between land and sea. But when it had reached dry land and its fear—and perhaps even that of its owner—had subsided, land and sea were again separated (at least in the horse's immediate vicinity), which spurred a cathartic outpouring of laughter and, doubtless, relief.

Colden, Helme, and Chandler then sailed into Narragansett Bay. Traveling east to Point Judith, they stopped and surveyed the Point and "the courses there" and then sailed around it and north to Boston Neck, the long peninsula constituting the eastern shore of the Pettaquamscutt River. They observed the island of Conanicut across Narragansett Bay's West Passage, as well as "other Parts too tedious . . . to tell." Sailing north toward North Kingstown, they sailed "round points of Lands and Coves / Thro' various Fields and most delightful Groves." Continuing in that direction, they passed on their way to Greenwich, "*Hope* and *Prudence* that most pleasant Isle / And *Patience* also, a most fruitful Soil." They were particularly taken with Warwick, which they described in glowing terms:

> And in that Town did of their Dainties eat
> And in soft Slumbers pass'd the Night with Sleep.
> Here neighbouring Orchards in their verdant Blooms
> The gentle Air Sweetens with their Perfumes;
> Which pleasing Prospect did attract our sight
> And charm'd our Sense of smelling with Delight.

The survey team ate a rich meal in this small but bustling port town, and lulled by apple and peach blossoms, slept the night before heading north to Pawtuxet, which divided Warwick and Providence. "Passing along still by the flowing Tide," they sailed to the "Famous Town of Providence." Upon approaching, they made the careful observation that "This Pleasant Town does border on the Flood," which suggests the extent of the tide marked an important point of transition in the debate over jurisdiction.[33] In fact, Rhode Island's colonial charter stipulated that Rhode Island held jurisdiction three miles east and northeast of Narragansett Bay, but that when the bay became a river, new rules applied. Once the surveyors reached "the mouth of the [Seekonk] river which runneth towards the towne of Providence," Rhode Island held jurisdiction only as far as its eastern bank.[34]

Chandler noted how the city's placement at the head of the tide gave it a distinct character. The survey team noted that the city was at once "Nature made" but "(with Art allied)" the people of Providence capitalized on their city's propitious location and built a successful "Place of Trade." Strong drink, including "Wine in Bowls," was readily available to "chear . . . hungry Souls." So liberal was this tidewater trade port that "Here Men may soon any Religion find." It was the cultural diversity and lax laws of Providence combined with the port's growing commercial success that brought Europe's great littoral nation, "brave *Holland* to my Mind."

Next, the survey team explored the east Bay. They sailed past Seekonk and Barrington and continued south past the mouth of the Warren River to Bristol, where they "turn'd a while to rove" Poppasquash, or Papoose-Squaw, Point, a long peninsula forming the western shore of Bristol Harbor and named for the place to which women and children fled during times of war. They walked through a shaded Black Cherry grove that was so impressive "Methinks young Lovers here with open Arms Need no young Cupids to inspire their Charms." Across the harbor, they explored the town of Bristol, where "Generous Hearts did give their liberal Treats" but met a women who shocked them with such "impious Talk" that "Her gravel'd Notes . . . made some of us smile." They sailed from the busy Bristol waterfront to Hog Island at the mouth of the harbor and northeast toward Mount Hope, where they climbed the hill toward "the Royal Spring / Which once belong'd unto an Indian King." There they surveyed the seat of King Philip and "saw the Place where quartered he did hang." They described the view from the top of Mt. Hope, the point of highest elevation along the Bay's eastern shore:

> Upon this Mount the wandering Eye may gaze
> On distant Floods, as well as neighbouring Bays
> Where with one Glance appears Ten Thousand Charms
> With fruitful Islands, and most fertile Farms.

Chandler made little distinction between New England's most prolific gardens and the bays, rivers, and creeks that surrounded them. From on high, the littoral formed an integrated matrix of water and land, flood and farm. So seamless were the bay and its shores that when the survey team sailed north to Assonet, which both Rhode Island and Massachusetts had agreed was the far extent of Narragansett Bay, they failed to comment. As such, they simply "turn'd about new Courses now to steer" and sailed south.[35]

Passing the eastern shore of Aquidneck Island, they brought their cruise to a close. Stopping at Sakonnet, they heard a tale of horse thieves, whereupon they traveled around Sakonnet Point and into the ocean "where Dreadful Billows roar." Chandler "survey[ed] the tossing Sea" and then "turn[ed] and view[ed] the Beach and Sands." Sailing east, they anchored in Dartmouth, which they described as "a most liberal Town / Whose liquid Treats their generous Actions crown." No doubt with a drink in hand, Chandler concluded his survey by noting that they sat down with their field book "to make a Plan."[36] Although the survey book has since been lost to history, the "plan" or map survives.

The Bay Borderlands

If Colden, Helme, and Chandler's map described the physical geography of the Bay, their poem highlighted at least some of the ways that borders and borderlands were deeply human and widely variable constructs. When shaped by politics and drawn by convention, borders often followed artificial lines imposed across a given space. But in the case of Narragansett Bay, those manmade lines often became entangled with the natural contours of the landscape. Along the Blackstone River, the border between Rhode Island and Massachusetts followed the flow of the river northwest to Pawtucket Falls, but then veered due north in a line that, ignoring the existence of hills and cliffs, was only discernible to the needle of a compass. As political scientists have shown, borders even constituted entire regions or zones. Beyond the control of any one group, these middle grounds or neutral zones shaped a permeable but mutually recognized borderland between distinct groups of people.[37] At once political and natural, porous and impenetrable, the marshes and broad bays that stretched from Massachusetts, across Rhode Island, and to the boundary with Connecticut became at various times all three types of boundaries.

During the seventeenth century, the Bay often served as a border. The Wampanoag and Narragansett Indians had long observed the Bay as a natural barrier between them, an arrangement that continued with the arrival of Europeans. "*Narragansett-bay* and river, which borders upon us," wrote John White in 1630 from a fishing station in Cape Ann, "is full of Inhabitants, who are quiet with us, and Trade with us willingly, while wee are their neighbours, but are very jealous of receiving either us or the *Dutch* into the bowels of their Country, for feare wee should become their Lords."[38] White described a border based on natural features and social differences. At the time, English jurisdiction in Plymouth Colony ran to the eastern shores of Narragansett Bay. Separated by a body of water, numerous islands, and myriad rivers and marshes, the Narragansett Indians and English settlers conducted trade and even lived on good terms, but their territory was clearly defined. Such "jealousy" no doubt strongly suggested to neighboring colonial powers that their presence was not welcome on Indian land. But the Bay's role as a boundary was short lived, and over the course of the seventeenth century it developed into a porous, uncontrolled region of movement and exchange.

The canoe played a vital role in the Bay's transition from boundary to borderland. It was by canoe that Roger Williams first landed on the place that would become Providence in 1636, and throughout the seventeenth century the canoe was integral to any form of travel on Narragansett Bay and throughout coastal New England. In nearby Massachusetts, William Wood was so struck by the sheer number of canoes he noted that "every household [had] a waterhorse or two."[39] By 1647 there were so many watercraft in Providence that the town government hired "Water-bailies" to patrol the shores.[40] An emphasis on the ways the English wielded guns and disease to wrest control of Indian lands has all but eclipsed the extent to which, at least for a short time, Indians, with their deft ability to navigate coastal waters by canoe, maintained the upper hand. When Roger Williams purchased Providence outright from Canonicus and Miantonomi, it was well understood that the Bay was still Indian territory. When the two sachems later sold Aquidneck Island to William Coddington in 1637, Canonicus and Miantonomi attributed their right to sell it "by virtue of our [the sachems'] general command of this Bay."[41] Although we can't assume that the sachems devised the contract's wording, the statement does highlight English acknowledgment of Indian rights concerning waterways.

One of the reasons Indians maintained control of the Bay during this period of early English settlement was that Native Americans produced most of the canoes and were usually more skilled in their use. In 1643 Roger Williams remarked, "It is wonderful to see how they [Indians] will venture in those Canoes, and how (being oft overset as I myself have been with them) they will swim a mile, yea two or more safe to Land."[42] In his second voyage to New England John Josselyn marveled in 1663 at the ways "the bold *Barbarians* in their light *Canows* rush down the swift and headlong stream with desperate speed, but with excellent dexterity, guiding his *Canow* that seldom or never it shoots under water or overturns."[43] Although Native Americans controlled Narragansett Bay, at least initially, the English quickly gained proficiency. And the canoe became the primary means by which Indians and the English alike navigated the political tensions that came to shape their estuarine borderland.

At times the canoe was even an important tool of diplomacy. In 1675 during King Philip's War, in a letter recounting a conversation with a Narragansett sachem, probably Canonchet, Williams was careful to note, "(being then in my Canow with his men with him) that Phillip was . . . deafe to all Advice and now was overset and Catcht at every Part of the Countrey to save himselfe but he shall never get ashoare."[44] From the safety of his canoe with Canochet and his men in theirs, Williams sought to win the allegiance of a potential hostile. The canoe made the Narragansett territory permeable. Under the power of a paddle, Europeans and Indians glided silently and at times even effortlessly through a contested space that a little over forty years earlier had been "jealously" guarded.

There were places among the swamps surrounding Narragansett Bay that even canoes could not go, and for borderland people these often became geographies of refuge. Throughout the seventeenth century, Narragansett Bay's impenetrable swamps often served as safe havens during times of conflict. Williams noted, "These thick Woods and Swamps (like the Boggs to the Irish) are refuges to the women and children in Warre, while'st the men fight."[45] But even fighting men used the swamps for tactical advantage. According to William Hubbard's 1677 account of

King Philip's War, when Indian forces retreated into the "great swamp upon *Pocasset-Neck*," Captain Henchman, who led the Plymouth forces, was "not willing to run into the Mire and Dirt after them in a dark Swamp, being taught by late Experience how dangerous it is to fight in such dismal Woods, when their eyes were muffled with the Leaves and their arms pinioned with the thick Boughs of the Trees, and their Feet were continually shackled with the Roots spreading every Way in those boggy Woods." Dark and difficult to navigate without considerable local knowledge, the swamps gave the Indians the upper hand. But true to his sinister surname, Henchman, instead of pursuing his foes on such difficult ground, surrounded Philip's forces and resolved to starve them out.

But if bogs and swamps served to conceal Philip's men, the Bay itself provided a means of escape. "The Swamp where they were lodged," recalled Hubbard, "not far from an Arm of the Sea, coming up to *Taunton*, they taking the Advantage of a low Tide, either waded over one Night in the End of *July*, or else wafted themselves over upon small Rafts of Timber very early before Break of Day."[46] For Benjamin Thompson, who chronicled King Philip's War in verse, the protean nature of swamps and marshes extended to Indians themselves, whom he described as supernatural "Elves" that united in "Swarmes" among the swamps to build their "nests."[47] Conjured in his rank retreat, Roger Williams noted, "Phillip['s] great Designe is . . . to drawe . . . forces . . . into such places as are full of long grasse, flags, Sedge etc., and then inviron them round with Fire, Smoke and Bullets."[48] In this sense, swamps were not only geographies of refuge but also spaces from which to mount attacks. As Jill Lepore has shown, the English were loath to enter New England swamps precisely because Indians were "entirely invisible in swamps, disembodied, indistinguishable from the vegetation around them." The fear of such wet, vegetative chaos was so strong among English colonists that in 1624 they coined the word "swamp" and even on occasion employed it as a verb, as in Indians "swamped themselves."[49] So important were these watery landscapes to Native American patterns of engagement that it was no coincidence that one of the war's most decisive battles took place within a "hideous swamp" in the Narragansett Country. In December 1675 amidst a "quagmiry-Wood," about 3,500 Indians took refuge in four or five acres of uplands surrounded by swampland and accessible only by a downed tree that led across the water.[50] Although the battle resulted in a swift English victory—a massacre, really—the results might have been different had the swamp not been frozen solid.

The English sought asylum among the swamps of the Bay as well. Although his identity and even much of his past were never revealed, Theophilus Whale—perhaps the regicide but perhaps not—nonetheless sought refuge from suspicious neighbors and royal authorities in the marshes at the head of Pettaquamscutt Pond. Whale was a source of conjecture for many, but by the early eighteenth century Narragansett Bay was known to harbor many men of questionable repute.[51] Rhode Island had acquired such a dubious reputation that in 1699 Lord Bellomont, the Royal Governor of New York, New Hampshire, and Massachusetts, described the colony's government as "the most irregular and illegal in their administration that ever any English government was."[52] When Bellomont apprehended and convicted the pirate William "Captain" Kidd, he then set out to find Kidd's accomplices, some of whom were hidden in and around Narragansett Bay.[53] Captain Thomas Paine, a notorious brigand who commanded the eight-gun frigate *Pearl*, was one of them. Sailing under a questionable commission, he attacked St. Augustine, which exacerbated tensions between Britain and Spain. As a gesture of peace, Charles II gave written orders that Massachusetts "give no succour nor assistance to any [pirates], and especially not to one called Thomas Paine." Instead, he demanded they "exterminate" any pirates "as a race of evildoers and enemies of mankind."[54] In response, and much to the consternation of Massachusetts and New Hampshire authorities, Paine the pirate went to Rhode Island, where he bought a home, married the daughter of Caleb Carr (who would later become Governor), and in 1689 was admitted a freeman. Like so many others, Paine sought refuge along the shores of Narragansett Bay. Despite his attempts at reform, Paine had run with the wrong crowd for far too long. When Captain Kidd sailed into New England, his first stop was Paine's house. Kidd anchored and fetched his former comrade, whom he asked to "secure some things" for him. Although Paine initially refused, he eventually hid some of Kidd's gold and as a result was later drawn into the trial that would send the latter to the gallows.[55]

Widely scorned as a warren of iniquity, Narragansett Bay was known to harbor more than a few buccaneers. When, in 1679, a group of English privateers sought permission to unload their loot in Jamaica after having attacked the Spanish in the Bay of Honduras, they told the Royal Governor that "unless they were permitted to bring it [their loot] into [the] harbour . . . they would . . . sail to Rhode Island and or to the Dutch, where they would be well entertained."[56] If Jamaica, the paragon of pirate lairs, barred access, the second choice was Narragansett Bay. Indeed, Rhode Island's large fleet of merchant vessels could so easily be converted into fighting ships that many captains had been tempted down the slippery slope from privateer to pirate. So commonplace had taking prizes become and so practiced were Rhode Island's seamen at the art of forceful seizure at sea that during the Seven Years' War Newport commissioned more than sixty privateers—more than any other colonial port.[57] Rhode Island's checkered reputation and the littoral's long history as a hideout even led some during the Seven Years War to accuse the colony of sheltering deserters. Rhode Island's colonial agent Richard Partridge vehemently denied "harbouring and protecting the Men of Wars Men," explaining that "those malicious Reports" had been "propagated by our Enemys to serve some sinister view."[58]

This watery maze of inlets, islands, and creeks was so attractive to the depraved and dissolute because it was nearly impossible to patrol. That distasteful elements sought refuge within the borderlands of Narragansett Bay certainly caused mouths to pucker. But the colony's southern border, which was open to the ocean and wholly unmanageable, also raised the specter of attack. In 1708 Governor Samuel Cranston wrote to the Board of Trade explaining that Rhode Island had, during Queen Anne's War, incurred "great charge and expense, in keeping watches and wards upon the sea coast." Although Cranston explained that Rhode Island's vessels had made "frequent expeditions by sea, in order to secure our coast from being infested with the enemy's privateers," it was, he lamented, "[*impossible . . . to forti*]fie ourselves so as to keep an enemy [*from entering into our Bay and rivers, or to obstruct*] there landing in most places."[59] No matter how prepared its militias and how vigilant its lookouts and patrols, the Rhode Island government could not possibly do enough to prevent the French from entering such a vast, chaotic network of tidal rivers and salt ponds.

For the same reasons that the Bay harbored pirates and deserters and lay open to marauding marines, it also developed, somewhat reluctantly at first, into the slave capital of the North. On May 30, 1696 the brigantine *Seaflower* under the command of Thomas Windsor arrived in Rhode Island with forty-seven African slaves, unloaded and sold fourteen for between £30 and £35 each, and then pressed on for Boston, where he sold the rest. But from 1698 to 1707, Governor Samuel Cranston attested to the Board of Trade, "[W]e have not had any negroes imported into this colony from the coast of Africa, neither on the Account of the Royal African Company, or by any of the separate traders." Cranston went on to explain that Rhode Island, barring those delivered by the *Seaflower*, had never had any slaves imported directly from Africa. Instead, the "whole and only supply of negroes" had come from Barbados. Having queried the "chiefest of . . . [Rhode Island's] planters," Cranston concluded that his colony would "find but small encouragement for the trade." The principal reason for their aversion to slave importation, the governor explained, was "the general dislike our planters have for them, by reason of their turbulent and unruly tempers."[60] So averse were some that they actively limited slavery within Rhode Island's borders. In 1708 the General Assembly enacted a duty of £3 per slave imported into the colony. Although that initial act stipulated that the duty would be returned if the slave was subsequently exported, in April of 1708 the Assembly passed another law that made the duty permanent, regardless of the slave's final destination. On February 27, 1711 the General Assembly affirmed the "act for laying a duty on Negro slaves that shall be imported." But this did not sit well with London. Keen on removing any impediments to the expansion of the Atlantic slave trade, the Crown mandated that Rhode Island repeal that legislation in 1732.[61]

If Rhode Island entered into the slave trade cautiously at first, by the mid-1720s it embraced it with verve. In the Narragansett Country, which held the densest slave population in New England for most of the eighteenth century, blacks composed roughly 22 percent of the population in 1730, 19 percent in 1748, and 15 percent in 1774.[62] In Newport, which had the second-highest slave population in New England behind Boston, blacks composed 14 percent in 1730, 17 percent in 1748, and 13.5 percent in 1774.[63] Although some historians have claimed that Northern slavery was more benign than that of the South, records from Rhode Island show that Narragansett planters, much like their Southern counterparts, were brutal and controlling, both physically and psychologically.[64] In addition to bodily abusing their slaves, masters also organized and shaped the landscape for the purposes of surveillance and restraint. Increasingly during the eighteenth century, plantation owners implemented "scientific" practices of management, which included measuring fields, timing workers, and partitioning land in ways that allowed them to monitor crops and the slaves who tended them.[65]

In Rhode Island, planters turned to stone walls as the preferred method of dividing property and containing slaves. Wall construction had begun in earnest during the mid-eighteenth century.[66] By the nineteenth century, 78 percent of Rhode Island's field borders had been constructed of stone, the highest concentration of stone walls in New England.[67] Built often by Indians and black slaves, walls and fences served as "symbolic barriers" of control.[68] For example, when on September 9, 1751 a "headstrong and Disobedient" slave named Hannibal belonging to the minister James MacSparran of South Kingstown ran away, the reverend had his workers build a "Rail Fence round the Field behind the orchard," presumably to prevent or at least deter any future attempts at escape.[69] He also directed his slaves to build more permanent barriers. In September and October 1743, MacSparran's and his neighbor's slaves "sledded stones" and built walls around the north end of his north orchard.[70] But for many slaves, particularly those on the biggest plantations, which often had tidal river, salt pond, or even deepwater Bay frontage, access to the marshes and creeks liberated them from the control that stone walls either intentionally or inadvertently created.

Just as the Bay provided a safe haven for slavery's expansion, it also became a waterscape of resistance for the oppressed. In the same ways slaves used forests as geographies of resistance in inland environments, slaves living in the littoral looked to the Bay and the marshes and swamps surrounding it for refuge.[71] On December 8, 1728 a fugitive slave belonging to Thomas Wickcom "was found dead in Dyre's swamp," her escape cut short when she likely succumbed to the cold.[72] On November 21,

1774 the *Newport Mercury* announced that "a Mulatto man, named PRIMUS" owned by Cornelius Harnet, Esq. of North Carolina had run from Newport. The announcement explained that Primus had "formerly belonged to Mr. BENJAMIN BRENTON . . . and is supposed to be somewhere in Narragansett at present."[73] Probably a sailor, Primus, having landed at Newport only a few miles from his former home in Narragansett, used the Bay and his knowledge of it to make his escape. Similarly, on August 29, 1751, upon realizing his slave Hannibal had "been out," James MacSparran "stript and gave him a few Lashes till he begged." In defiance, Hannibal, "naked as he was above ye waist" fled via the sea and was found near nightfall at Block Island. Returned by Henry Gardiner, Hannibal, MacSparran explained in his diary, "had wt is called Pothooks put about his neck."[74]

Providing slaves with considerable mobility in and around Narragansett Bay, ferries became an important tool of slave resistance. Often, "under pretence of being sent or employed by their masters or mistresses, upon some service," these slaves, the Rhode Island General Assembly lamented, boarded ferries and sometimes disappeared. So many slaves had taken to the Bay that in 1714 the Assembly passed a law "that no ferryman or boatman . . . shall carry or bring any slave . . . over their ferries, without a certificate under the hands of their masters or mistresses, or some person in authority."[75] In some cases, the unbounded nature of Narragansett Bay meant that ferries and privateers crossed paths. In 1757 the General Assembly noted that "it frequently happens that the commanders of privateers, or masters of any other vessels, do carry off slaves that are the property of inhabitants of this colony." In response, the Assembly passed the first law of its kind: if a commander or master "shall knowingly carry away from, or out of this colony, a slave or slaves," the vessel or merchant ship will be fined £500. The law also stipulated that if a slave was suspected to be on board a ship, the slave's owner had the right to inspect it.[76] It is difficult to tell whether slaves were impressed, kidnapped, or had looked to the sea as a means of escape. Doubtless, all three occurred at various times. But in any case, the unbounded, open-ended nature of the Bay borderlands provided opportunities for incredible mobility among slaves—opportunities that might not have existed farther inland.[77]

The most extreme form of resistance to which slaves were forced to resort was suicide.[78] The Bay provided a place for that too. In the case of one slave owned by Thomas Mumford of Kingstown, Rhode Island, who in 1707 had "committed the horrid and barbarous murder" of his master's wife, the slave, it was presumed after his body washed onto the shores of Little Compton, "threw himself into the sea and drowned himself, by reason he would not be taken alive." By taking his life, the waters of the Bay provided the ultimate freedom. Hauled off the beach, the dead slave's body was taken to Newport and, in a gruesome act of public display, the General Assembly ordered "that his head, legs, and arms be cut from his body, and hung up in some public place, near the town, to public view, and his body to be burnt to ashes." Their justification for dismembering the body was that the "terror" would dissuade other slaves from committing the like.[79] But it is possible that the grisly proceedings were precipitated by other factors as well. The body had drifted, presumably from Kingstown all the way to the shores of what was then Massachusetts—roughly fifteen miles—suggesting that even after death, the slave had attempted escape. In other words, the ocean at the mouth of Narragansett Bay had not only provided the slave with a means of suicide; it also, at least for a period of two weeks, facilitated the perpetrator's flight. To simply commit his body to the deep—to eternity—would acknowledge, if not condone, the sea's ability to obstruct justice. It might also concede to the corpse some glimmer of agency. As such, his body was hauled onto dry land, cut into pieces, and left for the gawkers, buzzards, and crows.

If the Bay held qualities that made it a permeable political borderland and a geography of refuge and resistance during the late seventeenth and early eighteenth centuries, stone walls, legislation restricting the movement of slaves, and attempts, though often futile, at patrolling Bay waters were small gestures toward ordering Rhode Island's coastal margin. But it was the monumental effort required to lay lines on paper that began to bring the Bay's blurry contours into focus. Colden, Helme, and Chandler's systematic survey clearly showed the lay of land and sea. But the true location of Narragansett Bay and the edges that would come to redefine it could only be discerned by the people who lived by and worked on its waters.

Placing Narragansett Bay

On June 10, 1741, a Wednesday afternoon, sixty-seven-year-old Thomas Church, the eldest son of Benjamin Church, who had earned fame for leading English forces during King Philip's War, took the stand on behalf of Massachusetts Bay. Church explained to the commissioners that he had lived in Little Compton and Tiverton for the last forty-five years and during that time had "followed the Sea bout twelve" and was "Master of a vessel ten years successively." But all told, he had known the waters of the Bay for fifty years. His father, he explained, "used often to take me with him in a boat about the said Bay and other places when I was a boy." For Church, the Bay was small, spanning only the stretch of water "between Narraganset Shoar and Beaver Tail" at the southern tip of Conanicut. He complained that until five years ago, when the controversy over Rhode Island's borders resurfaced, he had never heard of any other stretch of water called Narragansett Bay. The salt river that flowed from the ocean between Aquidneck Island and Sakonnet had, he avowed, always been called by sailors "The coming-in between Rhode Island and Seconet- and Seconet River, but by no other name that I know of." Farther north, the Sakonnet River widened into Mount Hope Bay, which again narrowed to become the Taunton River. But Narragansett Bay, Church testified, they certainly were not.[80] For Thomas Church, the Bay was a small body of water and the extent of Rhode Island was in turn equally limited.

But for others, the Bay stretched wide. Elisha Wing, seventy-two, of Wareham, Massachusetts, and formerly of Sandwich on Cape Cod, took the Quaker solemn affirmation and explained that he had known Buzzard's Bay for fifty-two years and had lived roughly a mile-and-a-half from its shores for the last twenty. Throughout his life, Wing explained, he had known the body of water to be called both Buzzard's Bay and Monument Bay and that "he had heard credible persons, inhabitants of Sandwich say that it had been formerly and in their time called the Narraganset Bay; and that they had heard the Sailors and Traders with the Indians on the South Shoar call it Narraganset Bay." Buzzard's Bay, he explained, was fed by a river known by some as the Herring River and others as the Monument River but "the first Setlers ancient People in

Sandwich," Wing attested, "say that in their time it had been called Narraganset River."[81] Although Wing did not identify the western bounds of Narragansett Bay, he placed its eastern shore roughly twenty-five miles east of that explained by Church.

Such discrepancies in their Bay descriptions were no doubt rooted in political allegiance, but confusion also swirled in the murky waters of historical memory. Wing, a Quaker, clearly sought to bolster Rhode Island's claims by placing Narragansett Bay almost to Cape Cod, and Church, whose father was a Plymouth Colony hero, was protective of those lands that he no doubt believed his father had secured. Wing further testified that the Monument, Herring, or Narragansett River, as it was variously called, flowed into the head of Buzzard's Bay. He noted that "said part of Plimouth Township did border on said Bay as long as he remembers and that he had heard that that part which borders on said Bay was an additional purchase to the town of Plimouth."[82] Wing recalled that the lands west of the river were later additions, suggesting they were illegitimate add-ons. For Church, the names attributed to bodies of water in southern New England had only been called into question since Rhode Island began an aggressive campaign for territorial expansion. "I have never heard of any other place called by the name of Narraganset Bay," he explained, "until within these five years past and since the controversy . . . when people talked of extending the Bounds of Rhode Island Government."[83]

Key to establishing the region's territorial history was understanding Native American place names and Indian modes of tribal jurisdiction. Europeans surely imposed their own names on the landscape, often omitting reference to natural features of the terrain. Similarly, they imposed property boundaries that often did not follow the natural contours of the land and watercourses. Native Americans, conversely, employed place names that composed a rhetorical map of natural features and available resources.[84] Far from supplanting or jettisoning Native American ways of understanding the landscape, the New Englanders who squared off over the boundary between Rhode Island and Massachusetts readily tapped traditional Indian knowledge to strengthen their claims. When Church was asked if he was familiar with "any tract of land anciently called the Pockenoket Country," he responded by explaining what his father had told him that "King Philip was chief Sachem of that country, which I take to be Mount Hope and places adjacent." He then explained that "Asamequin," or Massasoit, was King Philip's father.[85] By asserting that Massasoit and his son were the Wampanoag sachems of Pocanocket who presided over the lands that included Bristol, Warren, Barrington, Swansea, and Somerset, Church invoked Native American authority to situate lands long held by Europeans within the former bounds of Plymouth Colony.

But so bewildering was the winding, marshy terrain between Providence and Plymouth that confusion over place names often resulted. Church had situated Pocanocket near Mount Hope at the confluence of the Sakonnet and Taunton Rivers. But others suggested it was miles to the east. This discrepancy prompted Massachusetts Bay to call the Indians Benjamin Squinimo of Middleborough and John Simon of Little Compton to locate a place called "Assawampset" that several Englishmen had at times called Pocanocket. The sixty-one-year-old Squinimo had lived at Assawampset, he explained, for his entire life. He knew the place as "Assawampset" or "Sawampset," so called for trees the Indians knew as "sawamps" and the English called beech, and described it as a "neck of land about three or four miles long and in some places a mile wide and in others narrower." Although he admitted he knew of no other place called Pocanocket, his description of the landscape described multiple places—a stretch of land comprising Barrington, Bristol, and Warren southeast of Providence and that of Assawampset Pond twenty miles east in Middleborough.[86] Although the English desperately sought the mental map of ecological place names that Native Americans had developed over millennia, the widespread acceptance of English nomenclature combined with the watery nature of the landscape had smudged that map and rendered it illegible. Placing Narragansett Bay would require a more learned approach.

In turn, the people of the Bay constructed a new geographic understanding of the littoral by melding Native American knowledge with published English accounts, contemporary observations, and the opinions of resident "experts." To mine the historical record, the Commissioners called upon the minister, scholar, and historian Thomas Prince of Boston. Originally from Sandwich and decidedly well read, the fifty-four-year-old Prince was asked to provide a lay of the land as history—history in 1741—had revealed it. He explained that he had always known Buzzard's Bay as "Manamet Bay" and sometimes as "Bosworth's Bay," of which "Buzzard may be a corruption." He believed that the attribution of Narragansett Bay for Buzzard's Bay was rooted in a mistake printed in a pamphlet written by Edward Winslow and published in London in 1623.[87] Describing William Bradford's journey with the English-speaking Indian Hobomok to Manamet twenty miles south of Plymouth, Winslow noted that the Manamet River ran into Narragansett Bay.[88] "Bradford," Prince explained, "and others that went to Manamet labour'd under that mistake of Manamet Bay being part of Narraganset Bay 'til the year 1627 when they built a vessel at Manamet and . . . then found out the Narraganset Bay by sailing round into it, and trading with the Natives there."[89]

The mistake, however, was understandable. A tortuous network of tidal rivers and salt ponds that radiated inward from the surrounding sea nearly converged in the scrub oak mazes of southeastern New England, making the misrepresentation of this coastal space all but inevitable. "[I]t is my opinion," explained Prince, "that they by mistake apprehended that called Manamet Bay to be the Narraganset Bay, and that they took it to be part of the Bay which had been discover'd the year before at Poconoket . . . by Mr Winslow and W. Hopkins, in company with Squanto . . . , and shown to them by Masassawit the great Indian Poconoket Sachem, who told them that the Narragansets lived on the other side of that great Bay."[90] As Prince explained, in their westering wanderings, Winslow and Hopkins, guided by Squanto, had come upon waters that the Pocanockets shared, albeit at a distance of about four miles, with the Narragansetts. When they had wandered to the south, another bay with similar grassy fringes and strong afternoon winds opened before them, leading them to believe that these bays were the same. This instance, Prince attested, was the only time that Buzzard's Bay had been called Narragansett Bay.[91] Subsequent accounts simply repeated the mistake.[92] Prince concluded by saying that he believed Narragansett Bay was between Rhode Island (Aquidneck) and the Narragansett Country, "but the generality of Writers that I have met with," he testified, "take the Bay to be between Seconet Point and Point Judith."[93]

To Prince and others, southeastern New England was a meandering network of waterways that, after percolating through marshes and swamps and pulsing with the tide, coursed in every direction. So bewildering was the watery world of southeastern New England that geographic descriptions became muddled and place names confused.

Seventeenth-century graphic representations of southeastern New England reflect the littoral's indeterminate geography and the jumbled geographic knowledge it created. Citing William Wood's map printed in his 1634 *New England's Prospect*, Prince testified that the Pawtucket or Blackstone River that met Narragansett Bay at Providence had also been known as the Narragansett River. This map, Prince believed, "was the first Map that I know of that has been printed of these par[t]s."[94] Although Prince believed it was "full of errors" and "not to be depended on," he did note that the Narragansett River was "the same that is now called Pautucket River." This recognition suggests that Wood's map, regardless of the glaring differences it held with the map produced by the boundary commission's surveyors in 1741, still resonated with the geographic sensibilities of southeastern New England colonists more than a century after it had been created.[95] Mirroring the confusion over geographic boundaries and place names held by many of the boundary dispute deponents in 1741, Wood's map of 1634 showed not two distinct bays—Narragansett Bay to the west and Buzzard's Bay to the east—but a matrix of coastal channels that split the land of southeastern New England into numerous peninsulas. Although Wood failed to label the Taunton River, he did draw what looks to be the northeast-most tidal arm flowing into Narragansett Bay, which, as he conceived it, extended almost to "the great Baye," or Cape Cod Bay, nearly severing Plymouth Colony from the mainland. The Blackstone River, or what Wood labeled the "Narragansett River," extended almost all the way to the Charles River, and the "Merimock" River is not far beyond. Wood's map suggests that to seventeenth-century English sensibilities, coastal New England was a veritable labyrinth of waterways that, teeming with the same types of fish and bordered by similar vegetation, made erecting distinct territorial boundaries nearly impossible.

Although Prince had criticized the map as inaccurate, he took it upon himself (much to the commission's concern) to "correct" Wood's effort, which suggests he saw in it glimmers of accuracy. Prince noted that the

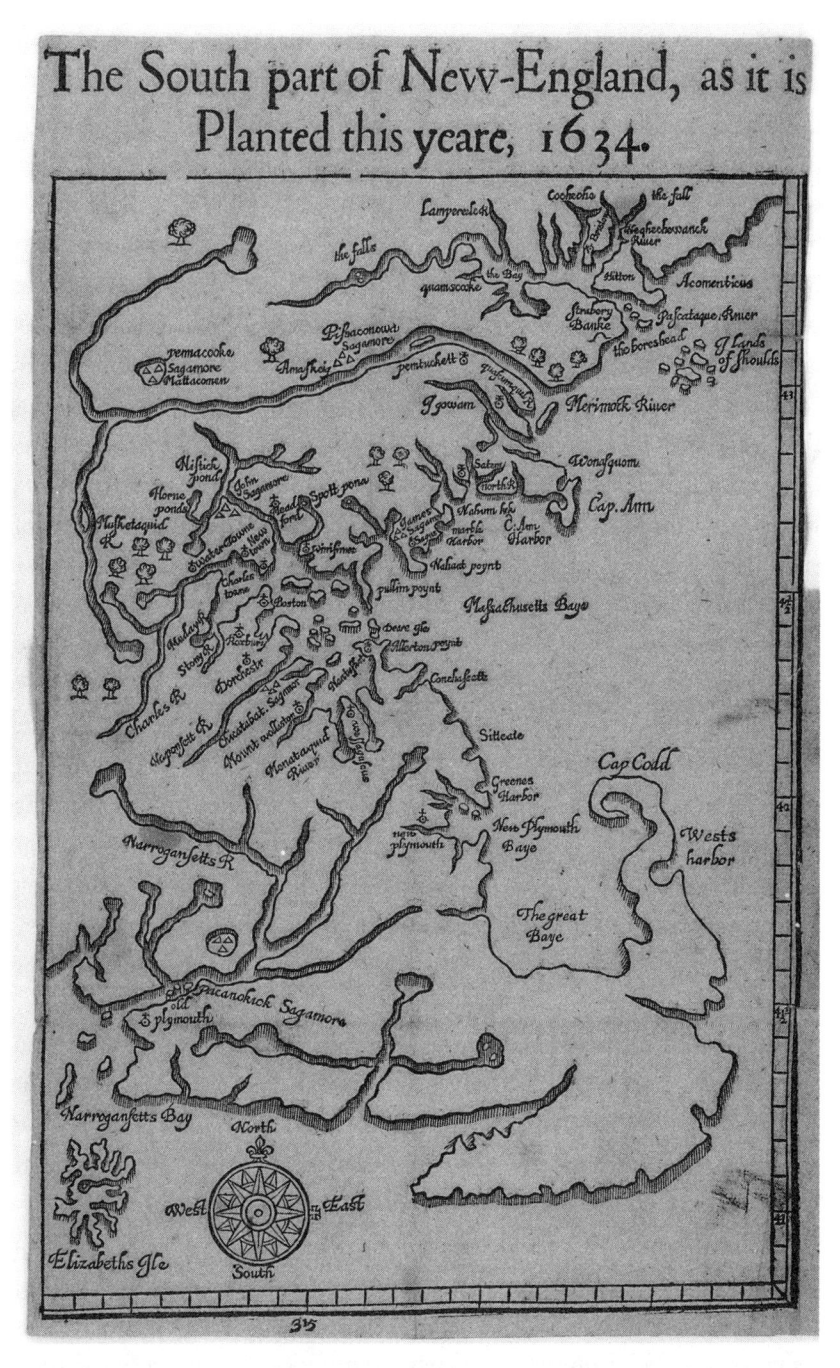

William Wood's watery 1634 map of southern New England was used as evidence in the 1741 boundary dispute between Rhode Island and Massachusetts. From William Wood, *New England's Prospects* (London: Tho. Cotes for John Bellamie, 1634). Courtesy of the John Carter Brown Library at Brown University.

island appearing south of Narragansett Bay on Wood's map had been labeled "Elizabeth Island" but that he had taken it upon himself to erase this designation from his copy. He explained that within the previous two months, "I apprehended them to be a mistake and erased them out of the print," because the "Elizabeth Islands . . . [are] twenty or thirty miles to the South of the head of Monamet Bay, and in this Map 'tis represented as twenty or thirty miles to the Westward."[96] Upon realizing the problematic nature of his changes, Prince restored the original name as it was written. But as he studied the map, he explained to the boundary commissioners that he subsequently added other place names, including the Seekonk River and Monamet Bay. "I have been these two months last past," Prince explained, "consulting said Map at Boston, and have wrote a great many words on it, as I grew satisfied of the situation of said places where I wrote said words."[97] He explained that some of the words had been added before the commission had summoned him to testify and others had been added after but was adamant that he had made the marks in his own handwriting only so that they "might more easily and readily form a truer idea whereabouts those places lay." For Prince, an eighteenth-century historian of early New England, it was important to interpret Wood's map in terms of the geographic assumptions of his seventeenth-century predecessors. If he had dismissed their ideas altogether, he would not have scrutinized the map so carefully. But in adding his own notations, he admitted—at least to himself—that Wood's watery and seemingly whimsical depiction of geographic space a century earlier was not that far off.

Although the depositions produced largely inconclusive results when eighteenth-century colonial authorities attempted to divine seventeenth-century notions of jurisdiction among the waterways of southern New England, an examination of the ways these documents were interpreted by deponents like Prince adds a foundational element to early New England historiography. Primary texts that have become the mainstay of modern historical analysis of seventeenth-century New England—Wood, Winslow, Bradford, Gookin, among others—had become, Prince's remarks revealed, bulwarks of New England identity among the learned elite. Prince, who in 1736 published his deeply researched but little-read *A Chronological History of New England, in the Form of Annals*, had carefully studied these documents often in manuscript form.[98] When referencing William Bradford's journals, he had studied the originals then in the possession of Bradford's grandson who, Prince explained, "lent it to me, together with several other Books in Manuscript." Prince read the material in Bradford's "own hand writing," which was, he explained, "very plain [and] fair."[99] Primarily a chronicler who saw New England history through a religious lens, Prince showed a deep reverence for his predecessors and the land they settled. "New England," he wrote, "open[ed] Her Arms to embrace them: they judged they now Ought to improve the offer and rather chuse a hideous *Wilderness Three Thousand Miles* across the Ocean."[100] For Prince, an untamed New England beckoned the brave. The uncontrolled terrain made their misguided sense of direction understandable. Prince himself expressed sympathy. Commenting on the bounds of Plymouth's patent and citing Bradford's manuscript and Winslow's journal, Prince wrote, "I was always uncertain whether Sowams was the same spot with that of Poconoket, but I understood them to be near one another, and Believe that it was in or near to Bristol, and . . . about forty miles from Plimouth." An important piece of the boundary deliberations, the exact location of Sowams and Pocanocket and their whereabouts had elicited much debate without producing consensus. In many ways, the studious Prince—and by proxy the commissioners who relied upon his expert testimony—imagined the landscape of southeastern New England in seventeenth-century terms. This was a country where meandering water routes and footpaths cut through marshes, swamps, and dense forests. Places like "Pocanocket," "Sowams," and "Assowampset" were but small outposts within this broad, confused waterscape. Their borders had never been measured, or even contemplated for that matter. But as tensions over jurisdiction rose, the exact location of these "island" outposts grew in importance.

If vague and sometimes conflicting historical understanding made placing Narragansett Bay difficult, human-induced environmental change further complicated the proceedings. On behalf of Rhode Island, the seventy-one-year old John Bowen of Rehoboth in Bristol County, Massachusetts Bay, testified that the contested town of Seekonk was originally named for the "the plenty of Geese that used formerly to be." Bowen explained that Seekonk derived from the word "Coank," representing "the noise they [the geese] were want to make."[101] It is likely that the first syllable derived from its proximity to salt water. Hunted from local existence, the geese that no longer flocked on the shores of Seekonk made placing this contested space more difficult. In short, an environmental transformation that removed the area's namesake further blurred colonial jurisdiction.

In other cases, environmental action, or the "improvement" of coastal space drew clear lines of demarcation. Of "Gold-Island," a two-acre dot of land in the Sakonnet River between Tiverton and Portsmouth, John Cook, a fifty-nine-year-old shopkeeper in Newport who had lived "the greatest part of my time" in Tiverton, asserted that although the island was closest to Tiverton in Massachusetts Bay, "the Portsmouth people have always improved it as belonging to Rhode Island." The "Rhode Island people," Cook explained, "have ever improved it by putting their rams on it and cutting wood off it." By grazing the island's pastures and taming its forests, Rhode Islanders had made it their own. Tiverton had never attempted to intervene, Cook explained, because "it was concluded that it lay in the Eastermost [sic] of the Narraganset Bay. And that I have, ever since I was a boy, heard the Ancient People say that Rhode Island lay in the Narraganset Bay."[102] For Cook, the line between Massachusetts and Rhode Island was drawn partially by the natural course of the water, but especially where the estuary—or at least the part of it that included Gold Island (now known as "little" Gould)—was improved.

Expert knowledge played an important role in placing Narragansett Bay but so too did political allegiance. Elisha Gibbs, a fifty-six-year-old mariner from Newport who had been master of a vessel for twenty-four years, explained that he had known Narragansett Bay since he was boy and had always understood it as filling the space between Sakonnet and Point Judith. Having learned from his father, who was "one of the first Setlers of Seconet," he also held intimate knowledge "all of the creeks up said Bay," explaining that the easternmost edge of the Bay ran north up the Sakonnet River to Assonet. On behalf of Rhode Island, Gibbs claimed that the myriad rivers and smaller Bays—Taunton River, Assonet River, Mount Hope Bay—were all piece of the larger Narragansett Bay.[103] Still others tried to extend Rhode Island's boundaries even farther. The sixty-nine-year old Benjamin Chase of Tiverton explained that his reading of the Rhode Island Charter and his "hearing what the Ancient People said" led him to conclude that "the Narraganset Bay was between Montauge Point [on Long Island] and Gay-Head" on Martha's Vineyard.[104] For Chase, Narragansett Bay was a vast stretch of water, and Rhode Island held claim to much of southern Massachusetts.

To counter Rhode Island's arguments, Massachusetts hinged its case on the location of King Philip, his kin, and their proximity to Plymouth Colony. Although the court records do not reveal the exact words and motivations of counsel for either Massachusetts Bay or Rhode Island, the evidence does suggest that the former believed that if it could establish the bounds of Pocanocket, which was known to rest within the Plymouth Colony's jurisdiction, then it could make a compelling case for ownership of that land. According to its colonial charter, Plymouth Colony extended roughly twenty-five miles north of Plymouth village to:

> a certain Rivulet or Runlet there commonly called Cohassett alias Conehassett toward the North and the River commonly called Narraganset River towards the South and the Great Western Ocean towards the East and between and within a straight line and directly extending up into the main land toward the West from the mouth of the said River called Narragansett River to the utmost limits and bounds of a country or place in New England commonly called Pokenocutt alias Sawamsett Westward and another like straight line extending it self directly from the mouth of the said River called Coahassett alias Conahasset toward the West so far up into the main land westward as the utmost limits of the said place or country commonly called Pokanacutt alias Sawamsett do extend.[105]

In many ways mirroring the distorted vision of William Wood's 1634 map, the descriptions in the Plymouth patent were imprecise. From Cohassett on its northern end, the charter explained that the colony extended westward from the ocean to the "utmost" or northernmost limits of "Pokenocutt alias Sawamsett" on the "Narragansett River," or Blackstone River. The colony then extended along the river south to its outlet with the ocean. But the bounds of Pocanocket were vague, especially since there were two Sowamsetts—one in Barrington and another roughly twenty miles east in Middleborough.

Joseph Titus, a seventy-six-year-old wheelwright from Rehoboth, a village in the contested territory, explained that he had seen King Philip when he was a boy and knew that he dwelt at Mount Hope and that his people "liv'd on the East side of Patucket River . . . because they were at enmity with all the Indians on the West side of said River, insomuch that if one of either side wounded a Deer he durst not cross the river to pursue and take him."[106] For Titus, King Philip's Pocanocket stretched from Bristol to Providence and perhaps farther north along the Blackstone River, or what Plymouth had called the Narragansett River. In addition, he explained (while admitting his memory was foggy) that that the people of Seekonk, next to Rehoboth, had purchased their land from Plymouth Colony and were therefore under its jurisdiction. Samuel Newman, seventy-eight, also of Rehoboth and who had also seen King Philip when he was young, confirmed Titus's description of Pocanocket's boundaries and his understanding of the colonial jurisdiction under which Rehoboth fell, explaining that Philip had sold the land to the town of Rehoboth and that "it was always under Plimouth."[107] If, however, the living memory of Philip and his men added credence to their country's provenance, the limits of Pocanocket—at once Philip's and his forbearer's village, home range, and region—remained elusive. For a boundary commission forced to rely heavily on anecdotal descriptions of who owned what, where, and when, the Bay borderlands defied any easy attempt to lay down a line.

Although the English came to rely on Native American history to define eighteenth-century jurisdiction, they found that forced or coerced Indian removal had further blurred their understanding of this contested space. Samuel Titticut, a seventy-six-year-old Indian laborer of Swansea in Massachusetts Bay, testified that he had known King Philip, who had lived at Bristol, and that a man named Tiask, one of King Philip's captains with whom he had often hunted, had told him that Philip's territory was between Bristol and Senichedeconet, presumably a place near Providence, which, Titticut explained, "comes from Shenicke, which is a Grey Squirrel."[108] He explained that upon visiting Senichedeconet, "there were no Indians there, only their planting fields . . . ," for Tiask explained to Titticut that the Indians there "were removed." At Barrington, Titticut explained, he had known Thomas and Peter Cheese, who "belonged to King Philip" and had heard presumably from them that there were "pretty many Wigwams there" but that they had never seen any Indian dwellings there. Titticut himself had also heard there were Indians living along the tidal Palmer River, which flowed through Barrington, Swansea, and Rehoboth, but that he had never actually seen them there.[109] The displacement of coastal Indians en masse by war, disease, and forced removal had permanently eroded Native Americans' loose but nevertheless cohesive conceptions of coastal space. A culturally informed geography divested of the inhabitants who had defined it became tangled and confused amidst the brambles and bittersweet of historical amnesia that had grown in their absence.

The loss of historical memory alone did not create such deep geographic confusion. Indian removal also changed the physical environment, further obfuscating earlier notions of jurisdiction. Samuel Titicutt explained that the Indians who had lived along the Palmer River once had numerous planting fields there but that those lands "have not been improved by the English."[110] Not only had Native American memory of the place been removed, but so too had any evidence of Indians' impact on the physical geography. The Native American sense of jurisdiction was integral to understanding the boundary between Rhode Island and Massachusetts, but then when the physical footprint of their labor on the land disappeared, the boundary, like a neglected stonewall, crumbled with the passage of time. That both colonies had neglected to improve the land further added to the confusion. When the people of Portsmouth, Rhode Island, improved Gold Island in the Sakonnet River, the people of Tiverton in Massachusetts Bay concluded that the property belonged to their colonial neighbors. When neither side improved the land, the boundary was far less clear.

The political infighting that resulted from an increasingly contentious boundary dispute on the colonial periphery had, since Rhode Island first received its charter, challenged metropolitan authority. Even after a commission was sent by the Crown to determine the boundary between Rhode Island and Massachusetts, the two colonies failed to reach an agreement. On March 11, 1663, the commission explained "we can make no final Judgment by consent of parties . . . till His Majesty's judjment and Determination of their Bounds be known." To keep the peace, they declared the boundary would follow the "Salt water betwixt" Aquidneck Island and Sakonnet, running north around the island, to the Providence River, up the Seekonk River, and north to the Massachusetts Line.[111] But such fluid borders left title to the Bay's terrestrial borderlands in abeyance. In 1663 and 1664 the towns of Rehoboth and Swansea maneuvered for the Sowamset lands in what would become Barrington, precipitating a heated dispute.[112] For the Crown, this rivalry warranted a stern rebuke. On April 23, 1664 King Charles II urged the Massachusetts governor "to suppress and utterly extinguish those unreasonable jealousies and malicious calumnies which wicked and unquiet spirits perpetualy labour to infuse into the minds of men." For the Crown, this confusion over jurisdiction threatened to undermine the very foundations of colonial authority. "[O]ur subjects in those parts," the King's correspondence declared, "do not submit to our government, but look upon themselves as independent upon Us and our laws, and that We have no confidence in their affection and obedience to us."[113]

Such rancor among Englishmen, the King explained, threatened not just the chain of command but also the welfare and security of his colonial possessions. He demanded that "lewd aspersions must vanish upon this our extraordinary and fatherly care" and that colonists in New England heed the "several instructions given to our Commissioners . . . which will exceedingly advance the refutation and security of our Plantation there."[114] If the fighting didn't stop, the threat of invasion was all too real, particularly from the Dutch. It was imperative, the King declared, that "We may protect our subjects of our several Plantations from the invasion of their neighbours and provide that no subjects of our neighbour nations how allied soever with Us may possess themselves of any lands or rivers within our territories and dominions."[115]

Tensions that escalated into outright violence threatened relations with the Indians. In 1665 the King lamented that "the Great Men and Natives of those countries . . . complain of breach of faith and of acts of violence and injustice which they have been forc'd to undergo from our Subjects, whereby not only our Government is traduced but the reputation and credit of Christian Religion brought into prejudice and reproach with the Gentiles and inhabitants of those countreys who know not God."[116] Such acrimony among countrymen, explained the king, soiled the fabric of English civilization and threatened to undermine Protestant Christendom's precarious foothold in the New World. But tacitly, he also suggested that internal struggle raised the specter of Indian attack.

The calamity that would come should the Indians attempt to capitalize on English infighting was well understood by Crown and country, center and periphery, alike. Hauled into court after having threatened Swansea with violence in March, 1671, Philip signed a treaty at Taunton on April 10 and another on September 28 in which he professed that he and his "Council and . . . Subjects do acknowledge ourselves subjects to His Majesty the King of England & the Government of New Plymouth and to their laws."[117] Philip promised to pay the government £100 and to deliver five wolf heads in tribute. Likely responding to a request spurred by mounting tensions between the southern New England colonies, he also promised "not to dispose of any of the lands that I have present" and promised "not to make war with any but with the Government Approbation of New Plymouth." Colonial authorities apparently sought similar assurance among other sachems who resided in the contested space between Massachusetts and Rhode Island, for on November 3 of the same year, they called to court Takamunna, a Sachem from Sakonnet, who agreed to "abide by the same engagement of subjection."[118]

Although Philip and Takamunna's deferential treaties in 1671 belie the mounting tensions that engulfed the entire region in total war in 1675, they nevertheless highlight the important role that the contested space between Rhode Island and Massachusetts played in sparking conflict. Historians have outlined many causes for the start of King Philip's War, which in relation to population killed more people than any war in American history.[119] Many English contemporaries believed the war began because the Indians had started it: the murder of John Sassamon, the Christian Indian minister from Namasket who had warned Massachusetts Bay Governor Josiah Winslow that Philip was planning an attack, the English believed, had warranted reprisals. Others believed the war was fought in the name of Christianity. Still other English people saw the war as divine punishment for their sins. Many Indians, conversely, believed they had been provoked when Massachusetts Bay tried and hanged three of Philip's closest advisors.[120] More recently, Virginia DeJohn Anderson has argued that English encroachment on Native American land and its subsequent destruction by domesticated animals exacerbated tensions between Indians and Europeans.[121] The only consensus that can be found, however, is that the causes of war were innumerable and complicated.[122]

Regardless of the war's ostensible origins and the dark machinations that led to the death of Sassamon and the later conviction and execution of Philip's men, the rhetoric surrounding King Philip's War has been dominated by a narrative that simply pits Indians versus the English.[123] Philip's ability to forge alliances across the region, but particularly among the rival Wampanoags, Narragansetts, and Nipmucks of southern New England, girded much of his success. As the flurry of letters between colonists across New England attests, cross-colony alliances played a key role in securing an English victory. Even Roger Williams, the great friend of the Algonquian, who spoke their language and had lived among them on peaceful terms for nearly fifty years, became an important English informant for Rhode Island, Connecticut, Massachusetts Bay, and Plymouth alike. So polarized had relations become during the war that the same bands of Indians with whom Williams had lived and traded for decades became "barbarous men of Bloud who are as justly to be repelld and subdued as Wolves that assault the sheepe."[124]

But the vagaries of cross-colony diplomacy—and more often the lack of it—destabilized the region in such profound ways that they, too, helped set the stage for war. The contest over land was not simply between the English and Indians.[125] Rather, it was often between the various English governments in southern New England whose shrill infighting sometimes included but often eclipsed Native American interests. And while the English were preoccupied with themselves, opportunity arose for Native Americans to reclaim their place within a contested landscape. In this sense, it was no coincidence that John Sassamon's body was found frozen beneath the ice in Assawampset Pond and that the first Indian attack upon the English in King Philip's War occurred at Swansea, in the heart of the Narragansett Bay borderlands. These were watery spaces where English control of the land was tenuous at best. For Indians, these coastal margins provided opportunity. If New England was, as the first wave of colonists had believed, a *vacuum domicilum*—empty, wild, and ripe for peopling and improvement—then, by the second generation of English settlement, those contested spaces where English ownership was cloudy, improvement was lacking, and lines of jurisdiction were blurred by the ebb and flood of the tide became geographies of Indian agency. Indians on Narragansett Bay took advantage of inter-colonial infighting, seeking and often gaining advantage amidst confusion.

The power struggle that precipitated King Philip's War surely included the breakdown of English-Indian relations, but the struggle was more often dominated by inter-English strife. Quite different from borderland disputes between multiple autonomous nation-states— England, France, or Spain, for example—this was a dispute between competing English powers in the region. Within the contested lands, particularly in the tidal marshes, even individual towns were no strangers to confrontation. In October 1670, Rehoboth and Swansea, both villages of Plymouth, went to court over tidal pastures. The court decided that although the "Hundred Acre Meadows," a broad swath of tidal pasture in the upper reaches of the Barrington River, lay within Swansea, it would be owned by Rehoboth. The "Five Ten Acre Lots" that belonged to Sowamset, or Barrington, but lay within Rehoboth, the court decided, would belong to Swansea.[126] Presumably these tidal lots were granted to the towns that had improved them, but the complexity with which this patchwork of ownership extended across the area provided interstices of opportunity for Native Americans. Within only six months, Philip and his men paraded their weapons through Swansea, terrifying Plymouth and Massachusetts Bay authorities. Much as Charles II feared, English infighting opened the door to Indian attack. Perhaps in waging war, Philip and his allies were not simply responding to English provocation. Rather, when placed in the context of years of English squabbling, Philip was seeking an advantage or at least a chance to negotiate with his neighbors on more equitable terms.[127] For an impatient and at times truculent sachem whose people had endured more than half a century of displacement, enslavement, disease, and famine, the circumstances were dire. And so it was in the borderlands at Swansea that on June 24, 1765 Philip and his men set the war that would bear his name in motion.

Laying the Line

Had Philip survived the war and lived to see the litany of boundary disputes that unfolded in his ancestral homeland, he would have been puzzled, if not exasperated, by the irony of English property politics. Sixty-five years after the English had united to remove him and his people from the southern New England landscape, they convened once again at great expense and commitment of time to piece together any scraps of historical documentation and traces of contemporary memory that could reveal the extent of Philip's territory. The Native American sachem who was once condemned as a bloodthirsty savage by the English became, posthumously, the star witness in a lawsuit that in many ways defined the greater colonial project. In many instances, the English had jettisoned Native American place names and imposed European patterns of land use, all in the name of civilization. But in this case, Indian knowledge was essential. The ancient bounds of Native American territories and the place names that defined them served as correcting spheres to the English surveyor's compass. At the very least, they served to mitigate any deviation caused by the forces of peripheral politics and metropolitan mandate.

The centrality of Native American history to the English understanding of colonial jurisdiction was not lost on the boundary commissioners. After about a month of hearing testimony and deliberations, the court reached a verdict on June 30, 1741. The commissioners explained, first, that no evidence had been produced that could prove "that the water between the main land on the East and Rhode Island on the West was ever at any time called Narraganset River." This ruling excluded Massachusetts Bay's claim that they have always enjoyed a boundary with Rhode Island that ran up the "middle of the Narraganset River between the End of Rhode Island and Little Compton where the said River runs into the Main Ocean."[128] The commissioners also explained that "no evidence has been produced of the extent of the Pauconoket Country to Seconk or Patucket River." They agreed that Pocanocket was the seat of Massasoit and his son Philip and that they and their land fell within the Plymouth patent. But the extent of the land remained unknown, as was the allegiance of Indians living nearby. The court explained:

[T]hough there be some evidence that the Indians at enmity with King Philip or with other Indians in amity with him, lived on the West side of the said River, and that the Indians subject to King Philip or in amity with him lived on the East side of said River, there is no evidence that all the Indians subject to or in amity with King Philip lived in the Poconoket Country.

Pocanocket, the Indian territory that had been ruled by Philip but which lacked finite boundaries, could only be determined on the grounds of political alliances. It was acknowledged that Indians west of the Blackstone River lived "at enmity with King Philip" but, the court concluded, there was no evidence that all of those with whom Philip lived in "amity" on the east side resided in Pocanocket. In short, there was no way to determine where Pocanocket began and where it ended. Such ambiguity precluded Massachusetts Bay from claiming title to the land. In addition, Massachusetts Bay had been unable to produce the Plymouth Patent letters, so their argument grew even more tenuous. Finally, under colonial law it was determined that "no jurisdiction within the King's Dominions in America can be held by Prescription or on the foot of Prescription."[129] In other words, land ownership could not be based on prolonged possession. This excluded any earlier rulings. Upon close examination of the 1664 determinations by the King's Boundary Commission, the 1741 commissioners determined that it was only meant to be temporary "for preserving the Peace on the Borders of both Colonies without determining the Rights and Titles of either."[130] In sum, the claims made by Massachusetts Bay were weak at best. The true test of jurisdiction lay in the colonial charters.

Yet Rhode Island's boundaries were still defined in relation to Narragansett Bay, so the commissioners had to place the Bay once and for all. "The Court is of the opinion," they wrote, "That the Narraganset Bay is and extendeth it self from Point Judith on the West to Seconet Point on the East, and including the Islands therein."[131] In line with Defoe's definition of a bay, Narragansett Bay comprised all the water and islands between two headlands and extended inland toward a source of fresh water and a sheltered harbor. At Narragansett Bay's northern end, however, flowed both the Taunton and Blackstone

Rivers. That Taunton lacked an extensive deep-water port, like that of Providence, perhaps led commissioners to exclude it from consideration, leaving it within the jurisdiction of Massachusetts Bay. Nevertheless, Rhode Island received the lands three miles inland from the Sakonnet River, which was now officially an arm of Narragansett Bay. From Tiverton, the Rhode Island boundary jogged northwest along a line measured three miles inland (in a northeasterly direction) from the northeast corner of Bristol Harbor and again at Rumstick Point, continuing northwest until it met a line extending three miles (also in a northeasterly direction) from Bullock's Point. North of Bullock's Point, the eastern edge of the Providence River formed the border with Massachusetts until Pawtucket Falls, where the commissioners drew a line north along the meridian to Massachusetts' southern boundary.[132] All told, Massachusetts ceded five towns to Rhode Island: Tiverton, Little Compton, Warren, Bristol, and Cumberland, a wedge of land north of Pawtucket Falls that had been known as the Attleboro Gore. The commission's decision was confirmed by King and Council in 1746 and the land officially transferred a year later.[133]

Both sides appealed the decision. Naturally, Massachusetts objected, claiming "every part of it, as grievous and injurious to us."[134] Although Rhode Island had gained a considerable amount of territory, it nevertheless felt that all of Narragansett Bay, as the commissioners defined it, should be included. Rhode Island felt that "the most North East part of the said Narraganset Bay is at a place called Assonet," a fork of the Taunton River, roughly twelve miles east of the boundary the commission had established.[135] Ultimately, the appeals were rejected and the boundaries remained largely as the commissioners had established them until 1862, when they were modified yet again.

At first blush, the story of the 1741 boundary dispute between Rhode Island and Massachusetts presents a familiar tale of eighteenth-century colonial administration. Colonies needed boundaries, and the people who lived near them often argued. But along the coast, where stable land overlapped with an impermanent sea, a much more complicated story is revealed. The documents and depositions compiled in the commission's report reveal the extent to which coastal space was shaped by human activities. Narragansett Bay was not simply a geologic formation. Rather, it was a cultural construct. The littoral was a place where people lived and worked and shaped their surroundings through aggressive action—burning forests, clearing fields, and hunting birds, for example—but it was also the reflection of ideas about nature and human memory. That the boundary commission did not grant Rhode Island the land extending three miles east of Assonet, a sheltered harbor clearly connected to Narragansett Bay, suggests that much more than the physical landscape was at play. The peregrinations of New Plymouth's first settlers, the ancestral homelands of the Wampanoags, Taunton's stands of commercially valuable timber, and even the strategic importance of a southern port for the Bay Colony no doubt played an important role in the boundary commission's decision to partition the greater estuary.

Although Narragansett Bay had existed geologically since the end of the Pleistocene, it wasn't until 1741 that its headlands, islands, harbors, shores, and shoals were fully defined. The commission's surveyors, Cadwallader Colden, James Helme, and William Chandler, had produced for the court an impressively accurate graphical representation of Narragansett Bay and its surroundings. But it was through careful assessment of seventeenth-century imperial documents and anecdotal descriptions alongside a survey of contemporary memory and local knowledge that the commissioners constructed a bay in southern New England. In doing so, they provided insight into the ways the early modern English understood nature at the edge of the sea. Although the commissioners declared that prescription could not establish the ownership of land, the people who worked on the bay and improved its shores (or lamented that others had failed to) clearly felt that prescription played an important part in defining ownership in the littoral. That northern Mount Hope Bay and the Taunton River remained in the hands of the Bay Colony suggests the Boundary Commission, despite their official rejection of the principle, at some level believed it too. But coastal work also implied ownership through a form of intellectual prescription. Landscape, after all, was not just a physical space. Rather, it constituted the demands and desires of the people who shaped it.[136] Europeans and Native Americans who hauled goods across bay waters and harvested fish, sand, and seaweed gained fluency in its myriad inlets and outlets, its tidal creeks, and salt rivers. For fishers, farmers, and coasters who had toiled on the Bay for years, who had amassed the skills and knowledge needed to navigate this demanding waterscape, ownership came from interaction with and interpretation of their surroundings. For the boatman who could feel the heat, eye the sky, and predict the southwesterly sea breeze to the minute, the brackish precincts in which he plied his trade were his and his alone. With knowledge of ledges and eddies, mudflats and marshes came ownership. With comfort in the silence of a salt creek and the familiar smells of his weirs came ownership. No imperial mandate or colonial tax collector could take that away. And in this sense, the inability to reconcile local conceptions of ownership with metropolitan designs denied progress. The littoral was not quite land and not quite sea; it was a tenuous space in between.

The ambiguity of ownership and the blurred lines of jurisdiction inherent in coastal space frequently made Narragansett Bay a favorite among those who sought sanctuary from authority. For some—at various times, smugglers, pirates, African slaves, Native Americans, and even secretive hermits—the watery nature of the Bay and its surrounding marshes and swamps provided geographies of refuge. For others, including King Philip and his people, the borderlands that formed where ecological systems converged and lines of colonial jurisdiction overlapped provided geographies of opportunity, spaces in which marginalized people (in a colonial context) asserted autonomy and maneuvered for advantage. As a coastal feature of southern New England, Narragansett Bay was ecologically complex and biologically productive. But as a space that was at times physically and intellectually malleable and at other times not, as a geography of work, war, and many things legally and culturally untoward, and as a space defined by the tension between local knowledge and imperial mandate, the Bay was a deeply human construct.

Scoured by the tides of time, the sands of historical memory had shifted. Material changes to and human dislocation from the waterlands of Narragansett Bay shaped the way people remembered coastal place names and boundaries. Subjecting recollection and documentation to the logic of the law, they produced a more clearly defined line between colonies, thereby adding new definition to the shores of southern New England. Even if this new coastline was only imagined in the minds and on the maps of men, ideas about the natural world shaped the way humans interacted with it. As the littoral people of Rhode Island reimagined coastal nature geographically, they began to reimagine it scientifically as well, adding intellectual order to the edge of the sea. And with new precision came progress.

CHAPTER 4

Natural Knowledge and a Bay in Transition

THE WINTER OF 1740, recalled William Greene of Warwick, on Narragansett Bay's western shore, "was the coldest known in New England since the memory of man." According to Greene's account, which was recorded as a "memorandum" and discovered by his great-grandson a century later, "extreme cold" hit in early November and continued with such "considerable force" that by December the Bay had frozen solid. The people of Providence, Bristol, and Newport, Greene recalled, ambled between the towns over the ice. A Boston paper reported that "Shays & Slays" passed to and fro over Newport Harbor. So deep was the freeze that ice connected the mainland with Block Island, extending "from there Southward, out to sea." In addition to the extreme cold, more than thirty storms slammed southern New England so that snow laid "knee deep" across the region. As icy drifts buried stone walls and fences, cattle roamed freely across the countryside. So much snow fell on the last day of January that some animals were "smothered." By April, Greene, who was soon to become Rhode Island's colonial governor, had lost nearly half of his sheep.[1]

Winter brought a new nature to Narragansett Bay. Brown, dry, and caked with frost, the marsh grasses lay dormant. Snow buried the beaches. And great slabs of ice twisted, shifted, and groaned with the rise and fall of the tide. Only infrequently did the Bay freeze clear across, as it did in 1740. But its brackish coves froze annually. In 1761 Samuel Tillinghast noted that Warwick's Apponaug inlet froze on December 18 and remained that way until January 13. After thawing, it "Froze Over hard" again in late February, the ice reaching as far as Warwick Neck roughly three miles to the east. When the temperature dropped the Bay was transformed, for apart from "skating" and "skimming" across the "the Cove," as the Apponaug villagers referred to their harbor, the ocean's edge stood still.[2]

For Tillinghast, who maintained a shop, owned a sloop, tilled a garden, kept cattle, and frequently harvested fish, crabs, eels, quahogs, and oysters from the Bay, the winter was a time for pause and reflection. Although he recorded the weather throughout the year, his winter observations seemed to revel in the dramatic changes to the Bay that came with the cold. Through one of his home's eighteen windows, each of which held eighteen panes, Tillinghast could see clearly across the Post Road and over the inlet.[3] He noted passing squalls and days when winds were particularly high or variable, observations that he surely made with an eye on the water. But as winter settled over the Northeast, he was most carefully attuned to the ways ice and snow affected Apponaug's tiny harbor. So impressed was he by these winter transformations that he chronicled every detail in his diary and then summarized them each spring. For example, in March 1763 he explained that the cove froze on December 19 and thawed a week later. At the beginning of January it froze again and remained that way until the end of the month. When a storm hit in February a "Very high Tide" caused coastal flooding that carried away dams and bridges and "Mov'd [an] Abundance of Wood fences and Other things." In the weeks following, the cove repeatedly froze and thawed until March 14, when the ice "all Went Away."[4]

Such careful observation of and near constant interaction with the estuary made it clear to Tillinghast that nature at the edge of the sea was anything but static. From his sloop in the summer and his window in the winter, he saw a Bay that cycled with the seasons. He saw the growing bustle of spring when oxcarts labored through muddy streets and the wharves of the west Bay sprang to life. He saw a summer harbor filled with sails and a shoreline crowded with neighbors seeking reprieve from the heat in their homes. He saw his autumn hamlet littered with leaves, which washed into the Bay with the cold rains of November. And finally, he felt and heard the deep cold and quiet of winter, when the graves were dug, the Bay grew gray, and somehow in the frosty air at the end of the pier the smoke from a pipe tasted different.

If the edge of the sea marked two different ways of knowing the natural world, then the dramatic changes that came with winter—namely, severe storms and the transformation of the Bay from a liquid to a solid—encouraged some coastal people to contemplate the "nature" of the place. Tillinghast pondered coastal space by dutifully recording atmospheric fluctuations and the ways they changed the Bay. Like so many eighteenth-century weather diarists, he came to know nature by observing it systematically and in its immediacy.[5] William Greene, conversely, imagined Narragansett Bay through the long lens of time. For him, the ice-bound estuary was less a subject of analysis and more a source of wonder, a phenomenon so awesome that he felt compelled to commit what he had seen to paper, presumably months or years later. For Tillinghast, the dutiful diarist, the Bay staged a world that was rational and regular. But for Greene, who saw men walk on water and a world of white extend clear to the Continental Shelf, the frozen Bay was nothing less than miraculous.

This tension between the empirical and extraordinary highlights some of the ways the "nature" of the estuary was shaped by the broader intellectual currents of the eighteenth century. When in 1784 the philosopher Immanuel Kant characterized the Enlightenment as "man's emergence from his self-incurred immaturity," he explained that Europe's epochal transition from an unqualified faith in religion, superstition, and traditional beliefs about social hierarchy toward science, reason, and a new era of human relations was anything but tidy and all-inclusive.[6] Change soaked in slowly, and opinion varied widely from place to place, person to person, and over time.[7] By the middle of the eighteenth century, the people of Narragansett Bay had begun to rethink their relationship with their climate, the weather, and, by extension, nature more broadly. Like other Englishman of the empire, the weather watchers of Rhode Island struggled to reconcile the acts of an angry God with those that operated in accordance with physical laws.[8] An experienced boatman who was not afraid to shove off in a storm, Tillinghast clearly felt confident in his ability to read regularity into the heavens. Steady rains in the southwesterly winds of early autumn, for instance, were little cause for delay.[9] But when the sea breeze subsided on a hot summer afternoon and the leaves of the trees curled before a cool northwesterly, it was time to steer for shore, lest his sloop succumb to the violent gusts and lightning strikes that were all but sure to follow.[10] Yet many of his neighbors and even Tillinghast himself sometimes felt the presence of an awful, unpredictable power. If the diarist from Apponaug came to know the natural world more rationally than others, there were times when the weather or its effects remained inexplicable, thereby rendering that understanding incomplete.

As atmospheric observers ordered the estuary above the waves, others, including Bay legislators, saw the need to systematize the space below. By mid-century, Rhode Islanders voiced mounting concern over the decline in Narragansett Bay oyster stocks. The seemingly endless oyster beds of the seventeenth century were being depleted at alarming rates.[11] But the search for a solution only led to confusion. That oysters were harvested with wagons and oxcarts like vegetables, pickled and barreled like animals, and their shells burned to produce lime, a mineral, caused lawmakers to question their place in nature.[12] The process of inquiry, classification, and debate—all attempts to impose intellectual order on conceptual chaos—represented important steps toward littoral development. At the same time that Rhode Islanders embroiled in the boundary dispute were organizing the Bay geographically, and scribbling weather watchers were organizing it meteorologically, fishermen and legislators had begun to organize it biologically. The ocean's edge still held a touch of mystery, but by the middle of the eighteenth century Rhode Islanders had begun to reimagine the littoral in ways that made it more responsive to human demands. Molded materially and ordered intellectually, the coast acquired new coherence in accordance with the scientific spirit of the times.

"Let there be a dome in the midst of the waters," commanded God on the second day, "and let it separate the waters from the waters." The Book of Genesis held that God thereby "separated the waters that were under the dome from the waters that were above the dome," which he called "Sky."[13] According to Judeo-Christian tradition, the process of ordering the oceans and atmosphere began with the dawn of time. Out of chaos, God thrust apart the fluid seas and fluid skies and held them in position. Then he gathered the waters on the surface of the earth to "let the dry land appear."[14] By consolidating the seas, God set the stage for civilization. When he later loosened the heavens in the time of Noah, he destroyed it.[15] For thousands of years, celestial fluids mingled with those that pooled upon the surface of the earth. The oceans and the winds that drove them were inextricably connected. The color of skies and the seas were mutually reflective. And in all cases it was God's hand that controlled them. When Jonah defied God's wishes and attempted to flee by sea, the Lord "hurled a great wind" over the ocean so that his ship "threatened to break up." When Jonah finally repented from within the belly of a whale, "the Lord spoke to the fish, and it spewed Jonah out upon the dry land."[16] The weather, the water, and all that lived in it bowed to God's command.

For one of Europe's most venerated thinkers, little had changed by the middle of the eighteenth century. In the sixth volume of his *Histoire Naturelle* (1756), Georges Louis Leclerc, Comte de Buffon echoed Genesis when he explained that the "limits of the waters are marked out by the finger of God." When the sun and moon acted in concert during the equinoxes, he continued, the tides grew highest, representing "the strongest mark of our connection with the heavens." Similarly, the air, "still lighter and more fluid than water," bowed to the push and pull of celestial bodies and "the immediate action of the sea." The winds, he explained, "push and collect the clouds. They produce meteors and transport to the arid surface of islands and continents the moist vapours of the ocean. They give rise to storms and diffuse and distribute fertile dew and rains." "Nature," he concluded, "is the external throne of the divine magnificence. Man, who contemplates her, rises gradually to the internal throne of the Almighty."[17] For Buffon, the ocean and atmosphere were created literally by the hand of God and their interaction dictated by his wishes. But only those who studied God's works would bear witness to his mysteries.

By bringing new purpose to old ideas, Buffon highlighted some of the ways that climate, a term that once included all aspects of the environment—temperature, wetness, dryness, the tides, winds, geology, and all the flora, fauna, and people it supported—was reevaluated during the age of enlightenment.[18] Buffon, like many adherents of "natural theology," sought to harmonize the melodies of piety and progress.[19] Weather observers in particular often melded traditional beliefs with newer systems of scientific thought.[20] But as eighteenth-century natural philosophers increasingly championed "curiosity," a term that came to imply careful, systematic inquiry, many abandoned their preoccupation with preternatural "wonders," thereby reimaging the climate and weather in more modern ways.[21] These new empirical conceptions of nature soaked in slowly. For those who searched for meaning in maritime disaster, for example, a rational weather that operated independent of God's will developed not through any discernable break with the past but through a more chaotic process of "wet fragmentation."[22] If, by the middle of the eighteenth century, enlightened thinkers had begun to add order to the atmosphere, their efforts were often irregular and discontinuous.

Accordingly, Buffon embraced his faith through science. And climatic differences, particularly those that distinguished Europe from America, piqued his curiosity to no end. Of special concern for Buffon was the amount of moisture that he believed polluted the American atmosphere. New World mountains, the "most lofty of any upon the globe," he surmised, had served to "condense the vapours of the air." This humidity, he explained, created springs, which in turn fed "the greatest rivers in the world." In proportion to its size, Buffon explained, there were "more running waters in the new continent than in the old." American Indians who had never "checked the torrents, directed the rivers, nor drained the marshes" had left "immense tracts of land covered by stagnant waters." Further perpetuating the wetness were the tangles of trees and "coarse weeds" that blocked out the sun. The resultant dampness had depressed temperatures to such an extent that the continent was consumed by "unwholesome exhalations." "In these gloomy regions," he concluded, "Nature remains concealed under her old garments, never having received a new attire from the cultivation of man, but totally neglected, her productions languish, become corrupted, and are prematurely destroyed."[23]

The enduring sogginess of the New World, Buffon believed, lay in its long submersion under the sea. "America," he explained, "has remained buried under the ocean longer than the rest of the globe." As proof, he identified the preponderance of "sea-shells in many places under the very first stratum of the vegetable earth." As a comparatively new continent, America was simply playing catch-up while its prospects for recovery were hampered by a climate that had enfeebled its people. He believed Native Americans were "cold" and their animals "diminutive," claiming that "the ardour of the former, and the largeness of the latter, depend on the heat and salubrity of the air." But, he continued, the conditions in America would improve once the "lands are cultivated, the forests cut down, the rivers confined with proper channels, and the marshes drained." When these measures were accomplished, he concluded, "this very country will become the most fruitful, healthy, and opulent in the world."[24] For Buffon, the sea and its lingering effects had postponed American progress. But if the lands and waterways were improved, the climate and all that inhabited it would benefit.

British commentators lodged similar complaints about the American wilderness and the climate it created. Most notably, in his *History of America* (1777), the Scottish historian William Robertson commended Europeans for having "cleared and cultivated a few spots along the coast." But, he lamented, among the "northern provinces of America, Nature continues to wear the same uncultivated aspect, and in proportion as the rigor of the climate increases, appears more desolate and horrid." He was particularly dismayed by the "exuberance of vegetation" and the "prodigious marshes" that had been left completely unattended. He found this failure to improve the place unconscionable, arguing that "When any region lies neglected and destitute of cultivation, the air stagnates in the woods, putrid exhalations arise from the waters; the surface of the earth, loaded with rank vegetation, feels not the purifying influence of the sun or of the wind; the malignity of the distempers natural to the climate increases, and the new maladies no less noxious are engendered." Calamity cascaded into a country where the land was left to fester. Foul gasses spread disease and even created new ones. As a result, Robertson concluded, America was "remarkably unhealthy."[25]

If, however, Buffon believed America's troubles lay in its long slumber beneath seas, Robertson saw in the ocean's edge the source of its redemption. In its undulating geographic "form," he explained, America was predisposed to prosperity. By contrast, Africa, which formed a "solid mass, unbroken by arms of the sea penetrating into its interior," lacked the natural features necessary for commercial exchange and would therefore "remain for ever uncivilized." When, conversely, continents were

> opened by inlets of the ocean of great extent, such as the Mediterranean or Baltic; or when, like Asia, its cast is broken by deep bays advancing far into the country, such as the Black Sea, the gulfs of Arabia, of Persia, of Bengal, of Siam, and of Leotang; when the surrounding seas are filled with large and fertile islands, and the continent itself watered with a variety of navigable rivers, those regions may be said to possess whatever can facilitate the progress of their inhabitants in commerce and improvement.[26]

For all of the environmental woes caused by the sea's wet legacy, its enduring presence would nevertheless open the doors to civilization. Health and wealth would percolate among America's estuaries, bringing vim to its people and vigor to their animals. Easily subdued but eternally renewed, the coast would bring America salvation.

An important point of reference for climatic inquiry in New England, Narragansett Bay proved Robertson at least partially right. In 1739 the Rhode Island historian John Callender praised the people of Aquidneck for "subduing and cultivating a Wilderness" to create their "*happy island,* as a safe Retreat form the stormy Winds." He felt compelled, however, to correct a "vulgar Error" perpetuated by some climate observers, who, presumably writing from afar, attributed New England's cold winters to the region's lakes. From Rhode Island, he explained, most of the larger bodies of water, apart from Lake Champlain, lay west of northwest, thereby clearing them as the cause of the cold. Having studied recent maps, he, in partial agreement with Buffon and Robertson, nevertheless concluded that the winter's icy air emanated from "the long Draft of Winds" that swept over the "uncultivated . . . perpetual Forest, which breaks the Rays of the Sun, and prevents their Reflection from the Earth."[27] A point of pride for Callender, improvements alongshore had transformed Rhode Island into a temperate "fruitful Land, the *Garden of New England,*" and it was only the distant moldering interior that continued to cause seasonal discomfort.[28]

Callender was not the only one who saw something special at the mouth of Narragansett Bay. Andrew Burnaby, the Vicar of Greenwich, who traveled through Rhode Island in 1759, claimed Aquidneck had "the most healthy climate of North America." With mild winters, cool summers, and "tolerably good" soil covered with corn and crawling with cows, Aquidneck, he explained, "enjoys many advantages."[29] Several years later, in his *Account of North America* (1765), the French and Indian War hero Robert Rogers explained that although the shores of Narragansett Bay had rocky soils, "when properly improved" they produced a variety of grains and "fruit common to the climate." The jewel of the Bay, he continued, was Aquidneck, "which, for its beauty and fertility is the garden of the colony, and is exceeded perhaps by no spot in New England."[30]

So suitable was the garden metaphor that others likened Aquidneck Island to *the* Garden. In his 1720 *History of New-England*, the English historian and Independent minister Daniel Neal explained that "for the Fruitfulness of the Soil, and the Temperateness of the Climate," Aquidneck Island "'tis deservedly called the "Paradise of *New-England*." He explained that Narragansett Bay had made Aquidneck a "Coat warmer in Winter, and . . . not so much affected in Summer with the hot Land Breezes, as the Towns on the Continent."[31] A generation later, Jean Palairet, an agent for the Dutch States-General, echoed Neal's sentiments when he explained that "by reason of its Fertility, and the Goodness of its Air," Aquidneck Island was "stiled the Paradise of New England."[32] A "garden," a "Paradise"—Aquidneck was at once burnished by human effort and blessed with divine endowments. Perhaps improvements to the land had softened the climate. Or maybe the proliferation of roads and rolling meadows just shaped the way people perceived it. But those who wrote about Rhode Island all acknowledged that Aquidneck, had, in stark contrast to the mainland, become an exemplar of environmental perfection for the region.

During the eighteenth century, islands and the climatic conditions that came with their cultivation became important subjects of philosophical inquiry. Partial to and proud of their island origins, some English writers claimed their oceanic climate had created ideal conditions for scientific advancement.[33] In his 1667 history of the Royal Society,

Thomas Sprat went so far as to explain that "[O]ur climate, the air, . . . as well as the embraces of the Ocean, seem to join with the labours of the *Royal Society,* to render our Country, a land of *Experimental knowledge.*" So favorable was Albion's islandness that Sprat believed "Nature will reveal more of its secrets to the English than to others."[34] If, however, the English held a home-island advantage, natural knowledge was also generated among islands on the periphery, some of which, moreover, were controlled by imperial rivals. On St. Helena and Mauritius, for example, natural historians, working within these small, semi-enclosed ecosystems, produced some of the first coordinated critiques of human-induced environmental change. Careful study of the Eastern Caribbean likewise contributed to climatic understanding.[35] That eighteenth-century observers of Aquidneck saw analogous connections between the island's climate and scientific progress suggests that they too had begun to draw contrasts between coastal spaces and the interior.

This process of corroborating climate speculation with historical documentation and direct observation placed local environment conditions in new theoretical frameworks. Once considered nearly synonymous with latitude, "climates," as they were imagined during the seventeenth century, divided the globe into horizontal temperature bands according to which many believed weather could be predicted by dragging a finger from Europe laterally to corresponding locations.[36] By this logic, Boston's climate was similar to that of Barcelona, and Newfoundland's to that of the Netherlands. When, however, early settlers were confronted with North America's extremes of heat and cold and the bitter truth that lemons would not grow in Virginia, they began to question those theories. While some colonial promoters simply denied climatic differences, other English writers contrived more "scientific" hypotheses. Some claimed that the sun lost its power after traveling over the ocean. Others avowed that fewer stars in the North American night made the continent colder. Still others, such as John Mason of Newfoundland, believed that the abundance of freshwater lakes in America depressed temperatures—a proposition John Callender of Rhode Island later vehemently rejected.[37]

Some proposed specific methods for effecting climatic improvements while others advertised the outcomes. In a paper presented to the American Philosophical Society, the Philadelphia physician Hugh

Williamson proposed "clearing and smoothing the face of the country," which he believed would make the winters warmer, the summers cooler, and the country more healthy and fruitful.[38] Thomas Jefferson, extolling the virtues of America to France (and addressing Buffon in particular), believed these types of efforts, if only incomplete, had nevertheless produced favorable results. "A change in our climate . . . is taking place very sensibly," he explained in his *Notes on the State of Virginia* (1787). "Both heats and colds are become much more moderate within the memory even of the middle-aged."[39]

For all their wild speculation, implausible schemes, and unflagging optimism, the one thing most colonists could agree on was that improving the land improved the climate. Human initiative brought order to broader environmental conditions.[40] Aquidneck Island, which had been sculpted into the finest garden and was bathed in the freshest air, became a shining example of New England's potential. It is unknown whether felling trees and draining marshes actually tempered the Bay climate or only made those who lived there feel better about it. But the process of scrutinizing climatic conditions and linking them to a growing corpus of imperial atmospheric observations placed the Bay prominently among broader Atlantic and global networks.[41] It also drew a clear distinction between the more moderate coastal climate of Narragansett Bay and the extremes found farther inland. Whether real or perceived, a climatic line had begun to form along the coast.

Organizing the Skies

If the people of Narragansett Bay believed they could shape broader climatic conditions through the "tedious, and laborious Business" of subduing their surroundings, as John Callender put it, it was the process of recording daily atmospheric fluctuations that, in effect, created Rhode Island's weather.[42] Long considered part of the environmental backdrop of a place, "climate" was something sporadically punctuated by extraordinary, divinely inspired events such as thunderstorms or meteors. But "weather," or atmospheric fluctuations that occurred continuously in relation to standardized time scales—hours, days, weeks, months, and years—was something culturally constructed through systematic observation and dedicated record-keeping. In other words, the temperature, wind direction, cloud cover, and precipitation, among many other variables, did not become "weather" per se until they were placed into temporal frameworks and then presented on paper in ways that allowed them to be organized and perhaps analyzed later. These new systems for normalizing daily atmospheric phenomena developed slowly and often in parallel to those that emphasized meteorological wonders.[43] For some, the skies were regular and rational, while for others, the celestial waters rippled before the breath of God. But by the middle of the eighteenth century, streaks of enlightenment sensibility had begun to coalesce in the skies over Narragansett Bay. The heavens still held a touch of the divine, but the people of Rhode Island had begun to order the atmosphere in ways that made the Bay more predictable.

Writing in 1724, the Baptist pastor John Comer of Newport firmly believed that God held dominion over the skies, the seas, and all who ventured between them. While sailing one evening in October of that year from New Haven to Boston, he was hit by an "extreme storm of wind and rain attended with thunder and lightning." So severe was the weather that none onboard "ever expected to see the light of another day." But, he explained, "through God's wonderful goodness" they arrived safely in Plymouth Harbor the next morning.[44] Several years later, Comer again reveled in the "wonderful salvation" God had granted a sinking ship from New Haven. While sailing from Newport to Antigua in September 1728, Captain Robert Gardner set a south by west course (225 degrees) but soon found his helmsmen could only steer south-southwest (202.5 degrees). Although the sailors were "faulted for this variation" in the ship's heading, Comer later reasoned in his diary that "God so ordered it in his Holy Providence." While the men on deck struggled to fill their sails, Captain Gardner, resting in his cabin, awoke repeatedly from "an uncommon dream of seeing strong men . . . so broken yt [that] he could scarce understand ym [them]." So distressing was this vision that he abandoned his bunk for some fresh air on deck, whereupon he saw something floating in the water—six people clutching the sinking remnants of a vessel from New Haven. "God who is a God working wonders," wrote Comer, "found out a way for their preservation." "The weather as God ordered it," he continued, "was calm till ye night after they were taken up."⁴⁵ The hand of heaven had not only steered Gardner's ship toward the wreck, but had also shaken him from his sleep and shielded him and his crew from the winds long enough to complete the rescue. For Comer, God controlled nature and cradled all those who were worthy to receive him.

God also announced his presence by posting signs in the skies. On October 22, 1730 Comer witnessed the "most terrifying awful and amazing Northern Light as ever was beheld in New England." Over the Bay and low along the horizon loomed "a very great brightness and over it an amazing red bow extending from North to East like a dreadful fire and many fiery spears." The haunting spectacle continued for several hours so that the people of Newport were "extremely terrified." "Words can't express ye awfulness of it," he trembled. "Wt God is about [to do] is only known to himself."⁴⁶ By focusing his commentary on "awful" weather events like auroras and extraordinary storms, Comer, like many other English eighteenth-century atmospheric observers, and especially men of the cloth, imagined the weather as symbols of God's presence and portents of his plans.⁴⁷

Extraordinary occurrences inspired awe among non-Christians as well. The Reverend Ezra Stiles, a dedicated diarist who often set his gaze aloft, noted in the summer of 1769 that the Jews of Newport "are wont in Thunder Storms to set open all their Doors & Windows for the coming of Messias." When one particularly violent storm swept over Aquidneck Island on July 31, he noted that they opened their homes and "employed themselves in Singing & repeating Prayers."⁴⁸ At other times, the God of Moses had to compete for attention. Of the earthquakes that frequently rumbled under south-central Connecticut, the Reverend Stephen Hosmore of East Haddam explained in 1729 that when he asked an "old Indian" about the occasional tremors there, he was told that "the Indian's God was very angry because Englishman's god was come here." Although Hosmore believed that the Christian God was most likely the source of the sounds, he admitted that he could not discern whether there was "anything diabolical," or Indian, at play.⁴⁹

The methodical, dispassionate weather observations of the farmer and fisherman Samuel Tillinghast, conversely, suggest that he gave little credence to the notion that preternatural powers played any role in daily life on Narragansett Bay. In December 1757 he noted that there was considerable snow on the ground and that "falling weather" had settled over Rhode Island. As the wind blew and the cove "broke up," Christmas Eve and Day saw a "Very Thawey Time" with driving rain and "Lowry, Cloudy" skies. He awoke at 5 a.m. on December 25, peered into the mist, and remembered that his wife had died exactly one year earlier. "[I]t Continues," he wrote, "a Very Melancholy Time & Season with me." His wife's untimely death—she was only forty—was clearly the source of his pain, but the unsavory weather surely aggravated his sense of loss. So low was Tillinghast that he wrote, "[W]hat a miserable thing, it is, to Expect . . . to be Cover'd up forever in Darkness."[50] That Tillinghast, who regularly attended Baptist meetings, denied his wife an afterlife—and on Christmas Day, no less—suggests that either he had little patience for Providence or had been thoroughly defeated by the gloom.

Tillinghast translated the weather into his diary with almost mechanical precision. Although he lacked weather instruments, he nevertheless developed a precise narrative system for describing atmospheric conditions. The temperature usually ranged from "warm," "sultry," or "Hot," to "Raw," "Sharp," or "Extreme" cold. The winds varied from "Fine Calm" and "Little" or "Small Breezes" to "fresh," "Blustering," or "Hard" when the winds blew strong. He described precipitation in terms of quality—rain, hail, sleet, or snow—but rarely in terms of its quantity. Instead he used more holistic descriptors that characterized the general feeling or mood at hand. Stormy days could be "Lowrey," "Darke," "Squally," "Smokey," or "Durty." Brighter days were "Moderate," "Clear," "Pleasant," or "Very Pleasant."[51] He employed these descriptions consistently and usually in the same order, providing first the date, then the clarity of the sky, the wind strength and direction, and temperature. He then described the day's overall disposition and added anything worthy of remark—"Much snow on the Ground" or "Considerable high Tide This M'g"—and then described his activities for the day, such as killing a calf, shoeing a horse, or attending a burial.[52]

Tillinghast's diary provides little evidence that he shared his notes with or was influenced by other weather observers, but his methods reflected broader British record-keeping trends. Most notably, his diary echoed the natural philosopher Robert Hooke's "Method of Making a

History of the Weather," which was published in Sprat's 1667 *History of the Royal Society*. Intended to encourage more systematic participation in weather observation across the empire, Hooke's methodology urged weather observers to record the wind speed and direction, the atmospheric pressure, humidity, and the "constitution and face of the Sky or Heavens," using words like, among others, "Cleer," "Hazy," "Thick," and "Lowring," all descriptors that Tillinghast employed at various times. He also recommended noting extreme weather events alongside any "Aches and Distempers" they caused. Finally, he instructed weather observers to make careful note of "Any thing extraordinary in the Tides." Consistent with Hook's final recommendation, Tillinghast, writing from his harbor-front home, made his observations "conversant in or neer [sic] the same place."[53] Although Tillinghast neither notated his observations on a grid nor measured the temperature or barometric pressure as Hook had recommended, his efforts nevertheless systematized the skies in ways that closely followed directives from London. Indeed, his well-informed qualitative observations of the weather and tides would have surely impressed even the most cultivated metropolitan minds.

The inclusion of published calendar pages in Tillinghast's diary added chronological structure to his records. Like other eighteenth-century New England weather watchers, Tillinghast bound leaves from a local almanac between the pages of his own observations.[54] Produced in nearby Newport, Nathaniel Ames's *Astronomical Diary*, the most popular almanac in New England, was a logical choice.[55] At times, Tillinghast jotted notes in the margins of the almanac pages. At other times, he wrote in his journal with satisfaction when astrological predictions failed to match the actual weather. On January 31, 1758, for instance, he noted that the month had ended with northwesterly winds and "Extream Cold," refuting the southerly winds and warm weather his almanac had predicted. "Mr. Almanack Maker," he scoffed, "Mist it Very Much."[56] Although Ames had called it wrong, his predictions encouraged Tillinghast and, presumably, other readers on and around Narragansett Bay to imagine the weather not just in terms of periodic spectacles or the slow cycles of the seasons, but daily and in accordance with the civic calendar.[57] Its astrological hocus-pocus aside, the almanac served as a monthly template within which systematic weather observations could be organized.

Tillinghast further structured his observations by keeping tabs on the time of day. In some cases, he noted only "AM" and "PM," but frequently he specified the hour when, say, snow turned to rain or the skies cleared.[58] So important was the time to Tillinghast that he took careful care of his clock. On November 8, 1759 he cleaned it and on the 13th made adjustments when he found it running "A little too fast." He cleaned it again on December 17 and made repairs on February 1 and 2. In August of 1760 he "Cleand & fixed" his clock and did the same again in October and November.[59] He also took pains to maintain its accuracy. On November 24, 1759 Tillinghast "Fixed [a] Meridian in ye Window" by drawing a line across the glass to fix the position of the sun, planets, or stars as a way of telling time.[60] Tillinghast's method of cross-referencing his weather observations with monthly predictions and then making periodic celestial observations to calibrate his clock, which, in turn, increased the accuracy of his weather chronicle, suggests that he saw among the skies something rational and regular. By weaving the activities of his daily life into this newly normalized weather, Tillinghast, furthermore, made even anomalous atmospheric events mundane. A trip to church with Uncle Samuel Greene, for instance, diffused even the "perfect Storm" that socked Apponaug in April 1758. Likewise, the stench and mess of making soap surely rendered the "thick" northeaster of February 1759 little more than routine.[61]

Tillinghast was one of many whose efforts served to domesticate the heavens. During the 1720s some of the first thermometers and barometers began to arrive in the colonies, encouraging, much as Hook had hoped, wider participation in and more systematic methods for weather recording.[62] When Benjamin Franklin began experimenting with electricity in the 1740s, he added yet another level of control over the atmosphere. As lighting rods were mounted on rooftops and steeples, storms succumbed to human ingenuity.[63] Among English settlers in the Caribbean, even hurricanes became more intelligble.[64] By about 1770, global weather networks expanded rapidly as meteorological observation became state policy across Europe. As officials looked for ways to improve public health, the "new science" of meteorology was born.[65] Over the course of the eighteenth century, this expanding global network of organized compilers had made every effort to gather the celestial waters.

But leaks still persisted. Even for the temperate Tillinghast, scientific explanations for the weather sometimes fell short. Whether it was his way of framing misfortune or a narration of harsh reality, his diary frequently coupled death with extraordinary weather. When on April 30, 1757 he witnessed "a great Cirkle about the sun," he noted that Joseph Stafford's four-year-old child, who had been "well at play the day before," died with the "Rattles." When almost a year later he saw another circle materialize, "Old Mrs. Hawksey" died, although he admitted she had been "out of order Sometime." When extreme cold hit just two weeks earlier, he noted that Governor William Greene, the same man who had chronicled the bitter winter of 1740, succumbed to an apoplectic fit. A month later, as dry, cold air blustered in from the northwest and Narragansett Bay "froze hard," Tillinghast's forty-two-year-old cousin Elisha was out smoking his pipe when he was "Taken Sudingly with Coughing, Sat to bleeding & Died instantly." "By this and many Such instances which has Lately happened," he concluded, "we may Discover ye uncertainty of Time."[66] Although he made no mention of God, Tillinghast's tendency to link the limits of life to atmospheric oddities hinted at a deep-seated Protestant fatalism. Although he would have surely eschewed any eschatological analysis of the local weather, his remarks suggest that his empiricism had its limits.

By the middle of the eighteenth century, the people of Narragansett Bay had begun to renegotiate their relationship with the natural world. For those who embraced the Enlightenment spirit of curiosity, Aquidneck Island, in stark contrast to the noxious interior, was celebrated for its healthy climate and the happiness it created. As progress came to the shores of Narragansett Bay, commentators began to imagine and emphasize distinct environmental and climatic differences between the coast and continent. Other empirically minded observers looked carefully at the Bay's daily weather. Tillinghast saw in the skies a subject of inquiry, a source of utility, and a place that humans could comprehend and control. Although the scrawled notes of a middling merchant provide little evidence of material transformations over or among the waters of Narragansett Bay, they nevertheless suggest that an important period of intellectual transition was underway. By observing the atmosphere systematically and placing his observations within precise temporal frameworks, Tillinghast effectively organized the skies. While some clung to more traditional beliefs, he imagined nature in more modern ways. And he was not alone. At about the same time, others with similar scientific sensibilities turned their attention to the oyster banks at the mouths of the Bay's tidal rivers.

Animal, Vegetable, or Mineral?

On November 7, 1745 an English sailor left his home of Weybread in Suffolk and signed onto the *Adventure* of London, a West India merchantman, which, after setting sail from Falmouth, was attacked by a French warship only a few days west of Lizard Point on England's southern coast. Taken captive, the sailor, whose name is lost to history, was hauled to Canada, where he spent the better part of a year. After gaining his freedom, he traveled south into New England and on October 18, 1748 arrived at Mr. Gideon Freeborn's house on Prudence Island at the center of Narragansett Bay, where he "was Treated with a great deal of respect and Esteem." Freeborn's visitor explained in his journal that the island was seven and a quarter miles long, comprising seven "fine farms," and that Freeborn's was "the best by far." Nearly two and a quarter miles long, the estate, he wrote, "produceth almost Every thing for the Support of Human Nature." The visitor praised Freeborn's hemp, flax, and wool, which, he avowed, was "the best in all the Colony." He also admired the island's grains and apple, pear, plum, and peach orchards. The vegetables, he explained, were "Exceeding Good as Surpasses Common Belief." The "Musk and water Mellon," he gushed, had "as fine a flavour as any I Ever Eate in Spain or Italy and of an uncommon Size," some weighing, he believed, upwards of forty-three pounds. He likewise marveled at Freeborn's peas, beans, cabbages, turnips, potatoes, beets, onions, radishes, peppers, and "all sorts of sallads." He explained that the corn, squash, and beans were, following the local Indian tradition, grown in mounds, whereby "when the Corn comes to Shoot up into Large Stalks the Kidney Beans Lay hold of them with their Claspers and run up with it while the Pumpkins Shoot out their Vines through all the Intermediate Spaces below." Once the beans reached the top of the corn stalks, he noted, they were gathered. The stalks were used for cattle fodder and the beans given to the hogs or sold to the Poor. The pumpkins, he noted, were gathered in September and stored in "Deep Cellars to prevent the frosts from coming at them." These, he explained, were used as the English used turnips, to fatten their cattle, which produced "a Great Quantity of Exelent Milk" and meat that was "Exceedingly Sweet." "Finally, the corn was mixed with rye, which made "very good bread" that fed Freeborn's family. The grains were also used as fodder for his cattle, hogs, and fowl.[67]

In its ability to produce order and productivity from the soil, Freeborn's Prudence Island farm represented the apotheosis of terrestrial improvement, but the quotidian workings of his operation, the visitor observed, extended well beyond his fields. "Here in the Creeks and Bays round this Farm," noted the traveler, "Vast Quantities of wild fowl in the winter Season Such as Wild Goose; Duck and Mallard; Teal, Widgeon, Smee, and . . . other Sorts which I never before Saw which are Easily Shot by Stalking too [sic] them with an Horse." An extension of the farm, shores and shallows were harvested using the same beasts that hauled wagons of apples and hay. From high in the saddle, Freeborn farmed his coastal marshes. Fish, too, he reaped in great numbers. Freeborn and his fellow husbandman readily tapped the Bay's vast resources of bass, "Tortogue" [tautog], sheepshead, mackerel, alewife, and menhaden, of which, the visitor explained, "they export many hundred Barrells to the West Indies." Of shellfish, "this farm (and no other on the Island)," the visitor explained, "has all Sorts . . . Such as Lobsters, Crabbs, Scollops, Muscells, Quahauggs, Clamms, and Oysters the best I Ever Eate and in Such Quantities that a man and Boy at Low water will take as many as a pair of Oxen can bring home in a Cart and that in Less then an hour's Time the beds not being a furlong from their Door." Like onions, squash, or melons, the oysters of Narragansett Bay could, at the lowest tides, be gathered, piled into carts, and hauled through the barn doors at day's end.[68]

If island farmers like Gideon Freeborn saw the Bay as a natural extension of their farms—a space easily tamed by oxcarts and horses—it was only natural that shellfish, which grew from the mud much like terrestrial plants, would cause some to question where they fit within the contemporary understanding of natural knowledge. In his final journal entry, the traveling English sailor noted of Rhode Island that "The Assembly in this Colony Sat 48 hours on a Quere wether oysters were Fish or vegitables Caried per the former per 4 votes."[69] Unfortunately, the records of this Assembly debate are scant. The journals of the House of Deputies and the House of Magistrates were silent on the subject. The judicial records from Newport County, which held jurisdiction over Prudence Island, likewise produced no evidence of any legal action rooted in the oyster question. However, careful examination of the published proceedings alongside the handwritten notebooks corresponding with the Acts and Resolutions of the Rhode Island General Assembly produce one shred. Scratched onto the back of a 1734 resolution concerning oyster preservation (and subsequently crossed out in ink) was a note that read: "Great Quere[?] being [illegible] in to Lime passed[?] pickling up[?] & carrying in to favr[?] countrey[?]"[70] The full meaning of this scribbled scrap is far from certain. But its use of similar language to that of the anonymous traveler, who in turn provided an exact tally of votes—four more assemblymen felt that oysters were fish rather than vegetables—suggests that the debate could well have occurred and still held such relevance that his host Gideon Freeborn mentioned it years later in conversation. If, in fact, the debate over whether oysters were fish or vegetables did occur, it asked, ostensibly, "What is the nature of Narragansett Bay?" Was it an arm of the sea or a field ripe for harvest? In asking such questions, the representatives of the people of Rhode Island attempted to define their relationship with the littoral. The protracted duration of the debate—two full days—suggests they were torn.

The members of the Rhode Island General Assembly were not the first to ask such questions. In 1726 Benjamin Franklin, while sailing from London to Philadelphia through the Gulf Stream, found within the yellow strands of sargassum weed floating on the ocean's surface, "a fruit of the animal kind, very surprising to see." Franklin explained that "Upon this one branch of the weed, there were near forty of these vegetable animals; the smallest of them, near the end, contained a substance somewhat like an oyster." Noticing a small crab nearby, Franklin surmised that the curious animals were crab embryos and believed that "all the rest of this odd kind of fruit might be crabs in due time."[71] For Franklin, what appeared to be tiny mollusks were both "fruits" and "vegetables" filled with contents that resembled those of oysters. He did not differentiate between the two until he was led astray by what was most likely a pea crab, which often lived inside some species of oysters and other bivalves.[72] The oyster's blurred taxonomic identity was understandable.

The language of oyster harvesting often emphasized their plant-like qualities. By the outset of the eighteenth century, at least one town in southern New England had attempted to sow "seed" oysters where they had not previously existed. In 1711, oysters were planted in Plymouth Harbor with a mind for their propagation, but because the bank on which they were deposited was exposed for too long at low tide, the plan did not take root.[73] In a 1791 letter from General Benjamin Lincoln of Hingham, Massachusetts, to the Reverend Jeremy Belknap of New Hampshire, Lincoln explained, "We have undoubtedly been criminally inattentive to the propagation of the oyster in different parts of our shores; we can probably fill our channels with these shellfish with much more care than we can fill our pastures with herds and flocks."[74] That Lincoln felt the town had been remiss in its failure to seed oysters suggests the practice was commonplace in other areas. His comparison to pasturing animals also suggests that by the end of the eighteenth century, the littoral had become increasingly open to cultivation.

Although evidence pertaining to the Rhode Island oyster industry is sparse for the eighteenth century, by the middle of the nineteenth much of the bottom of Narragansett Bay, particularly that of its tidal rivers, had been leased by oystermen. Widely known as "planters," these farmers of the littoral "bedded" their "seed" oysters by shoveling them overboard from "planting boats."[75] Invariably men and most often seasoned veterans of "water-work," observed the American naturalist Ernest Ingersoll in his US Fish Commission oyster report, the nineteenth-century planters of Narragansett Bay didn't forage for shellfish.[76] They sowed their seed in beds using the same tools with which their terrestrial counterparts tended beets and radishes. Narragansett Bay's oyster beds even required preparation. "Much ground that is not now suitable," observed Ingersoll, "might be made so, but needs to be carefully prepared." The Pawcatuck River estuary was a case in point. He believed it was or had become too muddy and polluted for oyster cultivation "unless," he suggested, "the ground should first be prepared by paving the mud and killing out the eel grass."[77] Although environmental conditions on and around Narragansett Bay had changed dramatically by the second half of the nineteenth century, the rich culture that had developed around Bay oyster cultivation—the community of watermen and their patterns of work as observed by Ingersoll—suggests oyster cultivation had existed on Narragansett Bay for considerable time. At the very least, it is likely that word of other coastal communities "planting" oysters had reached the Bay by the middle of the eighteenth century, for it was then that years of overharvesting finally required ameliorative legislation.

By the late 1720s the decimation of Bay oyster beds had become cause for alarm. On July 14, 1729 the Providence Town Council heard a complaint that the "Long oyster bed," which had been the main source of oysters for people of Providence and the "poore of our several Neighboring Precincts" was "Likely to be wholey destroyed by those that . . . have took up the trade of Lime burning." In response, the Council outlawed the practice and ordered that anyone who harvested oysters for the trade would forfeit the shells or lime and pay a fine.[78] The problem, however, was not isolated to a single bed. Mining oysters for lime had become so widespread that the Rhode Island General Assembly was forced to tackle the issue head on. The Assembly recognized that "sundry evil-minded Persons in several Towns" had been "Catching of great Quantities of Oysters to burn into Lime, whereby the same are greatly destroyed and diminished."[79] In response, on November 7, 1734 it passed an act that gave town councils throughout the colony full power to enact laws that would serve for the "Preservation of Oysters & all bottom shellfish within their respective towns."[80]

But if this law merely opened the door to potential conservation measures, more forceful legislation during the 1750s addressed additional causes of overfishing. On the first Monday in February, 1755, a petition was submitted to the General Assembly explaining that "the great Destruction of Oyster by pickling and exporting them in shells is a Publick Damage to the People of this Colony, and tends to the Utter extirpation of them." The petition, signed by members of some of Rhode Island's leading families, including the Whipples, Browns, Greenes, Olneys, Angels, and Watermans, among others, asked that the General Assembly pass some act for the "General preservation" of the oysters. That at least five copies of the petition were circulated and signed suggests this was no half-hearted effort and that Narragansett Bay's oysters were truly in peril.[81] Overfishing was not the only cause of decline. In October 1756, a group of Providence men, citing an act that prevented hogs from foraging along the "Puckassett River," submitted a new petition in response to the "Great Damage done . . . by the Rooting of Hogs" that would ban pigs from all Providence waters.[82] That swine had been fouling water and plundering shellfish beds since they were first left to wander the Bay's shores in the early seventeenth century didn't make them any less destructive, or for that matter, disgusting, more than a century later.

But perhaps the biggest threat to oyster stocks was dragging. In 1766 a petition was submitted to the General Assembly that explained that many inhabitants of Rhode Island had "supported themselves" and "greatly benefited by the Great Plenty of Oysters taken within the Bays, Coves, Rivers, and Harbours . . . ," but that recently some had begun to employ "Draggs fitted for that purpose, which rake over the Beds and hill and destroy more Oysters than are taken, and thereby have greatly damaged and almost destroyed many of the Beds of Oysters." The Assembly outlawed the practice and imposed a £10 fine on anyone harvesting oysters by any other "instruments" that injured oyster beds beyond the "usual method of taking them with Oyster Tongs."[83] When Rhode Island oyster harvesters began to employ the plow, the Bay's bottom-dwelling community underwent dramatic changes. Not only were oysters hit hard, but the method of extraction buried those left behind, which in many cases led to their destruction.

That oysters were planted and harvested like vegetables, pickled and eaten as animals, and often mined from the seabed as minerals (for producing lime) suggests that this estuarine creature defied the existing (but quite new) system of taxonomic classification introduced by the Swedish naturalist Carl Linnaeus. In his 1735 *Systema naturae,* Linnaeus famously divided the natural world into the *Regnum animale* (animal kingdom), *Regnum vegetabile* (vegetable kingdom) and the *Regnum lapideum* (mineral kingdom). Any given piece of nature, he avowed, could be classified an animal, vegetable, or mineral. Presaging future parlor games, the oyster, seemingly all three, was difficult to pin down. In a sense, the indeterminate nature of their estuarine home had been imposed on oysters themselves. But because mollusks were so conceptually malleable, they were, of all ocean creatures, the most likely candidates for cultivation.

Although Linnaeus explained in 1745 that nature's "laws are unchangeable . . . and they admit of no improvement," he nevertheless felt that nature itself could be and should be shaped by man.[84] Linnaeus believed that by tapping knowledge of the natural world, he could create a cameralist, or closed and self-sustaining, economic system in Sweden. Unable to compete with colonial powers like England, France, or the Netherlands, which had profited handsomely by extracting natural resources from their overseas holdings, Sweden would instead send students and explorers abroad to gather specimens with which they could create a "trans-oceanic empire" within its borders.[85] For Linnaeus, the natural world should be tamed for economic gain, a sentiment widely shared among his contemporaries. Adam Smith and physiocrat François Quesney, for example, sang improvement's praises among Europe's monarchs and other elites. As a result, agricultural, industrial, and even moral improvement had by the second half of the century become defining characteristics of Enlightened Europe.[86]

For Linnaeus, the potential for improving nature did not stop at the shore. One of his most ambitious schemes was to farm Lapland's freshwater mussels to produce pearls. After taking an academic position in Uppsala in 1740, Linnaeus began implanting mussels there with tiny flecks of chalk or gypsum, irritants around which they formed protective pearls. Although his efforts yielded only limited success, he earned himself a handsome monetary award and noble status from the Swedish crown.[87] For Linnaeus, the culture of improvement was not limited to industry, agriculture, and the science of man. Entire bodies of water could be subdued. His dominion over fluid spaces was guaranteed when in 1762 he was given the informal honorific "lord of all of Sweden's clams."[88]

Although oysters and other mollusks were surely causes of conjecture among eighteenth-century natural historians and the laity alike, it was the discovery of the aquatic polyp that caused the greatest stir among the Enlightened elite.[89] When in 1703 the Dutch natural historian Antony van Leeuwenhoeck noticed, while out for a walk, the thin, hollow, and branched body of a polyp floating in a pond, he assumed it was a plant.

But thirty-five years later, a young tutor named Abraham Trembley of Geneva stumbled upon a similar creature and noticed that it moved. "The shape of these polyps," Trembley wrote in his *First Memoir,* "their green color, and their immobility gave one the idea that they were plants." But upon careful examination, the creature seemed to pulse through the water, and a series of small tentacles moved food toward what appeared to be a mouth. "This contraction and all the movements I saw the polyps make as they extended," he noted, "once again roused sharply in my mind the image of an animal."[90] At first, he thought it was an insect, but upon dissecting the creature, Trembley noticed its parts regenerated into new individuals. This was likely an animal, but one very different from anything seen before.[91] In 1740 he wrote a letter to his patron René Ferchault de Réaumur in Paris explaining his findings, which Réaumur shared with colleagues, setting Paris's science circles abuzz. Four years later, Trembley published his discovery in a paper that earned him accolades and piqued the curiosity of natural historians throughout Europe.[92] Numerous others confirmed his experiments and began testing other creatures, including sea polyps and worms, to see if they would regenerate too.[93]

The novelty of the polyp garnered considerable enthusiasm among natural historians, but the philosophical implications of Trembley's discovery were staggering. Proponents of Cartesian philosophy, which sought to remove God from the normal workings of the natural world, saw in polyps proof that life could regenerate through material mechanisms without divine intervention. Those who subscribed to a Newtonian view, which proposed that the natural world marched under God's command (albeit according to rational and discernable laws), naturally viewed the polyp as a threat. Newtonians had advanced, and by the early eighteenth century largely won, the argument that animals had divinely inspired souls, but the possibility of regeneration—the possibility of creating new animal life through purely mechanistic means—cast doubt upon the entire Newtonian platform. Probably the most contentious interpreter of the polyp was Julien Offray de La Mettrie, who initially began writing in the Newtonian camp but whose opinions increasingly reflected a materialistic worldview. His ideas became so atheistic that La Mettrie and his ungodly polyps imperiled the divine right of kings.[94]

Perhaps a plant or—God forbid—an animal, this simple aquatic creature threatened to unravel the fabric of society. There were mysteries in the deep, Trembley's findings showed, that could alter the human connection to and understanding of the natural world.

In its "quere" on the classification of oysters, the Rhode Island General Assembly attempted to define its relationship to the estuary that sustained them. Although oysters lacked the ability to regenerate like the polyp, they, like the environment in which they formed, defied the contemporary conceptual boundaries that ordered the natural world. Oysters seemed at once fruits of the soil and denizens of the deep. And for some, this mixing of conceptual realms was abhorrent.[95] In a poem titled "The Fate of the Mouse," published in the *New-England Weekly Journal* in 1737, a mouse scavenging for food in a kitchen at an "ill hour" of the night came upon an oyster with "expanded Jaws, and gaping Shell." Whereupon:

> The greedy Mouse, [so] fond of some new Dish
> Enters the gloomy Mansion of the Fish
> With Beard exploring, and with luscious Lip,
> He begs the Pickle of the Seas to sip.

Tempted by the oyster, the mouse approached and prepared to eat, but in an instant the oyster's shell snapped closed upon his head.

> In vain the Victim labours to get free
> From Durance hard, and dread Captivity
> Lock'd in the close Embrace, strange Fate! He cries
> In Pillory safe, pants, struggles, speaks and dies.

Tempted by the open oyster, the mouse crossed a conceptual line between land and sea. After eating the forbidden fruit—"the Pickle of the Seas"—the mouse, ever the terrestrial scavenger, rightly met its doom when it succumbed to ocean temptations. To his readers the anonymous author implored, "This moral learn, to move within their Sphere." The mouse did not, which led to its demise. The penalty for such a transgression, the poem explained, was, apart from death, public ridicule.

Mounted high on the wall above the master's chair, "The Fish a Monument sublime" became the butt of his jokes.[96] Although the poem was likely a metaphor for maintaining gender roles or perhaps class order, the use of a mouse and an oyster—a terrestrial creature and one from the sea—to convey deep-seated differences to a wide audience suggests that the conceptual chasm between land and sea was wide. But the two did meet and when they did, ever emblematic of the littoral, confusion and even absurdity reigned. It is the act of contact, however, that belies the author's intent to differentiate spheres, for when the oyster ate the mouse (or a woman defied her husband or a mechanic dressed above his station), the ways society defined nature in the broadest sense of the term were called into question. In this sense, a parable that is meant to maintain the status quo exposes or at least suggests that a shift in ideas has or is about to occur. The conceptual chaos represented by an oyster eating a mouse was simply the initial stage of ordering, or improving, the confusing realities of the natural world.

This desire to create order from chaos was fundamental to eighteenth-century natural history. In the first of his thirty-six volume *Histoire Naturelle* (begun in 1749), the Comte de Buffon explained that among the earth's "heights, depths, plains, seas, marshes, rivers, caverns, gulfs, [and] volcano's . . . we can discover, in the disposition of these objects, neither order nor regularity." But through systematic examination and description, he wrote, "we shall perhaps discover an order of which we had no conception."[97] Although Buffon intended his *Histoire* to be a synoptic study of all aspects of the natural world, he expressed, alongside climate, a special interest in the natural history of the oceans, perhaps because of their inherent unruliness. He provided a detailed analysis of shell middens, particularly that of Turenne, which according to Reaumur consisted of more than 130 million cubic fathoms of shells.[98] He included chapters titled "Seas and Lakes," "Tides," and "Inequalities of the Bottom of the Sea, and Of Currents." He included other chapters that examined islands, marshes, and one titled "Of the Changes of Land into Sea, and of Sea into Land," in which he posited that the oceans were the most powerful force shaping the earth. "The motions of the sea," he wrote, "therefore, must be regarded as the principal cause of all those changes which have already happened, and of those which are daily produced upon the surface of the earth."[99] Through systematic analysis of the sea's depths and fringes, Buffon brought order to the ocean. Ever the proponent of Enlightenment "progress," Buffon compiled an inventory of natural knowledge that in many ways, at least conceptually, improved the ocean and its shores.

Systematizing the sea, however, was no simple task because for Buffon it was a space in perpetual flux. Although, he conceded, "the motion of the sea . . . has continued invariably the same in all the ages," it had, nevertheless, always been an agent of change, and often to dramatic effect. The tides "acting with violence," he explained, submerged isthmuses and eroded coastal lands.[100] More recently, he observed, the ocean had bowed to human influence. Of the massive marsh at Romney in the southeast of England, Buffon observed "no man, who has ever seen this plain, can possibly doubt of its having been formerly covered with the sea, as, without the intervention of the dikes at Dimchurch a great part of it would still be overflowed by the spring-tides." And the sea, he explained, had altered the human environment in return. At the "island of Okney," he noted, the sea had created land, that "in less than 60 years, has been considerably elevated by the accession of fresh matter brought in by every tide." He explained that the ocean had receded from the mouth of the Rhone, soil had accreted at the mouth of the Arno, and Ravenna had ceased to be a seaport. "The whole of Holland," he explained, "appears to be new land." In sum, Buffon described an agitated sea, one in which humans had effected environmental change in some cases and responded to it in others. And these dynamics were not isolated. "These changes of sea into land, and of land into sea," he wrote, "are not peculiar to Europe. The other parts of the globe, if properly investigated, would furnish more striking and numerous examples."[101]

Systematizing the motion of the ocean was but one small step in taming it, but classifying natural knowledge concerning the sea's creatures remained vexing. As it had the members of Rhode Island's General Assembly, the oyster stumped Buffon. That which differentiated vegetables from animals, he contended, was the latter's abilities of locomotion and faculties of sensation. The degree to which the oyster lacked both made its taxonomic designation as an animal questionable. "If we could give to oysters . . . ," he wrote, "the same faculty of sensation as to dogs, but in an inferior degree, why should we not allow it to vegetables in a still lesser degree: this difference between animals and vegetables is not only not general, but even not well decided."[102] In terms of sensation, the oyster's classification alongside a dog seemed just as plausible as a plant classified alongside the oyster. Buffon was not convinced that the boundaries of taxonomic identity had been clearly established.

In an effort to classify this slippery creature of the sea, Buffon conducted an experiment. In one bottle he added the "liquor" of an oyster; in the second, water in which pepper had been boiled; in the third, water in which pepper had been infused; and in the fourth, water in which he placed some "vegetable seed." Upon sealing the bottles, he waited two days and observed in the jar containing oyster juice "a greater quantity of oval and globulous substances, which seemed to swim like fish in a pond, and which had all the appearance of being animals." He noticed, however, that the creatures "had no limbs nor tails" and appeared "to change their forms . . . becom[ing] smaller for seven or eight days successively." These, he believed, were not real animals because he had previously observed similar creatures that grew in "an infusion of jelly of roast veal, which had been also very exactly corked." The infusion of seed produced "an innumerable multitude of moving globules," as did the samples in which pepper had been boiled and infused, although it took them longer to develop. He assumed that the globules were the result of fermentation, so in an attempt to differentiate that of his subjects and that of minerals, he added some aqua fortis, or nitric acid, to some powdered stone, which, he observed, bubbled violently. He noted only, however, that the mineral's response "had not the smallest resemblance to the other infusions."[103] Aside from the arbitrary nature of the final step in his experiment, Buffon's inquiry—an attempt to differentiate between plants (pepper) and the oyster—was inconclusive. He had certainly advanced efforts to systematize the sea, but the littoral and its creatures were nevertheless resistant.

The presence of the ocean invariably complicated questions about the nature of nature. In its constant motion, opaque surface, and dark depths, the sea blurred the precincts of natural knowledge. Even into the early nineteenth century, scientists had not settled on a single unifying system of classification. As a result, taxonomic decisions were often shaped by economics and politics.[104] The volley of arguments that took place in Rhode Island's two-day oyster debate simply do not exist, and as such, it is impossible to know the motives behind the General Assembly's decision to classify them as animals (if the debate mentioned by Gideon Freeborn's guest actually did occur). Perhaps the animal advocates were asking whether this creature of the sea had a soul. If it did, and it was imbued with the divine, perhaps its wanton destruction by avaricious lime burners and predatory picklers raised ethical concerns. But it is also possible that political and economic motives shaped the terms of the taxonomic—read legislative—debate.

By the 1730s lime production had become an important industry in Rhode Island, one undergoing a robust period of expansion. Integral to the process of making brick mortar and wall plaster, lime was also used to tan leather, bleach cloth, prepare whale spermaceti for candle production, refine raw sugar, and for flux used in iron production.[105] As manufacturing played an increasingly important role in Rhode Island's eighteenth-century commercial endeavors, the demand for lime, a key ingredient in almost all of them, increased. Although lime rocks did exist throughout New England, colonists had relied primarily on burning abundant and easily obtained seashells to produce lime. It was widely recognized, however, that shell lime was inferior to that mined from the earth.[106] There were scattered early attempts to burn quarried lime in Providence in 1661, in Newbury, Massachusetts, in 1697, and near the bowling green in Boston in 1723.[107] But when lime rocks were later discovered during the early eighteenth century in Smithfield, Rhode Island, on land owned by the distiller Richard Harris, his sons, David and Preserved, organized its production. They developed quarries, constructed kilns, and secured vast tracts of forest to fuel them. Lime, which for more than a century had been produced locally using knowledge often passed from father to son or master to apprentice, became, in the lands north of Providence, a thriving industry.[108]

In Rhode Island that industry was controlled by a wealthy few. From its outset in the 1730s, Rhode Island lime-burning was an "oligopolistic" enterprise.[109] It is conceivable that when the wealthy men who controlled Smithfield's lime quarries began investing in the construction of kilns, they also sought to remove any competition from shell burners on the Bay. Legislation that ostensibly sought to protect oyster stocks from those who would mine the beds for its living source of lime could also remove any competition from the merchants upstream and in the "countrey," as the scribbled trace read on the back of the 1734 Oyster Act.[110] Harris was not above maneuvering in ways that sewed up the market. By contracting numerous agreements, compacts, and a complex system of leases and subleases, he had by 1767 established an indomitable monopoly.[111] But there is no direct evidence that Harris actively sought the prohibition of oyster harvesting to bolster his own business. The ban on oyster harvesting for lime production could have simply been the result of growing alarm over the destruction of an important source of food on Narragansett Bay. By mid-century, the "oyster-catchers" of New York, reported Linnaeus's student, Peter Kalm, in 1748, "own that the number diminishes every year; the most natural cause of it is probably the immoderate catching of them at all times of the year."[112] Similarly, in 1765 towns across Cape Cod enacted laws to curb the oyster's decline.[113] But it is unlikely that the General Assembly would have spent two full days locked in philosophical debate over the identity of oysters if there hadn't been money on the line.

For all the acumen with which Harris navigated building a business and cornering the lime market, he lacked the ability to distribute his product widely. That changed when a young upstart named Welcome Arnold, another Smithfield man, entered the business.[114] And his timing was right. During the eighteenth century, growing cities built, in some cases, extensive municipal infrastructures. Many wealthy merchants constructed elaborate brick homes.[115] And decades of war saw numerous forts rising on prominent headlands at the mouths of important harbors. Rhode Island's lime barons profited handsomely during this period of growth. They also played an integral role in reshaping the littoral. If the space between land and sea had been physically and conceptually fluid, when Harris's and Arnold's lime was added, those muddy margins began to harden.

CHAPTER 5

Improving Coastal Space During a Century of War

PURCHASED FROM THE Narragansett Indians by a group of Newport planters in 1657, Conanicut Island forms, at its southernmost end, a convex, symmetrical point that, extending from the main body of the island, darts southward into the sea. Perhaps it was just coincidence that Dutch pelt traders had set up shop in the area during the first years of the seventeenth century. Or perhaps the trade in fur had fueled the imagination of those who bestowed English place names. But when the island's contours were laid on paper, there was no doubt that its southern tip, known as "Beavertail," clearly resembled the water-slapping end of New England's most valuable rodent.

If by the middle of the eighteenth century Beavertail's name and strangely representative topography echoed Rhode Island's commercial past, its location at the mouth of Narragansett Bay and orientation toward the sea augured the colony's future. Atlantic world trade had brought economic prosperity to Narragansett Bay, and Rhode Island's stalwart defense of "liberty of conscience" had attracted people of all stripes in droves. Between 1730 and 1749 the colony's population almost doubled from 17,935 to 31,778. Most of Rhode Island's inhabitants lived in the numerous towns scattered along the Bay's shores, but just under a fifth of the population was concentrated in its biggest city, Newport.[1] With its prosperous merchants and proximity to the sea, Newport was clearly a city on the move, rivaling, for at least a short period in the 1730s, Boston and Philadelphia for economic supremacy. At the very least, by mid-century Newport was firmly established as one of the five largest urban centers in America, alongside Philadelphia, Boston, New York, and Charlestown.[2]

The growth of Newport as an important center of trade and the concomitant increase in commercial traffic during the early decades of the eighteenth century precipitated a dramatic reconfiguration of coastal space. While members of the Rhode Island and Massachusetts Bay Boundary Commission were busy *locating* Narragansett Bay, the businessmen of Newport were working hard to enlarge it. On February 1, 1730 some of Rhode Island's most prominent merchants, including, among others, John Brown, Christopher and Job Almy, Godfrey Malbone, Gideon Wanton, and Peleg Carr, submitted a petition to the General Assembly "to have a Light house built, either upon Point Judith, Beaver Tail, or Castle Hill," leaving the final location to the assembly's discretion. They explained it was "highly necessary & requisite to prevent the Loss of any vessels that shall come from Foreign Parts upon this Coast." The lighthouse, they felt, would facilitate navigation as evidenced by "our Neighboring governments where it has proved of a very great Advantage."[3] In fact, Boston, which had erected the first lighthouse in America in 1716, had experienced such rapid growth that by 1720 its harbor, comprising a whopping fifty-eight piers and wharves, had become a veritable maze of ships and scaffolds.[4] The merchants who penned the petition certainly stood to gain by a beacon that bolstered ship traffic. But it was also their contention that a lighthouse, wherever it was erected, would serve the common good, for "Navigation . . . ," they explained, was "a Public Weal to this Colony in general in the Employment of the Inhabitants thereof."[5] Addressing fears of Spanish attack during King George's War, in February 1739/1740 the Rhode Island General Assembly laid out grand plans for a series of watch houses and beacons along Rhode Island's southern coast. They ordered that lights be erected on Block Island, Point Judith, Beavertail, Newport, and Portsmouth.[6] But the economic realities of war put their plans on hold. As hostilities with Spain came to a close in 1748, they resumed talks. And on March 3 of that year, colonial officials ordered Captain Joseph Harrison, Mr. Abel Franklin, Captain Josiah Arnold, and Captain George Brown "to build a Light House at Beaver Tail in Jamestown" using money from the General Treasury.[7]

Surely, the lighthouse would protect commercial traffic, but the beacon as it was proposed was much more than a simple aid to navigation. That the petitioners, all merchants and oceangoing ship masters, left the location of the lighthouse to the General Assembly suggests they were less interested in the avoidance of wrecks than they were in putting Narragansett Bay on the proverbial map. For Boston, its lighthouse firmed its position as America's premier port. Newport merchants undoubtedly wanted the same. A lighthouse at the mouth of Narragansett Bay—whether at Point Judith to the west, Beavertail at its center, or Castle Hill, just south of Newport Harbor—would give Newport political and commercial clout. For a shipmaster scanning a sea of darkness, a beacon at, say, Beavertail would suggest bustle and opportunity nearby. The port that boasted an illuminated sea-mark was no sleepy burgh flanked by dusky headlands, but rather a city with money to burn every night from sunset to sunrise. There was no more fulgent form of conspicuous consumption than a towering lamp high above the sea replete with an officer to maintain it. In 1749, atop the surf-drenched crags at Beavertail Point, workers erected a fifty-eight-foot wooden tower capped by an eleven-foot lantern. It was only the third lighthouse built in America. (A light at Brant Point on Nantucket had been built three years earlier.)[8] When the oil-soaked wicks of Beavertail Light were lit, Narragansett Bay, this dazzling shingle at its door made clear, was open for business.

Thrusting its light across the waters of Rhode Island Sound, the beacon at Beavertail projected Narragansett Bay, literally, as far as the eye could see. The Bay was no longer confined to the headlands between Point Judith and Sakonnet, as the 1741 Boundary Commission defined it. Rather, the light as a point of reference became a solitary ocean signpost.[9] Doubtless, the whale oil lamp was dim, but its presence on the horizon, however faint, sliced a ship's highway toward shore. In this sense, the beacon, swifter than any saw, scythe, or shovel in its ability to impose a human presence on untamed nature, subdued the sea. John Stilgoe has shown that the construction of lighthouses along America's coast during the Early National period "announced the end of the locally controlled complex of structures and man-altered spaces."[10] His point was to highlight the shift from local to federal control of territorial space at the end of the eighteenth century, but his observations also expose, albeit tacitly, the length to which lighthouses extended "man-altered" space—regardless of who managed it—well beyond the shore.

Although flickering lamps do not transform an estuary, they are emblematic of broader changes that occurred in the littoral by the middle of the eighteenth century. Expanded local-resource use, an increase in oceangoing trade, and mounting political hostilities among European nations imposed a growing human presence on the sea. Although many devout weather observers saw among the arms of the ocean proof of God's eternal embrace, the steady decline in Narragansett Bay oyster stocks made it all too clear that the hands of man could shape the world beneath the waves. This spurred colonial lawmakers to contemplate, classify, and even pass legislation to protect this valuable resource. But if imposing rational order through knowledge of nature was little more than an intellectual endeavor—conservation measures only slowed the pace of biological decline—advances in technology, communication, and Rhode Island's commercial infrastructure ushered in sweeping material changes to littoral space. Atlantic trade spurred the growth of ports (and the proliferation of beacons), which, like the light at Beavertail, extended the human presence well beyond Rhode Island's shores. King George's War, the French and Indian War, and the American Revolution prompted the construction of numerous watch houses and forts along Rhode Island's southern coast which, with their lookouts and cannons, had similar effects. And finally, in the name of profit and power, the littoral was mapped to an unprecedented degree of accuracy by imperial authorities. In sum, the coastal landscape was remade to meet the demands of an expanding Atlantic world.

These changes surely had an ecological impact, but more profoundly, they transformed the relationship between people and the ocean's edge. If, during the seventeenth and early eighteenth centuries, the littoral had been a space neither "natural" nor "civilized," one that more often than not precluded progress, by the second half of the eighteenth century, technology, trade, and maritime war ushered in an era of littoral improvement and development. In their ambition to facilitate safe passage through their waters and discourage attack on their towns, the people of Rhode Island imposed ever more order to their estuary.

Securing Narragansett Bay

On May 8, 1723, Edward Low and Charles Harris, masters of the *Ranger* and *Fortune*, overcame the *Amsterdam Merchant* somewhere off the Atlantic coast. Working in tandem, the two pirate ships forced the *Merchant*'s captain, John Welland, to surrender. Boarding, Low and Harris's men ransacked the ship, taking money and valuable stores. For good measure, they sliced off one of Welland's ears and then sent the ship to the bottom. A month later, the pirates plundered a merchant ship from Virginia and then set sail for Block Island. Upon receiving the report, the twenty-gun *H.M.S. Greyhound*, under the command of Captain Solgard, took pursuit. After three days of hard sailing, the *Greyhound* approached the *Ranger* and *Fortune* near the east end of Long Island. Believing the *Greyhound* was a merchant ship, the pirates attacked, whereupon the three ships bombarded each other for an hour, the *Greyhound* taking the upper hand. Upon realizing they were battling a British warship, the pirates attempted to flee. But as the winds lightened, Solgard's men, manning the *Greyhound*'s oars, maneuvered between the two ships and engaged them once again. The *Fortune* escaped, but Harris and his *Ranger*, having experienced numerous injuries and several fatalities, surrendered.[11]

All told, thirty-six pirates were hauled into Newport. So unruly was the bunch that the Rhode Island General Assembly ordered the island militia to set up a "military watch" to "secure the said pirates for making their escape."[12] But in short order three "got off their Irons," and when the jailor, his daughter, and his "lusty young" male servant opened the door, the pirates overpowered them. The fleeing buccaneers, however, were soon apprehended and committed to the jail's "dungeon."[13] Soon after, Governor William Dumman of Massachusetts convened a Royal Admiralty Court hearing at Newport, where twenty-six of the pirates were condemned to death.[14] The court granted Rhode Island's Governor, Samuel Cranston, the choice of the "place within the seamark, and manner of execution." He chose Gravelly Point, at the northern end of Newport Harbor. There, on July 19, 1723 the pirates were given the opportunity to speak. The *Boston News-Letter* reported, that many advised "all People, and especially Young Persons, to beware of the Sins which they had been guilty of . . . ," including, "Disobedience to Parents, profaning the Lord's Day, Swearing, Drinking, Gaming, [and] Unchastity."[15] These lapses, they avowed, had led them astray. Upon the scaffold, constructed "within the flux and reflux of the sea," they prayed.[16] Nooses snug around their necks, the sound of the sea washing below them, the twenty-six pirates dangled to death in one of the largest public executions in American history. Fluttering in the sea breeze above their bodies was their own black flag emblazoned with "the Pourtrature of Death" holding an hourglass in one hand and a bleeding heart in the other.[17] "[T]hey [the pirates] . . . often us'd to say," noted the *New-England Courant*, "they would live and die under it."[18]

The mass execution of twenty-six pirates at Gravelly Point in Newport in many ways symbolized Rhode Island's commitment to patrolling its shores. During the 1720s Rhode Island made a concerted effort to clean up its act. In June 1726 and again in 1729, the General Assembly allocated funds to protect the colony from enemies and pirates. The Assembly ordered "one hundred pistols, one hundred cutlasses and so many muskets as will make up one hundred fifty." They also allocated funds for "forty half pikes and twelve good guns with carriages fitting and suitable for a sloop or other vessel."[19] In November 1738, four pirates were executed in Newport.[20] On August 21, 1760, two more were executed nearby on Easton's Beach.[21] In and of themselves, defense spending and executions do not evidence a wholesale shift in Rhode Island's relationship with freebooters. After all, the commitment with which royal authorities prosecuted pirates was notoriously fickle, and their willingness to condemn some while pardoning or even empowering others appeared at times even farcical.[22] Nevertheless, in the decades following Queen Anne's War, piracy on and around the Bay waned.[23]

Just as the lighthouse at Beavertail and a two-day oyster debate served as visual and intellectual methods of mastering the elements, the massive public execution symbolized an assertion of legal jurisdiction over Bay waters. That the pirates were hanged in the Bay's intertidal zone according to English custom was powerfully representative of littoral law and order. The Admiralty court at Newport specifically granted Governor Cranston the power to decide where "within the seamark" the execution would be conducted. The word "seamark" has had many meanings, including an elevated navigational aid on land. But among the British, it was typically used to describe the farthest reach of the tide.[24] For the purpose of pirate executions, scaffolds were erected at the edge of the sea. Between high and low water, where the Admiralty courts still held jurisdiction, pirates were hanged and their corpses left submerged or in the mud for three cycles of the tide. Upon the Wapping mudflats along the Thames River in London, the tethered bodies of pirates hanged at Execution Dock were typically hauled, smeared with pitch, and hung in gibbets to deter others from piratical behavior.[25] Although the New England papers did not report how long the dead Newport pirates steeped in Narragansett Bay, the ceremony with which their execution was conducted—the public address, the prayers, the ironic display of the pirate flag above the gallows—all suggest that officials in Rhode Island were serious about asserting control over Bay waters and those who misused them. With ropes likely spaced the length of two outstretched arms, twenty-six men would have spanned 130 feet of shoreline. The sprawling, gruesome display of bodies swinging above Newport Harbor asserted authority wherever the coasters and shipmasters who witnessed it went. As a final monument to Bay control, authorities dumped the bodies into a grave at the northern end of Goat Island between the high and low water mark. If the pirates hadn't spent the customary thirty-six hours soaking atop the muck, they spent an eternity buried just below it.

Integral to the assertion of legal control was military might. In direct and explicit ways, forts and naval ships extended authority over Rhode Island waters. Coastal defense and imperial policing transformed Rhode Island's littoral from something unmanageable and amorphous into something defined, patrolled, and controlled. The British military, Stephen Saunders Webb has argued, shaped the patterns of imperial expansion as much as commerce ever did.[26] Others disagreed, claiming that, "the colonial militias . . . were the sword of civil power."[27] Regardless of whether the military presence followed metropolitan directive or addressed specific defense needs of people on the periphery (or represented a combination of both), the militarization of Narragansett Bay played an important role in its development. The construction of numerous forts, the ubiquity of naval ships and, later, the completion of some of the most detailed maps of the area were potent assertions of military power in the littoral. If one extends Webb's contention that "garrison government," as he called it, defined the relationship between people on the periphery and the metropole, then it is reasonable to assume that a military presence in many ways shaped their understanding of coastal space as well.[28]

A few hundred yards away from the twenty-six-pirate Goat Island grave stood Fort Anne, which, guarding the mouth of Newport Harbor, was a "regular and beautiful fortification of stone with a battery."[29] The fort itself had been constructed largely with bonds issued by the General Assembly, but royal coffers supplied guns and other provisions. Little more than an earthen embankment when it was first constructed around 1700, the fort soon expanded. On May 6, 1702 the Rhode Island General Assembly voted "for the better defense of His Majesty's interests and good subjects . . . there shall be a fortification or battery built . . . near the harbor of Newport." The fort was of sufficient size "to mount therein twelve pieces of ordinance, or cannon."[30] By 1715, however, the fort had fallen into disrepair, requiring the colony to take £30,000 in bills of credit for maintenance.[31] Equipped with an eighteen-foot boat, a wharf, and causeway, it was also manned with gunners.[32] In 1730, three years after the accession of George II, the fort's name was changed to Fort George.[33]

Like the lighthouse that would be built within a few years at Beavertail, the fort served as an important navigational aid at the entrance of Rhode Island's busiest harbor. The cost of maintaining it, however, had saddled the colony with debt. Requiring extensive renovations, the fort needed caulking, sealing, and filling around its foundation. In addition, "to keep the guns from the weather," sheds had to be built.[34] Because the General Assembly felt Fort George had been of great utility "for the security of navigation," in 1732 Newport began requiring shipmasters to pay six pence per ton of freight that passed through its harbor or one-sixth of a pound of powder "for the use of said fort on Goat island."[35] For merchant ships and coasters plying Rhode Island's busiest commercial thoroughfare, the fort with its guns trained over the water was a beacon of security whose maintenance required communal sacrifice, at least according to the logic of legislators. Although Samuel Cranston had reported to the Board of Trade in 1708 that it was "impossible for us to fortify ourselves so as to keep an enemy from entering into our bay and rivers," by mid-century the brick, stone, and mortared edifice that was Fort George loomed formidably over Narragansett Bay's commercial entrance.[36] Military lookouts scanned the horizon. And cannons asserted coastal control. For all intents and purposes, military improvements had marshaled littoral space.

The ownership of the oceans and the assertion of jurisdiction among their waters had been hotly contested topics since the early seventeenth century. The Dutch jurist Hugo Grotius, widely considered the father of international law, argued in 1608 for a *Mare Liberum,* or for freedom of the seas.[37] Where the waters of the ocean flowed, he believed, no nation could claim absolute sovereignty. Responding in 1616 or 1617 but left unpublished until 1635, John Selden argued in *Mare Clausum* that the oceans could indeed be owned. "[T]he Sea, by the Law of Nature or Nations," he wrote, "is not common to allmen, but capable of private Dominion or proprietie as well as the Land."[38] For Selden, the King of England ruled the seas and the tidal lands adjacent. One important tenet of his argument was a belief that the sea, like rivers and the land, could be subdued. Grotius, conversely, believed that the seas were inexhaustible and therefore should be naturally *liberum*.[39] Rejecting that proposition, Selden explained that "truly wee often see, that the Sea it self, by reason of other men's Fishing, Navigation, and Commerce, becomes the wors for him that own's it, and others that enjoie it in his right." It was evident, he explained, that the activities of man on the sea changed it and often to its detriment. Those changes were most glaring, he explained, where pearls and corals were harvested. "Yea, the plenty of such seas is lessned every hour, no otherwise then that of Mines of Metal, Quarries of stone, or of Gardens, when the Treasures and Fruites are taken away."[40] Acutely aware that the seas were not eternal, that they could be changed in dramatic ways by the hands of man, he argued that they could be owned, just like any piece of dry land. "The Sea (I suppose) is not more inexhaustible then the whole world . . . ," he concluded, "And therefore a Dominion of the Sea is not to be opposed."[41] For Selden, it was perfectly acceptable to erect jurisdiction among the waves. Like a mine, quarry, or garden, the ocean could be subdued. Therefore, the seas and its arms could be owned.

Yet the question of how to control such a vast, boundless space had still to be answered. Ever the pragmatist, Cornelius van Bynkershoek of the Netherlands struck a compromise in *De dominio maris dissertatio* (1702), arguing that the oceans should be free but that a state could maintain sovereignty along its coasts. Sovereignty, explained Bynkershoek and others during the eighteenth century, could be extended as far as a cannon could fire. This "cannon shot" doctrine was widely accepted among European nations, barring the Scandinavian states, which asserted their right to a fixed distance of one Scandinavian league, or four miles.[42] The two systems operated in tandem throughout most of the eighteenth century until 1782, when Ferdinando Galiani proposed in *The Duties of Neutral Princes towards Belligerent Princes* a fixed distance of three miles, which was the "utmost range that a shell might be projected with hitherto known gunpowder." His proposal incorporated the "cannon shot" spirit while obviating the need to lace one's shores with artillery batteries. All nations would simply observe a three-mile zone, he explained, "Without waiting to see if the territorial sovereign actually erects some fortifications, and what caliber of guns he might mount therein."[43] Although much of Narragansett Bay and Rhode Island Sound were still well beyond the firing range of Fort George and the few other batteries scattered along the Bay's shores, the heart of its commercial enterprise was covered—and with vigilance. In 1756 the General Assembly announced it would levy fines on any vessels "to enter the harbor without having first obtained liberty from the [Fort George] captain, or gunner." They codified a fee schedule of "£4 for the first shot, £8 for the second, and for every shot after, £12," suggesting approaching ships at Newport were frequently targets of Fort George fire.[44] The "cannon shot" doctrine was no theoretical convention. Just as the lighthouse at Beavertail certainly extended the Bay to commercial traffic, the Fort George cannons promised the same for marauding navies and would-be brigands.

If colonial expansion on the periphery was, as Webb suggested, primarily a military endeavor, then the extent to which government spending for defense (and other infrastructure projects) relied heavily upon trade-derived bills of credit must be acknowledged. Between 1700 and 1740, Fort George's construction and upkeep and the requisition and

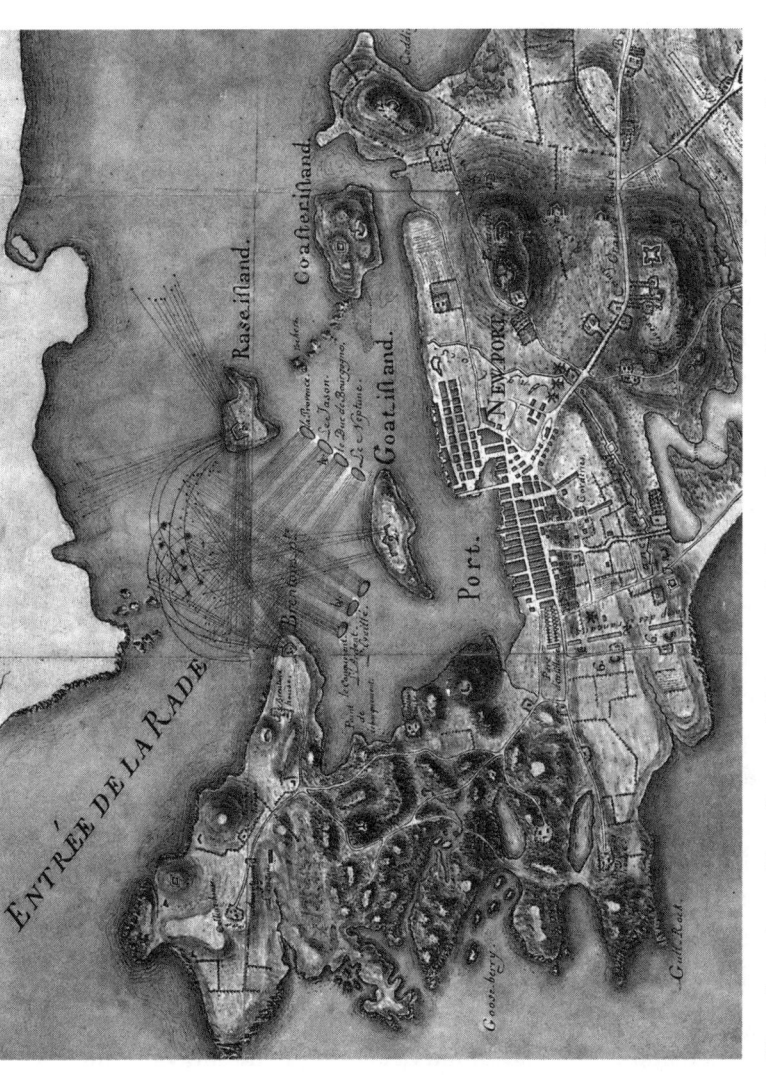

Shore batteries and heavily armed ships organized coastal space near Newport Harbor. *Plan de la ville, port, et rade de Newport, avec une partie de Rhode-Island occupée par l'armée française aux ordres de Mr. Le comte de Rochambeau, et de l'escadre française commandée.* (1780?) 1:26, 200. Library of Congress, Geography and Map Division.

purchase of a 115-ton privateer equipped with "twelve carriage and twelve swivel guns" as well as "small arms, pistols, [and] cutlasses" had cost Rhode Island more than £200,000, much of which was public debt drawn on commercial shipping.[45] In this sense, the improvement of Narragansett Bay floated on the currents of Atlantic world trade. The shrewd determination with which Newport merchants facilitated the Atlantic world exchange of hardwood from Honduras and Surinam, sugar from Barbados and Jamaica, slaves from the coast of Africa, and rum, candles, fish, and lime from their own backyard had given them, if not the means, then the leverage to invest in major infrastructure projects at home.

Fort George was just one of many capital improvements. In 1733 a harbor and commercial pier were constructed at Block Island on credit.[46] Rhode Island built a "large brick state house" and appropriated funds for its first lighthouse—on credit. Money was allocated for bridge construction and repair.[47] The colony also invested in highways, one of the most important of which connected Newport and South Kingstown by way of two government-funded ferries that extended the road across Conanicut Island over Bay waters.[48] Expanding infrastructure had improved Narragansett Bay so that it was safe and passable for ships and ponies alike. But Rhode Island's mounting debt and continued reliance on paper money had caused considerable grumbling and even outright condemnation from Whitehall. Rhode Island's boosters, however, believed the expenditures had been well worth it. "[W]e are become," Governor Ward wrote to imperial officials in 1740, "the barrier and best security of the New England trade. . . . [I]f this colony be in any respect happy and flourishing," he continued, "it is paper money, and a right application of it, that hath rendered us so."[49] That "right application" had in many ways added structure to an otherwise indefinite space. But if imperial officials remained incredulous to Ward's claims concerning bills of credit—by 1740 Rhode Island's debt had climbed to £340,000, the value of paper bills was plummeting, and counterfeiters were running rampant—Newport's commercial and strategic value was clear.[50]

As Whitehall witnessed the changes that unfolded on Narragansett Bay during the middle of the eighteenth century, it became increasingly evident how limited their knowledge was of the Bay that surrounded one of America's largest cities. Coastal space, after all, was shaped by social life. Economics, politics, religion, work, and technology, among myriad other forces, developed in tension with one another, thereby changing the "nature" of the place.[51] Over the course of the eighteenth century, Rhode Islanders had engaged in intellectual debate over the nature of the Bay's creatures. They had erected navigational aids on the Bay's shores, which not only secured safe passage into Bay waters but also marked the colony's entrance to the world economic stage. They had tapped new natural resources. They had fortified their urban center with guns on shore and roving ships at sea. And in the process, they built for Narragansett Bay a deeply human "second nature."[52]

Mapping and Maritime Militarization

During the French and Indian War, Whitehall realized it needed more detailed information on its changing empire. Despite Newport's growing commercial importance, there were few graphic representations of its harbor and the surrounding coastline. Through the 1730s, Rhode Island did not even appear on all British colonial maps.[53] Eager to win the financial support of the Board of Trade, Newport citizens had commissioned Peter Harrison, who had redesigned Fort George in 1745, to survey Newport Harbor and its defenses in 1755.[54] But it was at the behest of Dutch military engineer Samuel Holland, who had served in the British Royal American Regiment during the French and Indian War, that the Board of Trade and the King committed to a detailed survey of Britain's North American holdings. Holland was hired to survey the northern sections of continental North America and a German, William De Braham, the southern section. Concomitantly, the Board of Admiralty commissioned Joseph Frederick Wallet Des Barres, who was probably Swiss-born but underwent training at the Royal Military Academy at Woolwich, to survey the coasts for what would become the four-volume *Atlantic Neptune*. It was one of Holland's deputies, the Prussian-born Charles Blaskowitz, who ultimately conducted the most detailed surveys of Narragansett Bay that would become the basis for both the Board of Trade's and Navy's maps and charts of the area.[55]

In 1764 Blaskowitz began work in Rhode Island that would redefine the way imperial authorities understood the North American coast and the key role that Narragansett Bay would play on it. Over the course of two months, Blaskowitz completed a survey of southern Narragansett Bay that included Conanicut, Prudence, and Aquidneck Islands, as well as a detailed map of Newport that was meant to explore the possibilities of a naval base there or elsewhere on Narragansett Bay.[56] A written report of the survey possibly drafted by Grenada's colonial governor Robert Melville, who was in Newport for a time overseeing the survey work, explained that the Bay "is an excellent man-of-war harbour—affording good anchorage, sheltered in every direction, and capacious enough for the whole of his majesty's navy, were it increased four fold." Narragansett Bay, the report explained, had unique advantages for supporting a large military presence, "which cannot be found elsewhere in America."[57]

And those advantages were legion. First, the report explained, there were no "dangerous ledges or shouls" within the Bay, the entrance of which was "easy with all winds." Newport Harbor only rarely iced over, the report explaining that it "has not been frozen up so as to prevent ships coming in to safe anchorage since the year 1740." The Bay and its center entrance between Newport and Beavertail were also sufficiently deep for warships, and its islands were "admirably situated" for the construction of marine hospitals. But probably the Bay's greatest advantage was that a "whole fleet may go out under way, and sail from three to five leagues on a tack; get the trim of the ships, and exercise the men within the bay, secure from attack by an enemy."[58] Narragansett Bay, Melville believed, was perfectly suited to become an important hub of British naval activity in North America. Later surveys would concur, showing that the lack of tidal current and the Bay's numerous anchorages for large ships made it one of the most strategic ports north of the Chesapeake.[59] If Peter Harrison's chart of Newport was a grass-roots effort to win imperial financial support (which never came), Blaskowitz's chart was an assertion of imperial control.

The survey, Melville's hopeful report suggested, would transform Narragansett Bay into a naval stronghold. On the maps (that were referred to in the report but that no longer exist) Melville marked "several excellent sites for docks, ship yards &c." as well as locations for hospitals and defense structures, which would not only provide security against enemy navies but also "of the men against desertion." The final maps produced from the Blaskowitz survey, the report suggested, would extend the scope of surveillance over Narragansett Bay. No one would enter or leave without a nod from the British Navy. The land surrounding Narragansett Bay would also support the military cause. The map, Melville explained, included all of Rhode Island's roads and showed the "seats of the principal farmers." He noted that Aquidneck Island had "excellent soil, and [was] under the highest state of cultivation." Rhode Island society was even well suited to host the naval elite. The large population of West Indian and European "families of fortune" that summered in Rhode Island's healthy surroundings and the "many men of science and erudition" combined with a "very extensive and well selected" library, has, Melville explained, "rendered the whole mass of society much better informed in general literature, than any I have met with in any part of the world." The establishment of a large-scale naval presence in Rhode Island would be well received, and naval officers would be sufficiently entertained, for the people of Newport were "celebrated for their hospitality to strangers, and [were] extremely genteel and courtly in their manners."[60]

Rhode Island's reputation as a place of refinement and science had been carefully cultivated. And it was the Harrison brothers, men of great ambition and many talents who had emigrated from England to Newport during the 1740s, who had done much to polish the colony's presentation. Peter Harrison, widely considered America's first "professional" architect, had designed the Redwood Library in 1748 and 1749 and the Touro Synagogue in 1759, thereby endowing Newport with America's second lending library and one of its oldest synagogues.[61] His older brother, Captain Joseph Harrison, had built the lighthouse at Beavertail and played a central role in surveying Rhode Island's borders, which continued to foment discontent through the 1760s. Resolved to mend (and defend) Rhode Island's fences once and for all, the elder Harrison pressed the General Assembly to purchase a new piece of surveying equipment called simply a "Sea Quadrant" (actually an octant), which had been recently invented by London insurance agent Caleb Smith for the purpose of establishing latitude with "greater Certainty, and more frequently, than by . . . any other of the Common Instruments."[62] But instead of ordering the device from London, the General Assembly agreed in 1756 to pay Newport instrument maker Benjamin King £300 to produce one locally.[63] Made of dark, burnished wood and equipped with carefully calibrated mirrors and lenses, King's quadrant, which followed Smith's designs and Harrison's advice, was one of the most sophisticated surveying tools of the day.[64]

King's creation allowed Rhode Island to stake its territorial claims with mathematical precision, but by crafting a device of such complexity a world away from London, King also bestowed upon his colony the broader authority that came with the possession of specialized knowledge. The quadrant was so valuable to Rhode Island's physical and conceptual identity that the General Assembly carefully noted that although it would be stored at the Redwood Library, "the property thereof, shall stand and remain vested in the colony."[65] Commissioned for the common weal, the instrument placed the people of Narragansett Bay in direct communication with the European centers of science. At no time was this prestige more evident than in 1769, when stargazers in Providence pulled King's quadrant out of storage to observe the transit of Venus. According to the American Philosophical Society, their work held significance, for the weather had obscured visibility in Russia and Sweden, making Providence important for the purpose of "comparison with ... Greenwich."[66] In sum, King's collaboration with the Harrison brothers added a new level of order to the Bay and its surroundings. In Abraham Redwood's library, Peter Harrison had shaped a space for systematizing knowledge. In Touro Synagogue, he created a sense of stability and permanence for one of America's oldest Jewish communities. Joseph Harrison, working closely with King, had organized Rhode Island geographically over land and at sea and placed it with mathematical precision beneath the heavens. All of these efforts, in no small way, served to usher an imperial backwater onto the world stage.

As Narragansett Bay grew in importance, colonial authorities increasingly sought to control it, but their lack of tact with the people and unfamiliarity with the place often torched their efforts—quite literally. After the particularly combative Captain William Reid of the armed sloop *Liberty* took custody of two vessels he suspected of smuggling in July 1769, a Newport mob seized Reid and his men, dumped the *Liberty*'s guns overboard, ran the vessel aground, and set it ablaze.[67] When during the summer of 1772 Lieutenant William Dudingston aboard the H.M.S. *Gaspee* ordered the *Hannah* of Providence to submit to inspection, the *Hannah*'s captain, Benjamin Lindsey, fled in defiance. Perhaps due to a lack of local knowledge, Dudingston, while pursuing Lindsey, ran the *Gaspee* aground at Namquit Point in Warwick. That night a group of irate Rhode Islanders rowed longboats to the stranded vessel, shot Dudingston, seized the rest of his crew, and, upon delivering them to shore, burned the *Gaspee* to the waterline.[68] What was arguably the first armed confrontation of the American Revolution was reason for pause at Whitehall. That a British naval vessel was destroyed after having run aground in what for them was uncharted territory likely informed their decision to have Blaskowitz return.

Keenly aware of mounting political tensions within its own ranks and in Rhode Island specifically, the Board of Trade sought even more detailed surveys of the area. In October 1774, Blaskowitz began a second, more comprehensive survey of Narragansett Bay, this time including the waters of the Providence and Taunton Rivers. The chart added soundings throughout the Bay as well as detailed topographical information, such as marshes, ridges, stands of trees, and, as Melville had stipulated a decade earlier, planter property boundaries. A year later, in December 1775, British troops occupied Newport, and Blaskowitz's work was printed for the navy, no doubt with great haste, by Joseph Des Barres in 1776. In 1777 William Faden printed another version based on Blaskowitz's survey. Faden's title, *A Topographical Chart of the Bay of Narraganset in the Province of New England . . .* , expressed his hybrid goal of depicting both land and sea. By contrast, Des Barres's 1776 chart, which was based on the Blaskowitz survey but commissioned by the Royal Navy, included less topographic information.[69]

Faden's "topographical chart" relayed more information about Narragansett Bay than had ever been compiled in a single space before. In addition to detailed soundings of Bay waters and topographical contours, Faden included maps of Newport, Providence, and several smaller towns scattered around the Bay. Important ferry landings and gun batteries were also marked, including the number of guns in each battery and their size. One such battery perched on the eastern shore of the Providence River was annotated: "A Breast Work Commanding the

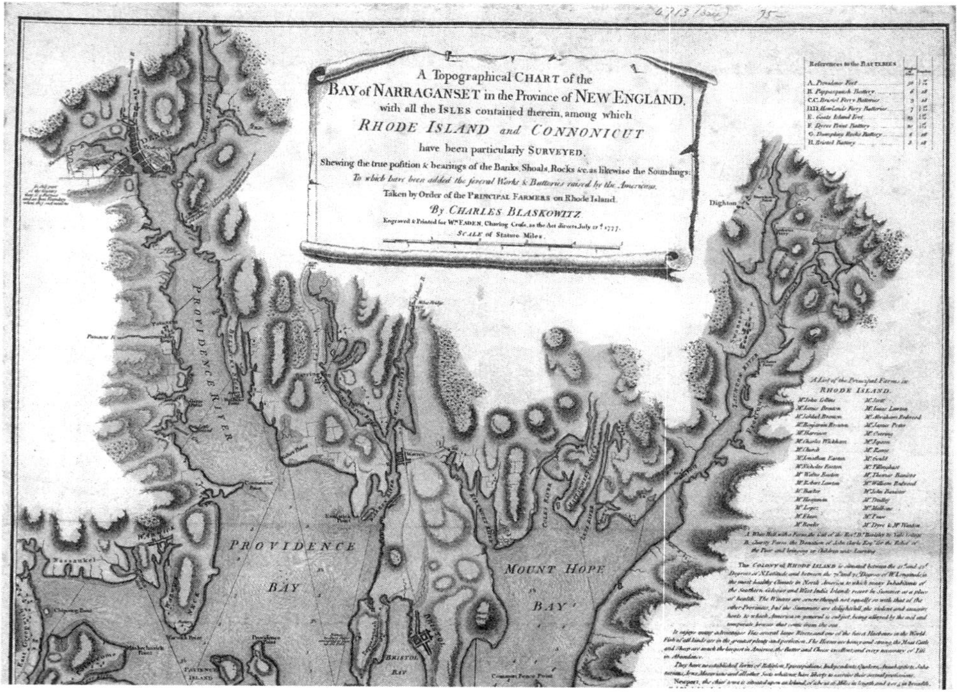

A Topographical CHART of the
BAY of NARRAGANSET in the Province of NEW ENGLAND,
with all the ISLES contained therein, among which
RHODE ISLAND and CONNONICUT
have been particularly SURVEYED,
Shewing the true position & bearings of the Banks, Shoals, Rocks &c. as likewise the Soundings:
To which have been added the several Works & Batteries raised by the Americans.
Taken by Order of the PRINCIPAL FARMERS on Rhode Island.
By CHARLES BLASKOWITZ
SCALE of Statute Miles.
References to the BATTERIES
A List of the Principal Farms in
RHODE ISLAND.
PROVIDENCE RIVER
PROVIDENCE
BAY
MOUNT HOPE
BAY
BRISTOL
Dighton

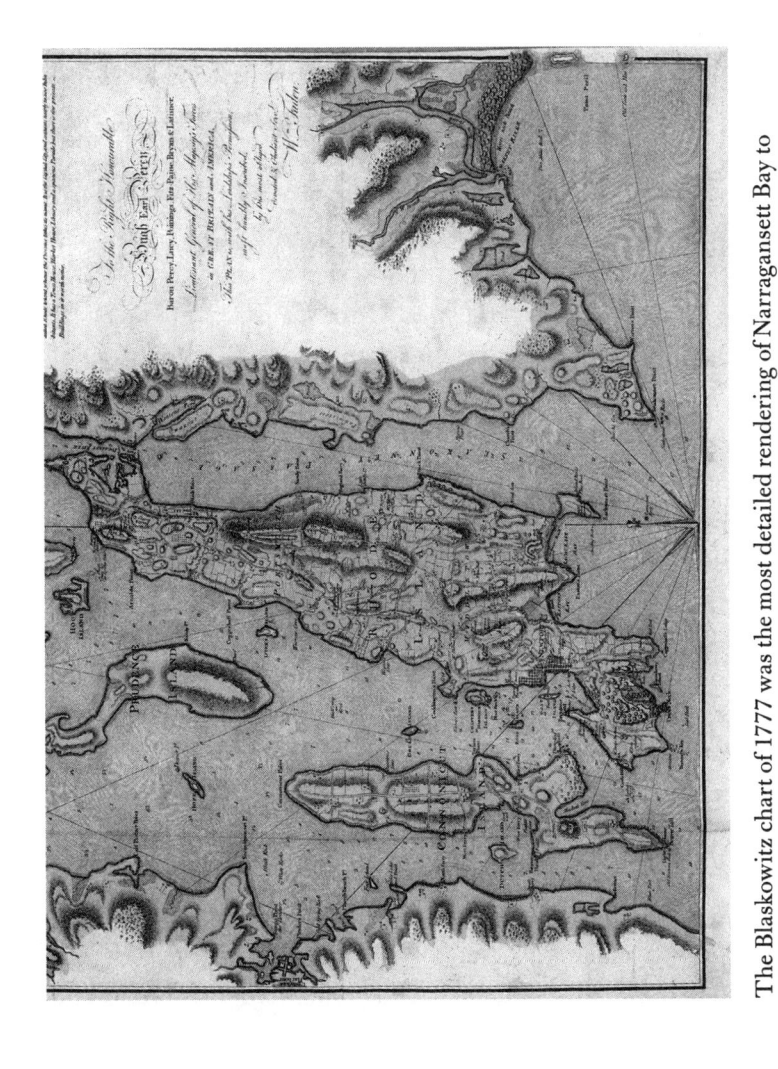

The Blaskowitz chart of 1777 was the most detailed rendering of Narragansett Bay to date, providing imperial authorities with unprecedented control over the coast. Charles Blaskowitz, *A Topographical Chart of the Bay of Narragansett in the Province of New England . . .* (London: Engraved & Printed for William Faden, Charing Cross, 1777). Courtesy of the John Carter Brown Library at Brown University.

Navigation up to Providence and Calculated for a Shelter for men with small arms but without cannon."[70] Even an "Iron Foundery where they cast cannons" was included outside of Providence. Faden also accompanied his chart with a narrative description of Narragansett Bay and its surroundings. He noted in the right margin Rhode Island's latitude and longitude, explaining that the colony was situated in the "most healthy Climate in North America." He explained that the winters were "severe, though not equally so with that of other Provinces" and that the summers were "delightfull . . . being allayed by the cool and temperate breezes that that come from the sea." Home to "one of the finest Harbours in the World," Narragansett Bay had "Fish of all kinds . . . in the greatest plenty and perfection." So too were its animals on shore: "Horses are boney and strong, the Meat Cattle and sheep are much the largest in America, [and] the butter and cheese excellent." Narragansett Bay, Faden's chart avowed, provided "every necessary of Life in Abundance."[71] As his colorful visual and textual descriptions revealed, Blaskowitz's survey and its 1776 and 1777 renderings provided much more than military reconnaissance.

The completion, publication, and widespread circulation of detailed charts and maps marked a wholesale shift in the ways this littoral space was understood. Blaskowitz's surveys as interpreted by Des Barres and Faden added coherence to the coast. Chandler's 1741 boundary commission surveys had produced a map that clearly displayed the general contours of Narragansett Bay, but it was the Blaskowitz chart that provided a truly comprehensive depiction of Bay waters. For coastal navigators, there is no substitute for local knowledge, but Blaskowitz's work placed navigational power in many more hands. Most striking in the case of Narragansett Bay was the extent to which those hands included imperial authorities for whom this stretch of coastal space had been largely ungovernable since English settlers had first inhabited its shores. Using Des Barres's chart, a naval shipmaster could, from almost any point in the Bay, triangulate his position. A drop of the lead line added bathymetric knowledge to his calculations. The map made it possible to estimate distance and time and reduce broad stretches of the coast into ordered, discernable units. Once it had been systematized, ocean space became pliable and more easily manipulated.[72] Surveyed and sounded, divided into latitude and longitude, the eternal sea, or at least one corner of it, had been rationalized and subdued.

Human efforts to order the oceans had been long in development. The introduction of the nautical compass in Europe during the thirteenth century had ordered oceanic space and patterns of human movement in dramatic ways.[73] According to Sir Francis Bacon, the compass, alongside the printing press and gunpowder, ushered in such sweeping changes that "no empire or sect or star seems to have exercised a greater power and influence on human affairs."[74] The highly detailed medieval portolan charts of the Mediterranean and Atlantic coasts likewise ordered the ocean's edge.[75] When in 1569 Gerardus Mercator published his world map, the first to scale lines of latitude as they moved closer to the poles, he further marshaled oceanic space by enabling navigators to plot courses across vast distances in straight lines.[76] Although mariners had long been able to determine latitude by measuring the angle between their ship and the sun and stars, longitude remained elusive until Englishman John Harrison introduced his H-4 marine chronometer in 1759, thereby allowing shipmasters to plot their position with reasonable accuracy anywhere on the globe.[77]

The promise of this new navigational technology, however, was slow to manifest. Chronometers proved so costly that few captains could afford one, let alone the two or three that were customarily used to ensure redundancy. As late as 1802, only 7 percent of British warships sailed with a marine chronometer. Instead, British naval officers were trained to perform lunar observations, the math for which was so complex that errors were frequent. When Admiral Lord Howe sailed for America in 1776, he nearly ran aground on Nantucket Shoals when his calculations placed him 300 miles off course.[78] If, however, blue-water navigation was still fraught with imprecision, recent advancements in tidal theory and piloting had asserted a new level of mastery over coastal spaces.[79] This new knowledge, combined with the vivid graphic representations produced by the Blaskowitz survey, had in many ways tamed Narragansett Bay.

Once commerce, science, and the specter of conflict had made Narragansett Bay more accessible, the grim realities of war—military occupation, forced migration, and outright violence—changed the place physically. By the time the H.M.S. *Rose,* a twenty-four-gun Man O' War entered Newport Harbor on the morning of December 11, 1774, a Rhode Island vessel, observed Ezra Stiles, had already set sail for Providence with a few remaining cannons from Aquidneck Island, lest they fall into the hands of the Royal Navy.[80] Gripped by panic, the people of Rhode Island were preparing, if not for war, then at least to defend their interests. Their preparations proved prescient, for in the following months the *Rose* hobbled Rhode Island's commercial interests and brought its principal city, Newport, to its knees. By the next summer, Deputy Governor Nicholas Cooke, writing on behalf of the General Assembly, demanded of the *Rose*'s captain, James Wallace, a reason for "interrupting . . . lawful trade, and preventing the importation of the provisions necessary for . . . subsistence."[81] Without addressing Cooke's questions, which the ever-condescending Wallace assumed had been written "on behalf of some body of people," his response demanded only to know "*whether* or *not,* you, or the people on whose behalf you write, are not in open rebellion to your lawful sovereign."[82]

Cooke deigned no reply, but the answer was an unequivocal yes. The General Assembly, observing the large and growing British fleet in Boston, had resolved to raise a Rhode Island army "for the defence and security of our lives, liberties and estates."[83] Responding to the "very dangerous crisis," Rhode Island commissioned 1,500 soldiers and ordered 2,488 lbs. of powder, 3,978 lbs. of lead, and 15,678 flints to be distributed among its towns in April.[84] These preparations did not go without protest. Rhode Island Governor Joseph Wanton anticipated that the move would threatened the colony's charter and "involve the country in all the horrors of a civil war." He also felt that Rhode Island simply could not afford such an extravagant display.[85] But the General Assembly would have none of it, refusing to administer their newly elected Governor the oath of office. Wanton's defiance rendered him powerless. And so the military mobilization continued.

Rhode Island had begun its build-up on land but quickly realized the necessity of securing itself at sea. In June 1775, the Assembly voted to equip two vessels to defend the colony, the larger of which would hold eighty men, ten four-pound guns, fourteen swivel guns, as well as a number of small arms and "all necessary warlike stores."[86] Rhode Island authorities also ordered the construction of two "row-gallies" capable of carrying fifteen sets of oars, swivel guns, an eighteen-pound cannon, and sixty fighting men.[87] In an effort to assert control over their shores, legislators passed laws restricting the dissemination of specialized Bay knowledge. Anyone who acted as a pilot on any armed ship, taking them "in or out of any of the harbors, rivers, or bays" faced stiff penalties.[88] In addition, new systems of communication consolidated the coast. Job Watson was stationed at Tower Hill in Kingston with the express purpose of maintaining a lookout, "in case any squadron of ships should be seen." On a hill east of Providence, a beacon was erected "to alarm the country [to the north], in case of invasion."[89] If removing a few cannons from coastal forts had left southern Bay waters in temporary disarray, the collective vigilance of the colony's Whig majority added cohesion, particularly among the brackish waters to the north.

Recognizing that southern Rhode Island had effectively already fallen to its foes, legislators maneuvered to cut their losses by ordering a massive evacuation of the colony's most valuable commodity: its animals. In August 1775, the General Assembly ordered 250 men to Block Island to remove "all the . . . cattle and sheep upon New Shoreham, excepting a sufficiency for the inhabitants." The best animals were immediately sent to the army, and the remainder was "disposed of to the best advantage."[90] Towns across the colony prohibited more than two neat cattle and five sheep from being ferried to Newport at a time, lest they be "taken by our enemies." With the locus of imperial power centered on the Bay, the islands were particularly susceptible to enemy control. As a result, Rhode Island legislators mandated that all animals be removed to the mainland under the protection of two armed vessels. If colonial record keepers were accurate, on August 30, 1775, 35 oxen, 48 cows, and 344 sheep were ferried from Conanicut to South Kingstown. Three days later, 1,908 sheep and lambs were removed from Block Island. Although their precise numbers are unknown, the animals on Prudence and Hog Islands were floated to Bristol.[91] According to Ezra Stiles, dispatches to George Washington reported that, in addition to sheep and other livestock, there were 5,000 head of cattle on Aquidneck Island that would certainly fall into enemy hands if Rhode Island forces there were withdrawn.[92] It had become increasingly clear that the vast herds that had fueled Britain's Atlantic world expansion were becoming important cogs in the engines of war.

People, too, were on the move. Lawmakers lamented that Wallace and his crew had prevented all "ferry boats, market boats, fish boats and wood vessels" from entering Newport with food and fuel. As a result, Newport was "exposed to the dreadful consequences that must inevitably arise through the want of the common and usual necessaries of life."[93] The city was in disarray, its people frantically packing their possessions onto wagons, horse carts, and boats.[94] By October 1775, Stiles noted that only thirty of 130 families in his congregation remained and that two-thirds of the inhabitants at the northern end of Aquidneck had fled.[95] Stiles himself was preparing to leave as well. Having removed a load of his own books and furniture on October 9, he fumed that the "infernal Wallace" had been met by several other warships, which were sailing north to "spread Terror thro' the Bay." When the people of Bristol chose to remove their dead and dying (largely from dysentery) instead of delivering 300 sheep to British forces, Wallace blasted the port for an hour. The next morning he fired on Portsmouth while another ship targeted Jamestown, his fleet "firing as if they would fill the Heavens with Thunder."[96] The situation had become so dire by January 1776 that Rhode Island legislators "strongly recommended" that the elderly, women, and children of Newport "remove to some place of safety."[97]

Years of coastal reconnaissance, careful mapping, and then outright aggression had, by the end of 1776, tamed the Bay in ways that opened the doors to an even wider British occupation. When army officer Frederick Mackenzie observed the massive mobilization of British troops on Long Island, he posited that they were bound for Rhode Island because "the harbour is a very fine one for our large ships, the Island [Aquidneck] very defensible."[98] His intuition was correct, for on December 6 and 7 the British fleet departed the East River, sailed up the Sound, and into the Bay. After conveying several regiments of British and Hessian soldiers onto Aquidneck Island, where they took control of abandoned American batteries, the Navy closed Rhode Island's principal points of entry. The frigate *Emerald* anchored between Aquidneck and Prudence Island, thereby closing the Bay's East Passage. Another frigate anchored at the northern end of Prudence Island to close the Bay's westernmost entrance. Two more frigates controlled the Sakonnet River.[99] If neither the *Rose* nor rebel forces had fully controlled the Bay during late 1775 and early 1776, the arrival and calculated positioning of so many warships had by year's end tipped the balance of power toward Britain.

But for a coastal people who knew the Bay's sandbars and tidal rips, who deftly navigated its undulating edges and the broad, deep passages in between, the unruly nature of the littoral provided numerous opportunities. On June 4, 1777, Mackenzie noted that a rebel sloop sailing under a "favorable wind" left Howland's Ferry, a narrow, quarter-mile passage between Portsmouth and Tiverton, and sailed south down the Sakonnet River toward Rhode Island Sound. The British battery at Fogland Ferry, a sand-and-cobble cape jutting from the Sakonnet's eastern shore, fired seven shots at the escaping Americans "without effect." On June 13, an American frigate, brig, and several other vessels, sailed out of the Providence River and anchored. Mackenzie surmised that they would "attempt to go to Sea the first fair wind." And these were not isolated cases. On that same day several boats filled with patriot troops passed between Mount Hope and the northern end of Portsmouth unmolested. Less than a week later a rebel schooner sailed out of Bristol and through the narrows at Howland's Ferry. Sustained damage from British fire, Mackenzie recalled, "did not prevent her from passing on."[100] That American ships frequently chanced escape with little compunction and in broad daylight suggests that British control of the Bay was tenuous at best.

Under the cover of darkness, rebel forces often held the upper hand. On the night of June 10, roughly fifty patriot fighters veiled in "thick fog" slipped across a small bay at the northern end of Aquidneck Island and surrounded and attacked a military post along the road to Common Fence Point. British soldiers repelled and then pursued their attackers, but, as Mackenzie explained, "they could discover nothing of the Enemy, except the noise of their Oars in going off in their boats."[101] A month later, a band of Rhode Islanders in small craft landed on Aquidneck Island, about five miles north of Newport, and, picking their way through the darkness to a house on the West Road, kidnapped the commander of British Forces on the Bay, General Richard Prescott, who was then rowed into rebel territory. Astonished, Mackenzie noted that "It is certainly a most extraordinary circumstance, that a General Commanding a body of 4000 men, encamped on an Island surrounded by a Squadron of Ships of War, should be carried off from his quarters in the night by a small party of the Enemy ... without a Shot being fired." So skilled were Rhode Island watermen at using the Bay to their advantage that they caused at least one British official to wring his hands in dismay. After an American brig slipped out of Bristol one night, it sailed carefully outside the reach of British cannons and anchored in plain view near Mount Hope. Mackenzie, the frustrated officer in charge, moaned that the "whole Rebel fleet may get out if they have only the spirit to risque a few shot from our Batteries as they pass, for as our Frigates are now stationed they cannot prevent them, if they take the proper advantage of Winds & Tides."[102]

Despite their efforts to patrol the channels and fortify the shores, British military authorities recognized that even the smallest American watercraft made their floating pickets permeable. In June 1777, patriot diarist Fleet Greene noted that British commanders in Newport placed "great restraints" on fishermen, thereby requiring them to "haul up their boats." Greene also noted that when British soldiers sacked and burned Warren and Bristol in May 1778, they torched 120 flatboats, took control of one galley, and burned several others before returning to Portsmouth.[103] His mind at ease following the same events, Mackenzie, the British officer in charge of the northern end of Aquidneck Island, concluded that the "destruction of the Armed vessels and so many boats, must undoubtedly prevent the Rebels from making an attempt on this Island for a considerable time." For good measure, several days later British troops descended upon Fall River and destroyed a "quantity of Plank for building boats" as well as the sawmill that had produced it. The British campaign to control water travel also served to suppress discontent within its own ranks. When the prodigious early summer runs of migrating fish reached Narragansett Bay in June 1778, Mackenzie explained that Rhode Islanders were limited to fishing with "Seines" because they had "neither boats nor Nets." "Indeed," he rued, "they are not to be trusted with the former, on account of their inclination to carry off [British] deserters."[104]

This watery borderland in which neither British forces nor Rhode Island rebels held full control was transformed once again when the

French fleet sailed into Narragansett Bay. Following the British surrender to American forces at Saratoga in October 1777, American diplomats brokered the Franco-American Treaty of 1778. This new alliance transformed both the war and the largely littoral theater in which it was fought. On the morning of July 29, 1778, the French Fleet under the command of Jean-Baptiste, comte d'Estaing was spotted sailing south of Narragansett Bay. The Newport town crier spread the alarm and, as Greene explained, the Tories fell into the "greatest confusion." Mary Almy, a Tory observer, explained that "distressed" shopkeepers quickly placed "locks and bars" on their doors. "Heavens!" she declared in her diary. "With what spirit the [British] army undertook the old batteries; with what amazing quickness did they throw up new ones . . . so earnest were they to give the Count a proper reception."[105] But their efforts were too little, too late. Soon after French warships entered the Bay's westernmost passage, British soldiers on Conanicut Island were forced to "quit the works and leave the Island to the Provincials."[106] Mackenzie noted that French warships streamed into the Bay and quickly secured all of its exits. Rather than surrender vessels to the superior firepower of the French, many British ships were set ablaze or run aground. The *Lark* was torched and its magazine of seventy-six barrels of gunpowder exploded, blasting burning embers clear across Aquidneck Island. When the British warship *Orpheus* exploded, books from its library landed three miles away.[107] Others were scuttled at the northern and southern ends of Goat Island to block entry into Newport.[108] "It was a most mortifying sight . . . ," Mackenzie wrote, "to see so many fine Frigates destroyed in so short a time, without any loss on the part of the enemy."[109] Any houses that could potentially provide cover for a French invasion were also fired. Mirroring the patriot flight two years earlier, British soldiers and American loyalists packed their belongings in fear. "[T]he dismay and distress so strongly impressed upon the Countenances of the Inhabitants . . . ," Mackenzie wrote, "formed altogether a very extraordinary Scene."[110] As the French fleet and rebel troops surrounded the island, clouds of smoke gave way to impenetrable fog. Exhaustion soon set in. Throughout Newport, Mary Almy recalled, "A solemn silence reign[ed]."[111]

This wrenching realignment of power utterly transformed the Bay and its islands. Not only was Newport Harbor filled with sunken obstructions,

The American War for Independence laid waste to Aquidneck Island. "The Siege of Rhode Island, taken from Mr. Brindley's House, on the 25th of August, 1778" in *The Gentleman's Magazine* (London: February 1779). Courtesy of the John Carter Brown Library at Brown University.

but the thousands of displaced marines forced ashore exacted a heavy toll on resources that were already severely diminished. Stray soldiers combed the countryside for food, seizing animals and stripping gardens clean. "The army," wrote Greene, "continues to lay waste the Island, cutting down orchards and laying open fields." Mackenzie noted that British soldiers systematically razed trees near Newport to build an abatis, or wooden embattlement, behind which they constructed an earthen breastwork upwards of six feet thick and four-and-a-half feet high. American forces, too, built embankments behind their lines, which extended as far south as Honeyman's orchard in Middletown.[112] When roughly 8,000 American soldiers led by General John Sullivan poured onto the north end of Aquidneck Island while the French bombarded the south, the island shuddered. Fires raged down its length as troops torched houses, fields, and fences. Bombs rained down and bodies piled up in the roads. "[Newport] Harbor," wrote Greene on August 8, "is in one continual blaze."[113]

Even nature itself played a role and took a toll. When the French pulled out of Narragansett Bay to meet British reinforcements under the command of General Howe, American Tories and British soldiers, Greene explained, expressed an "excess of joy and grief."[114] Such mixed emotion, however, was soon supplanted by certain misery when on August 12 a fierce gale, perhaps even a hurricane, blasted Aquidneck Island and the ships vying to control it. Severely disabled by the storm, the French fleet limped into the Bay several days later while the bodies of its sailors washed ashore.[115] The calm weather that followed, however, belied the increasing fury with which the war for Aquidneck continued. The incessant hammer of guns made anxiety, hunger, and death ever more conspicuous. When Hessian soldiers pursued rebel forces on August 22, Almy recorded that the East Road was "strewn with dead bodies." This was, she avowed, "a scene of blood and slaughter."[116] Fighting culminated on August 29 in what became known as the Battle of Rhode Island, which saw American patriots inflict heavy casualties on British troops and which marked the last major siege of the northern campaign.[117] But when rebel troops retreated from the island's northern end the following night, British and Hessian troops took no quarter. "[T]he inhabitants of Portsmouth and Middletown are plundered," wrote Greene. "Some families are destitute of a bed to lie on."[118] With the French refitting in Boston, British reinforcements reoccupied Newport, which remained in their hands for another year. Nine months after the English departed in October 1779, the French, under the command of the Comte de Rochambeau, returned to Narragansett Bay in July 1780.[119] If the wind, weather, and ebb and flood of an estuary could shape the rhythms of war, it surely had at the mouth of Narragansett Bay, where its waters and islands swapped hands time and time again. Fighting in the mud and muck of a marsh was messy business. Outcomes were rarely conclusive. Progress was often met with regress. And these patterns were repeated along the entire length of the Atlantic coast.

The American War for Independence was, after all, largely a littoral war. The "shock troops" of resistance, as one historian characterized waterfront people, played a central role in inciting protest up and down the Atlantic coast.[120] In response, in 1776, Britain amassed the largest display of marine military power ever by a European state.[121] But when the world's great blue-water empire waded chest-deep onto the shoals of its brown-water brethren, the results were less decisive than the Crown had hoped.[122] When British Regulars commenced their march on Lexington and Concord in the early morning hours of April 19, 1775, they first traversed the Charles River and landed at Lechmere Point, where they spent several hours laboring through the icy, muddy waters of the Cambridge marshes. In many ways, the bloody outcome of their mission was conditioned by, if not in some small part contingent upon, the delay caused by their tidal swamp traverse.[123] Similarly, the high casualties at the Battle of Bunker (Breed's) Hill were conditioned by the tide, which delayed the British landing on the Charleston shore, where the Mystic and Charles Rivers met the waters of Boston Harbor.[124] Control of New York was contested in the brackish borderlands of the Hudson River and Long Island Sound estuaries, and that of Philadelphia among the waters of the Delaware River estuary and Chesapeake Bay.[125] Even the movement of troops to interior theaters of war was affected by coastal dynamics. General James Robertson, according to Ambrose Serle, Lord Howe's secretary, explained "that he has known a Vessel reach Albany from New York in one Tide, which reaches the former Place in about 24 Hours."[126] When Howe decided not to ride the tide up the Hudson during the summer of 1777, turning instead to the Chesapeake, he opened the door to Burgoyne's defeat at Saratoga, which did much to forge a French-American alliance. Farther south, Charleston and Savannah were clearly coastal contests. And the Siege of Yorktown in 1781, which brought an end to hostilities, was an entirely amphibious effort, one whose outcome was conditioned in part by British susceptibility to malaria, which ran rampant among the swamps and marshes of the southern coast.[127]

The coastal dimensions of the war, moreover, were global. Up and down the Eastern Seaboard, naval convoys and patriot privateers clashed as they ferried food, firewood, and personnel between America's principal ports and among the countless creeks and rivers in between.[128] So obviously salty were the American theaters of war that they spurred Congress to create its own navy, handing command to Rhode Islander Esek Hopkins and appropriating nearly a million dollars to convert four vessels into warships and build thirteen new ones. Although only half were ever completed and Hopkins was removed from his post after only a year and a half of service, American sea fighters caused a stir, particularly when John Paul Jones and his crew sailed across the Atlantic and began attacking British towns and taking prizes.[129] Similarly, French ships wrought havoc across the West Indies. In September 1778, Dominica fell to French forces, as did St. Vincent and Grenada less than a year later.[130] American privateers also harassed British West Indian shipping, the islands, and even the Royal Navy, thereby requiring British strategists to redirect ships that could have otherwise been used to blockade the American mainland.[131] Once Spain and Holland entered the war in 1779 and 1780, fighting spread from the West Indies to the East.

All told, the waves of war slammed the American coast for six years. At times, they shifted the sands of sovereignty toward the continent—toward progress and the transformation of human relations. And at others, they drew them out to sea—toward tradition, the old order, and stasis. And in this sense, the American Revolution, as an ideological transformation, as something connected to but nevertheless distinct from the war for Independence, was littoral as well.

While its proximity to the sea had ushered in a golden age of growth in Newport before the Revolution, years of exposure to the storms of war had left it heavily and permanently eroded in the years that followed. Soon after arriving in Newport in December 1776, Frederick Mackenzie had climbed Quaker Hill roughly seven miles north of Newport and gazed over Narragansett Bay. He saw "many fine and well cultivated Islands, and the beautiful bays and inlets, with the distant view of towns, farms, and cultivated lands intermixed with Woods." The Rhode Island that unfolded before him was "the finest, most diversified, and extensive prospect I have seen in America." So striking was the scenery that he thought "The Ships of War . . . stationed in the different passages . . . appear as if they were placed there only to add to the beauty of the Picture."[132] But when Ezra Stiles, who had since become the President of Yale, returned to Newport in 1780, he described a very different landscape. Riding his horse across Aquidneck, he noticed that trees that had once lined the roads had all been felled. The "sundry Coppices or Groves & Orchards" had been, he wrote, "cut down and laid waste." He estimated that in

Newport about 300 houses had been destroyed. Although he still praised the island for its "natural Beauties," the city, he avowed, was "in Ruins."[133] Similarly, the geographer and Congregational minister Jedidiah Morse explained that before the war, 30,000 sheep had grazed the grasses of Aquidneck Island but by 1787 only 3,000 remained. So dramatic were the war's effects on Newport that he described it as "truly melancholy and distressing."[134] The city's waterfront had been hit particularly hard. Littered with sunken debris, the harbor was impassible with few options for ships to land, for during the frigid winter of 1778 and 1779 British troops had torn up the wharves for firewood.[135]

The War for Independence had made its mark on Narragansett Bay. When Francisco Miranda, a Spanish traveler, landed in Newport in 1784, he wrote that Aquidneck Island "justly deserves the title of the Paradise of New England," but that its lack of "trees and woods which the armies entirely destroyed during the war . . . gives an arid look to all the pastures and cultivated lands like that of the Mancha in Castille."[136] The remnants of "destroyed" military escarpments likewise defaced the country. A network of trenches and embattlements surrounded Newport, thereby "cutting the Island into two pieces." Even the Redwood Library had lost its luster, since, he explained, British troops had pilfered the collections. The buildings of Newport were equally decrepit, most resembling "hovels" and with no more than three dozen houses of "decent appearance." "It is a pity," Miranda concluded, "that such a beautiful location and one so advantageous to commerce should be in such a miserable and decadent state."[137] When Jacques-Pierre Brissot de Warville visited Newport four years later, little had changed. He described "idle men, standing with folded arms at the corners of the streets." With "rags stuffed in the windows," houses were "falling to ruin." Newport's "miserable shops," he explained, "present nothing but a few coarse stuffs . . . and other articles of little value." "Every thing," he wrote, "announces misery."[138] When the Newport native and celebrated physician Benjamin Waterhouse sent a letter to Thomas Jefferson in 1822, he regaled his longtime friend and colleague with stories of Rhode Island's golden age but admitted that war had changed the place. "The British destroyed for fewel about 900 buildings," he wrote, "[and Newport] has never recovered the delapidation [sic]." In subsequent years, Waterhouse explained, "The town of Providence has risen to riches & elegance from the ruins of the once beautiful spot; while Newport resembles an old battered shield—its scars and bruises are deep and indelible."[139]

As Waterhouse acknowledged, northern Narragansett Bay was recovering. When the Duke of La Rochefoucauld-Liancourt, François-Alexandre-Frédéric stopped in Rhode Island during his mid-1790s North American tour, he echoed many sentiments of earlier postwar observers. He praised Newport's superb harbor, but like Brissot de Warville, believed the small, unpainted houses surrounding it were "miserably bad." The entire island, he lamented, had been denuded "when the great trees were destroyed by the English during the war." In consequence, islanders were plagued by "an extreme scarcity" of fuel wood, and the land was "too much exposed to the winds, which often blow over it with a very troublesome violence."[140] But as he traveled from the north end of Aquidneck to Bristol, he admitted, "The prospect of the bay, of the islets with which it is interspersed, and the main-land contiguous to the bay, is extremely pleasing." He noted that although Bristol had taken considerable fire during the war, the houses there had been rebuilt and were "now more numerous than before." Nearby Warren, he explained, was building between eight and ten ships per year. Barrington, separated from Warren "only by a [tidal] river of its own name," comprised 150 "good houses." All three seaports, he avowed, were "much better built than Newport," and their town land values were appreciating. Passing north and west to Providence, the Duke came to rest at the home of the Quaker merchant Moses Brown. "Lofty, well-built and well-furnished houses," he observed, "are numerous in this town, which is becoming continually larger." For the Duke, the city and its surroundings at the northern end of the Bay were simply "more interesting than those of Newport." Providence had a distillery, "perhaps the greatest in the American States," as well as "extensive manufactures of nails and other forged iron-work." In addition, he noted that within the last year, "endeavors have been made to introduce the manufacture of cotton-yarn and stuffs into Providence." Since he was lodging with Moses Brown, the "author of this undertaking," the Duke no doubt received a detailed description of the enterprise, which he was told was "already profitable." The incredulous Duke, however, believed "this to be rather the boast of sanguine expectation, and of self-conceit, than the actual truth of the facts."[141] For all his ability to describe Rhode Island's postwar littoral landscape, this noble wayfarer was unable to see the new revolution that was about to reshape the Bay and America's future.

The proliferation of coastal beacons and forts changed Narragansett Bay for the people who plied its waters. War and the graphic representations of the coast that facilitated it transformed littoral space in ways that reached much farther. During the War for Independence, the French, who sought charts of Narragansett Bay, borrowed heavily from both Faden's and Des Barres's, thereby producing still more maps from Blaskowitz's work.[142] As military generals and naval shipmasters unfurled these plans, weighted their corners, and then positioned their troops and ships on paper, they added order to Narragansett Bay. In some cases, water and overland routes were closed or at least monitored with cannon batteries and lookouts. In other cases, they took control over resources such as food and fuel. In their war rooms and aft cabins, imperial strategists took these steps with a bird's-eye view of an otherwise geographically complex and highly variable space. But even the best-laid plans could not exceed the power of local knowledge. In all size of watercraft, by day and night, Patriot fighters often eluded British patrols. And in some cases, rebel forces used the Bay to take the offensive. If progress had come to Narragansett Bay with imperial efforts to map it, the trackless nature of the sea provided local people with watery avenues of opportunity. But by war's end, the push and pull of imperial control and local resistance had torn Aquidneck Island into pieces. Treeless, windswept, and wasted by fire and the spade, the island at the mouth of Narragansett Bay was destroyed.

Like the thousands of streams that fed the rivers that flooded the estuary each day, progress trickled into Narragansett Bay during the eighteenth century. Rhode Island's vast herds of animals had changed its tidal rivers and lagoons. Weather diarists had begun to organize the atmosphere. Inquiry into the natural history of Bay creatures, most significantly the oyster, had set the groundwork for understanding biological production in Bay waters. And mapping and war fundamentally reshaped the contours of political sovereignty and local economies. Although Rhode Island, true to its reputation as an outlier, had been loath to ratify the US Constitution, its legislators, facing commercial alienation and even threats of invasion, reluctantly agreed in 1790, making it the last of the thirteen former colonies to join the union.[143] Rhode Island's coastal margin was integrated into a new continuous American "coastline," one that mapmakers penned with the goal of forging a more unified national identity.[144] And with a growing sense of geographic cohesion, Americans increasingly imagined themselves as a "continental people."[145] Once an unmanageable borderland whose aqueous, indeterminate nature created geographies of iniquity for some and refuge for others, Narragansett Bay had by the end of the eighteenth century been improved physically and organized intellectually in ways that opened new possibilities for coastal environmental change during the nineteenth century. A space that had once denied human progress, that had stemmed the tides of history, was breached by the storms of technological improvement that at century's end gripped Narragansett Bay's northern reaches and transformed the rest of America in the process.

CHAPTER 6

Carving the Industrial Coastline

JUST ABOVE THE tidewater at Mill Bridge in Providence, a large crowd gathered on the morning of July 1, 1828. Dressed in his finest, Rhode Island Governor James Fenner joined fifty notable citizens aboard what could only be described as a "gentleman's barge." Seventy feet long, nine-and-a-half feet wide, and drawing only eight or nine inches of water, the *Lady Carrington* was fitted with a low-lying cabin top, "conveniently and neatly arranged," explained a local newspaper, which extended almost the full length of its hull. Hundreds had gathered along the river's banks to inspect, and perhaps even scoff at, a boat the likes of which had never been seen in Rhode Island waters.

At around 10 a.m. a cannon was fired, and the *Lady Carrington*'s guests erupted into applause. The "enlivening notes of a band of music" met the clop of hooves as the first boat to ply the waters of the newly built Blackstone Canal lurched northward from Narragansett Bay toward Worcester at the headwaters of the Blackstone River in central Massachusetts. Steadily, the horses pulled the *Lady Carrington* at a pace of four or five miles per hour on the levels, while pausing occasionally to pass through granite locks "of most substantial masonry," which was "particularly gratifying to those who had never before witnessed this operation." Apart from the "many hard thumps" the *Lady Carrington* received inside the locks, the canal boat's journey was, unlike those of its oceangoing counterparts, eerily silent. Passing mill buildings and tree-lined banks made of "good stone wall," the *Lady Carrington* ghosted through Horton Grove, "a most romantic spot directly on the Bank of the Canal," and then stopped at Bishop's Tavern near Scott's Pond where they were "most hospitably entertained." Echoing the chatter of so many conversations, the crowded canal boat was such a spectacle that "scores of neatly dressed females . . . thronged the windows" of nearby factories. At the village of Albion, where lock gates had not yet been installed, the *Lady Carrington* reversed direction, arriving in Providence that evening "amid a large concourse of spectators who had assembled to greet her arrival."[1]

The fanfare with which the *Lady Carrington* commenced and concluded its maiden voyage reflected the widely held belief that the Blackstone Canal would usher in a new age of prosperity. In the same newspaper account, its author announced that beyond providing a "delightful jaunt," the canal promised "immense benefits that must soon result from connecting the Atlantic with the most fertile portion of Massachusetts, the line of communication abounding with sites occupied by flourishing manufacturing establishments."[2] The forty-five-mile canal between Providence and Worcester, to the north, would, he explained, connect the port of Providence with a new agricultural and manufacturing hinterland. A link with the sea would connect Worcester to distant markets and would provide Providence merchants with new sources of agricultural production and new outlets for selling European goods and West Indian commodities at home. This, many hoped, would bring economic growth to the entire region.

Port cities in particular were looking for new avenues of commercial expansion amidst the bustle of early nineteenth-century America. Boston, whose population had ballooned from 18,000 in 1790 to 33,000 in 1810, was filling its marshes to make room.[3] Similarly, Providence had, in the years following the American Revolution, undergone dramatic growth, eclipsing Newport to become Rhode Island's most powerful city. The reasons for this shift were many. The British occupation of Newport during the war had taken its toll. In addition, in the years following, the British closed West Indian ports to American merchants, and London levied duties on American imports, particularly whale oil, which hit Newport hard. As international trade, especially that with Asia, required increasing amounts of capital, commercial activity became concentrated in larger cities like New York and Philadelphia, which benefitted from ready access to continental resources. Newport, whose exposure to and rigid commercial orientation toward the ocean had left it broken and nearly bankrupt, was outmatched.[4] After "her downfall," noted one contemporary chronicler, Newport had become but a "mere resort of strangers for a few weeks" every summer.[5] Responsive to the entrepreneurial maneuvering of a few influential merchants, most notably those of the Brown family, Providence invested in infrastructure by expanding its port and building roads and turnpikes.[6]

And the city expanded. Between 1790 and 1800 the population of Providence grew by 19 percent, while Newport's grew by only 4 percent.[7] Although the port of Providence was nearly destroyed in a massive gale that slammed the Bay in September 1815, the city soon rebuilt and expanded.[8] Between 1800 and 1820, the population of Providence grew by 54 percent, from 7,614 inhabitants to 11,745.[9] In that same time period, Newport's fell by 320 people to 7,319.[10] What had been one of the largest cities in America on the eve of the Revolution had failed to make the top twenty by the second decade of the nineteenth century. Although Providence merchants continued to look south over Narragansett Bay toward the sea, many looked inland as well. It was its ability to tap a terrestrial hinterland that made Providence prominent. Boston was similarly endowed, but whereas New England's largest city had embarked on an ambitious program of "gaining ground," Providence sought, rather, to extend the sea.[11] A forty-five-mile canal would effectively project Narragansett Bay deep into the New England interior.

If industrialization fundamentally transformed human interactions with nature, the Blackstone Canal made its most visible mark in the littoral. The establishment of an industrial market economy at the end of the eighteenth century marked a major shift in the ways that people related to the natural world. Capitalists sought to quantify, control, and assert mastery over nature by imposing new regimes of scientific management.[12] In New England, this shift reconfigured the region's major inland waterways.[13] Historical actors, however, typically set their gaze upstream where historical action was bound by the banks of a river. A look at the Blackstone Canal and its interaction with the broader estuary on which the American Industrial Revolution was born, conversely, reframes this "ecological revolution" in terms of how it affected the sea.

For so long, the estuary's tangle of rivers and the brackish bay into which they flowed had been shaped by the vagaries of wind, weather, and the tide. Flanked by thick walls of stone and its waters managed by a corporation, the canal, by contrast, sought absolute control. When the Blackstone Canal Company broke ground in 1825, mill owners along the Blackstone River protested, fearing the canal would divert water from their operations. Granting mill owners the "natural run" of the river, the courts compelled the company to build a complex network of holding ponds to manage the flow of water. From that point on, the "natural run" of the river was no longer a product of nature but the concession of a corporation. Narragansett Bay, or at least the upper reaches of its watershed, had been transformed beyond recognition.[14] The technological advancements and culture of scientific improvement that accompanied industrialization enabled the physical manipulation of littoral space, or at least the freshwater portion of it, as never before. Forty-five miles of ditches, trenches, locks, ponds, and dams extended coastal space deep into New England's interior. Moreover, the corporatization of this inland sea placed management of the littoral in private hands. But when the corporation and the system it had created failed, the canals and rivers became sewers for human and industrial waste, which, flowing downstream, subsequently posed the single most important environmental challenge for Bay waters.

The Canal Idea

In 1796 the fabulously wealthy Providence merchant John Brown, desirous of a navigable inland trade route from Narragansett Bay, hired an engineer to assess the feasibility of a canal from Providence north to Worcester and perhaps beyond to the Connecticut River. Brown's plan was to redirect the economic activity of western Massachusetts away from Springfield and that of Worcester away from Boston and channel it into his hometown, where he and his ancestors had, over the course of the eighteenth century, established a veritable commercial empire. Responding to a positive surveyor's report, John Brown prepared to invest $40,000; his brother Moses, his son Obadiah, Nicholas Brown, and Thomas Ives, a business associate, also agreed to contribute.[15] Brown applied to the Rhode Island legislature for an act of incorporation, which was soon granted, but Massachusetts, threatened by such an ambitious scheme hatched by one of New England's most powerful businessmen, denied his request. The plans for a canal were placed on hold.[16]

But by the early 1820s, the success of canal construction up and down the East Coast spurred renewed interest in developing the Blackstone Valley, which wound south-southeast from Worcester to Providence. The Middlesex Canal, which connected Boston and East Chelmsford, Massachusetts, was completed in 1803 and had, at least in its initial years, earned a profit. The Santee Canal in South Carolina, the first in America, had been operating since 1800. Other canals were under construction in Pennsylvania.[17] And the middle portion of the Erie Canal was open to navigation, while construction on the western and eastern portions was well under way.[18] Two groups of interested investors, one from Providence and another from Worcester, gathered during the spring of 1822 and commissioned Benjamin Wright, the Chief Engineer on the Erie Canal's middle section, to conduct a feasibility survey.[19] The Worcester contingent then penned a petition to the Massachusetts General Court asking for permission to proceed with its plans.

The petition highlighted the dramatic economic, social, and environmental changes promised by the construction of a canal. "[A]mong the many improvements which the enterprise of our countrymen is constantly projecting and multiplying," the subscribers explained, "none, at the present time, receive so great a share of publick attention, and promise such important results as the establishment of Canals with a view to enlarge and extend the natural channels of Internal Navigation." In its enthusiasm for the project, Rhode Island, the Massachusetts petitioners explained, "has given an impulse to the dormant energies of the country, and has opened to us new views of our present powers and our future destiny."[20] They explained to the General Court that those who would protest the construction of a canal "cannot any longer shut their eyes to the advantages with which nature seems to have peculiarly favored them." Establishing a connection between the "seaboard and the very heart and centre of our country," they emphasized, was akin to opening a "great artery" from which "new life may be diffused, as it were by a thousand veins, to every part of an extensive, fertile and

The Blackstone Canal extended Narragansett Bay forty-five miles north to the town of Worcester in central Massachusetts. Anonymous. n.d. Oil on canvas. RHi X3 3307. Courtesy Rhode Island Historical Society.

populous territory." Ultimately, the Blackstone Canal, avowed the petitioners, promised to "develop the immense resources of this territory, give a home to its native population," and showcase "the ingenuity of our mechanics."[21] In its promise of progress, the language of the petition suggests that the people of the Rhode Island coast had opened their eyes to the interior's true potential. The untapped energy of the countryside, in other words, would be released by establishing a connection to the sea. The storm of industrial progress would breach a guzzle through the dunes that would allow the once-stagnant waters of Worcester to at last flow downhill to the tidewater. The energy of industry would just have to be harnessed.

It was widely acknowledged that the canal would benefit not only its wealthy investors but also the broader community. Even before the petition for incorporation had been filed with the Massachusetts General Court, the Worcester-based *Massachusetts Spy* opined in April 1822 that the canal would "be highly beneficial . . . when it is considered what are . . . the wants of the County, as a manufacturing as well as an agricultural district."[22] The canal promised dividends to small farmers and manufacturers alike. On May 10, 1822 citizens crowded into Sikes's Coffee House in Worcester to weigh the pros and cons of a canal.[23] Among the people of Worcester, the sentiment was clear: a canal connecting the breadbasket of Massachusetts with the industrial heart of America was "hailed as highly auspicious."[24]

Nevertheless, there were others who believed improvements to the Blackstone Valley would hamper economic growth elsewhere in Massachusetts. Boston feared that communication between Worcester and Providence would lead to its own stagnation. One Boston paper concluded, "It would be well, if the citizens of Boston should endeavor to take measures, to prevent the loss of an important internal trade, which now centres at this city."[25] The *Boston Patriot* implored its "enterprising citizens . . . not to suffer our important internal trade to be diverted from our new and growing 'city.'"[26] The *Boston Centinel* protested the canal's construction, claiming that "any project which tended to decrease . . . prosperity [in Boston], though it might benefit a particular section, would be injurious to the general interests, and therefore on principles of 'sound public policy,' ought to be *rejected*."[27]

Completed in September 1822, the survey indicated that the construction of a canal was eminently achievable. Benjamin Wright's report, submitted on October 2, 1822, explained that the canal would stretch from the tidal limits of Narragansett Bay in Providence to Thomas Street in Worcester, climbing 451½ feet over forty-five miles. "The ground is remarkably favorable," he wrote. "The soil generally easy to excavate—the embankments neither large nor extensive—very little solid rock to be removed—the aqueducts and culverts are not numerous or expensive."[28] But one concern of the engineers was the availability of water not only to fill the canal throughout the year, but also to provide a sufficient supply to "the valuable hydraulic works now erected on Blackstone River and its branches." Unlike the Erie Canal, a canal along the Blackstone required engineers to consider the water needs of the numerous manufacturers whose livelihoods relied on a steady run of water. The smaller, trickling streams that formed the Blackstone's northern reaches, or the "summit level," would require considerable hydraulic engineering to make the canal system work. Wright noted that the 100-acre North Pond, two miles north of the Worcester Court House, could be dammed, which would flood the pond to 180 acres, allowing spring floodwaters to be saved for summer use. If the pond was eight feet deep and covered 140 acres, and the canal locks seven miles south of Worcester at Dority Pond in Millbury were designed to use 4,200 cubic feet of water each time they were filled, North Pond could provide the upper canal system with 11,616 locks full of water.[29] Wright's calculations considered almost every pond along the canal's route. Through intensive engineering, Wright's canal and the adjacent river would remain filled with water throughout the year. No longer tethered to the natural cycles of flood and drought, the canal, in Wright's estimation, would herein draw upon a heavily managed hydrologic system.

The projected benefits of the canal and Wright's propitious report garnered crucial support from the Massachusetts legislature. On January 15, 1823, the Blackstone Canal Company was granted a charter to build its canal to the limits of Massachusetts' jurisdiction. The charter also gave it the right to tap water from North Pond, Quinsigamond (or Long Pond), and Dority Pond "with such other ponds as lie upon or near said route." It also gave the company the right to construct reservoirs and

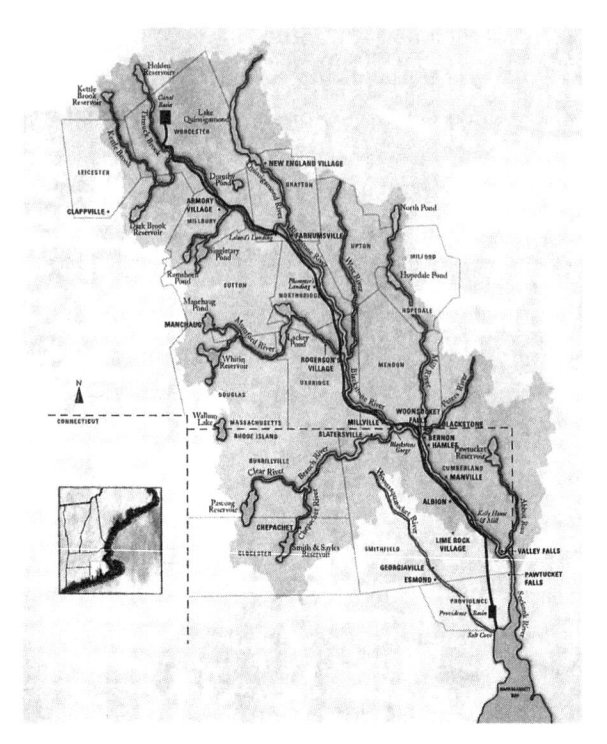

Map of the Blackstone Canal (c. 1830). Courtesy of the John H. Chafee Blackstone River Valley National Heritage Corridor/National Park Service. From *Landscape of Industry: An Industrial History of the Blackstone Valley* (Hanover, NH: University Press of New England, 2009), 9.

the power to "connect with said canal by feeders or navigable canals, any or all said ponds and reservoirs." The charter did, however, stipulate that the company would be liable for damages.[30] Upon receiving the approbation of Rhode Island legislators, the company would then open its book of subscriptions to investors. "We hope their exertions," wrote the *National Aegis,* "will not be intermitted [delayed], until the completion of this great and stupendous work."[31] On June 25, 1823 Rhode Island granted a charter to the Blackstone Canal Company in that state in a vote "nearly unanimous."[32] Although the Rhode Island charter was largely similar to that of the company's Massachusetts branch, it specifically addressed the canal company's obligations to maintaining the water level in the Blackstone River. "[I]t shall be the duty of said corporation," the charter mandated, "to allow the same quantities of water to pass from said ponds, brooks and streams of water, constituting parts of the sources of Blackstone River, whenever the same shall be necessary for the use of the factories and mills." The charter further stipulated that the factories and mills shall have "the benefit of the natural run of water."[33] It also, however, provided a number of contingencies when the "natural run" could be changed and set fines that would be levied should the company be found negligent, suggesting that legislators anticipated or had already entertained arguments over the canal's impact on water availability.

Mill owners had clearly voiced their objections. In a letter to Massachusetts Canal Commissioner John W. Lincoln, one canal advocate in Providence noted, "The only objection to granting a charter by our legislature will arise from the fears of mill owners that their water will be diminished by the proposed Canal."[34] On June 10, 1823 Rhode Island Canal Commissioner Stephen Smith wrote to Lincoln that "We had a very full meeting at Pawtucket of all those who are concerned in the waters of the Blackstone River in this state. The result of which was in no respect satisfactory to the friends of the canal."[35] Mill owners were so resistant that they maneuvered to stall the project. By the following March, little work on the canal had been completed, and Smith suggested to Lincoln that political posturing might force concessions from mill owners. "To bring them to terms," wrote Smith, "it had been thought best to talk the project down and in despair, appear[ing] to give up all hope of success when there is so much opposition. The fear of losing the water of Long Pond by a Canal to Boston will produce good fruit."[36]

Exasperation in Worcester at what was seen as commercial myopia among Rhode Island manufacturers had raised the possibility of digging east toward Boston. "[I]t becomes our duty, in justice to ourselves," wrote the *Massachusetts Spy*, "to improve the advantages we possess in the next best mode." That mode, of course, was a canal to the metropolis. "There can be no doubt," the papers continued, "that the Legislature of this State will grant leave to construct a Canal from this town to Boston on the same conditions that it was to have been done to Providence, as this diversion would confine the business within the State, which by the original plan, would have been carried to Providence." The cause of the stalemate between Rhode Island manufacturers and the Blackstone Canal Company, the papers argued, was that the Rhode Island charter was "so overloaded and embarrassed with unnecessary and vexatious restrictions, that it still remains, and probably forever will remain, a dead letter on the statute books of the State." Rhode Island's insistence on maintaining the "natural run" of the Blackstone River favored the rights of individuals over those of the Canal Company. To the business interests in Worcester, "The Corporation" in Rhode Island was "wholly at the mercy of every one who might fancy his individual rights trenched upon."[37]

The debate over who owned the rights to the upper reaches of the Narragansett Bay watershed had been percolating for decades. Rhode Island lawmakers had built provisions into the Canal Company's charter to protect the flow of water in the Blackstone River. These restrictive clauses reflected not only the economic importance of manufacturing to the area but also a keen awareness of the incendiary nature of any decision that pitted community rights against the desires of commercial entities. Throughout the eighteenth century, laws favored the interests of small farmers over commercial endeavors, such as the millers, fullers, and blast furnaces that depended on the rivers for energy.[38] For instance, the Hope Furnace Company, which built iron hollowware and cannons, entered arbitration in 1765 when its Scituate dam flooded the property of James Matheson, a local farmer. Siding with Matheson, the arbitrators ordered that Hope Furnace pay him an annual $110 in rent plus another $4 per year for flooding the land.[39]

But it was the river fisheries that most clearly highlighted tensions between the common good and commercial development. Important sources of food and income for Rhode Island farmers, the annual runs of alewife, blueback herring, shad, and salmon were threatened by river damming, which, accordingly, sparked protest.[40] At the insistence of farmers, Rhode Island passed laws in 1719 and 1735 that required clear passage for migrating river fish during their spring runs.[41] In 1742, 1743, and 1750, laws were passed to prevent fish from "being stopped in their course" on the Pawcatuck River.[42] When earlier fines of 40 shillings were considered insufficient to prevent seine fishing during the spawning run in Point Judith Pond, they were raised to £50 in 1755. In 1756, Job Randall and Benoni Waterman were appointed to draft a bill "for regulating mill dams with respect to the passage of fish up the rivers of this colony."[43] Time and time again, lawmakers had decided on behalf of the fish, which, tied closely to traditional ideas about common property, represented the greater good of the community.

So important were Narragansett Bay's river fish to Blackstone River communities north of Pawtucket Falls that people there physically transformed the environment to accommodate them. Scattered with boulders and pools, Pawtucket Falls had been an important fishing site for Native Americans before Europeans had settled the area.[44] Although some fish would have climbed the falls, others traveled around through a small channel known as "Little River."[45] When a new bridge was built across the Blackstone in 1714, Little River became clogged. When a dam was built across the river in 1718, there was no way for fish to travel upstream. On behalf of the fish, William Sergeant cleared Little River, which subsequently became known as "Sergeant's Trench." It provided fish with clear passage. That changed, however, when the banks of the fish-way became important mill sites. Over the course of the eighteenth century, dams were built in the trench to power the mills that sprung up along its banks. In 1730 a dam for an anchor mill was erected, and in the following decades several more were constructed to power ever-more-extensive manufacturing operations.[46] These projects elicited numerous petitions of protest from fishermen upstream but also garnered support from the more general population.[47]

The General Assembly largely found in their favor, requiring mill owners to provide clear passage for fish during spring spawning runs. But the trends on the trench were clear: an ambitious program of hydraulic engineering designed to protect traditional fishing rights was losing ground to industrial economic expansion. In its transition from a "little river," to a "trench," to a walled power canal, the Pawtucket Falls bypass had come to represent a society in transition.

As Providence grew and manufacturing along the Blackstone and Pawtuxet Rivers proliferated, the rights of rural fishermen steadily eroded. In 1769 the owners of Hope Furnace, who only a few years earlier had been ordered to pay restitution to a neighboring farmer, met a

Pawtucket Falls on the Blackstone River played host to the birth of the American Industrial Revolution, which fundamentally reshaped the upper reaches of the Narragansett Bay watershed. D. B. (Anonymous), [*Pawtucket Bridge and Falls with Slater Mill*]. Pawtucket, RI c. 1810. Water and ink on paper. RHi X5 22. Courtesy Rhode Island Historical Society.

more supportive response when they explained to the General Assembly that the Fish Act of 1735 was hindering their business. "[T]he inconsiderable fishery above the furnace," their petition claimed, "at a season when the labor of the people is most needful, will not, in any measure, equal the advantages which must be derived to the community, by the carrying on so large and so useful a manufacture."[48] For the owners of Hope Furnace, which included members of the venerable Hopkins and Brown families, the greater good would be realized not through access to fish but by the success of their blast furnaces. It was better for everyone, they believed, if farmers gave up fishing the paltry remaining stocks and joined the ranks of modernity in their employ. Playing a similar hand on the Pawtuxet's south branch a year later, Nathaniel, John, Griffin, and Christopher Greene explained that the fish laws had hampered their business, which had "been at a very great expense in erecting and building dams, forges, anchor works and saw mills . . . and employing a great number of hands." They, like their neighbors on the north branch, asked to be "totally exempted from preparing and providing fish-ways."[49] In both cases, the General Assembly agreed. Among the rivers of the broader Narragansett Bay watershed, industry had trumped tradition.

But it was Samuel Slater's mill and the dam that supplied it with power that marked a shift in the way people and subsequently the courts understood personal property and access to natural resources. After partnering with Moses Brown and Brown's son-in-law, William Almy, Slater built the first water-powered textile mill on the Blackstone River in 1790.[50] Shortly after, Slater built another larger mill 200 yards upstream of Pawtucket Falls and began constructing its dam, which, at 6 to 7 feet tall and roughly 200 feet wide, was likely the largest dam that had ever been built in America. But while the dam was under construction, the disgruntled blacksmiths Stephen and Eleazar Jenks and the miller John Bucklin, both of whom used Blackstone River water farther downstream, tore the dam down.[51] The Jenkses and Bucklin were angry not simply because the dam's construction had temporarily disrupted their operations but because Slater and Brown were outsiders—Slater was English and Brown was from Providence—and their business, the largest yet established along the Blackstone, was perceived as a threat. In retaliation, the Jenkses and Bucklin raised their dam by two feet in an attempt to push water back on Brown and Slater's wheel. Brown, Slater, and several other mill owners who were also affected and farmers who argued the taller dam threatened clear passage of fish filed a protest, which was upheld by the Rhode Island General Assembly in 1792. Brown used his powers of persuasion to convince the Assembly to remove from its purview decision-making powers over water rights for cotton mills. Subsequently, large mill owners like Brown and Slater were left unmolested. No longer was the fight over water rights one that pitted the public good against economic interests. By the turn of the nineteenth century, mill owners were vying for power amongst themselves.[52]

The ensuing battles over water rights fell to the courts. These legal decisions over water allocation imbued American common law with a penchant for progress. Specifically, it was Judge Joseph Story's 1827 circuit court decision in *Tyler v. Wilkinson,* which ruled on the apportionment of water among mills on the Blackstone River and Sergeant's

Trench, that set the precedent for interpreting water rights in terms of "valuable use." Although Story's decision acknowledged that "no proprietor has the right to use the water, to the prejudice of another," it was nevertheless used in subsequent cases to support the interests of mill owners.[53] Exempt from politics and favored by the law, cotton mills became not only the engines of economic progress, but also the arbiters of environmental control.

The Blackstone Canal Company challenged that status. In most cases, a cotton mill controlled the adjacent waters along the extent of its property and to the middle of the river, but the water itself could not be owned. In his 1824 treatise on the law of water, Joseph K. Angell, a Providence legal scholar and arguably America's leading authority on riparian rights at the time, explained that because water was "moving and transient by nature, it can admit of no permanent property. The only property of which it is susceptible is temporary and usufructuary."[54] Angell explained, however, that the use of water was not without limits. Citing the eighteenth-century English jurist Lord Blackstone, Angell noted that a mill owner could detain water only in a manner that did not injure his neighbors, particularly those who had established mills before him. "[E]very subsequent occupant," he wrote, "is obliged to exercise his right as not to incommode the first occupant." In the United States, Angell continued, the law allowed that when a disturbance was "*trivial*," and the use of water was "*reasonable*," there were no grounds for redress.[55] But the Blackstone Canal Company, a newcomer, promised to radically restructure the way water moved through the river. It would surely change water levels, perhaps even beyond "trivial" and "reasonable." And, moreover, half of the company operated outside of Rhode Island's legal jurisdiction. It is no wonder that mill owners balked.

Although the mills in Rhode Island largely controlled economic production on the Blackstone, Massachusetts controlled the river's headwaters. Diverting the canal to Boston surely jeopardized the economic benefits of connecting the port of Providence to a rich hinterland, but more importantly, it also threatened to turn off the Blackstone spigot. Angered over political gridlock in Rhode Island, the *Massachusetts Spy* outlined the advantages of digging to Boston. "The State may, if necessary," it explained, "divert the whole water of the Blackstone River by paying the damage which would accrue to the establishments *within the State*."[56] Not only would the entire river be diverted east, desiccating the economic engine of Rhode Island and dramatically altering Narragansett Bay by turning off one of the major sources of freshwater that drove its saltwater circulation, but Massachusetts would only pay damages to riverfront proprietors within its own borders. The Federal Government had also become involved, adding further doubt and confusion. A corps of engineers under the authority of Congress was investigating the possibility of a canal between Barnstable on the north shore of Cape Cod and Buzzard's Bay on the south, and another connecting Boston Harbor with the Taunton River on Narragansett Bay.[57] So long had Rhode Island stalled on the subject of the Blackstone canal that the *Providence Gazette* reported, "[W]e are almost led to believe that the project is abandoned."[58]

But the impending doom reported in the papers did not necessarily reflect the steady advances being made by canal authorities behind the scenes. In November 1824, Rhode Island Canal Commissioner Stephen Smith explained in a letter to Massachusetts Commissioner John Lincoln, "There has been continual warfare since I last saw you between the Manufacturers and the friends of the Canal." Smith assured Lincoln, however, that his interest in the canal had "never for a moment abated," that sentiment had turned in their favor, and that "there never was a time when the general impression was so . . . favorable as it is now."[59] By early spring of 1825, it appeared that Blackstone mill owners would allow the canal to proceed, but not without first securing their interests. "They agree with the importance of having the Canal," wrote T. Beckwith in a letter to William Lincoln, a Worcester attorney, "and I think will agree to something that will be satisfactory but they are all so busy securing mill privileges [water rights] that they have not much time to do anything about the canal."[60]

By April the canal had been approved. "The long agitated project . . . ," glowed the *Providence Gazette,* "has been resumed in good earnest." The subscription books were opened to investors and the Canal Company began purchasing property along the canal route.[61] They offered 5,000 shares at $100 each to be sold on Wednesday, April 27 between 10 a.m. and 1 p.m. at Franklin Hall in Providence. According to the papers, however, they sold 11,297 shares, amassing promises for $1,127,900, three-quarters of a million dollars more than the original estimates required.[62] Many of those promises never came to fruition, but to the public, the Blackstone Canal Company was flush with cash. Success seemed ensured. According to the papers, even protests among mill owners were a thing of the past. Both pistons of progress, the canal and mills now thumped rhythmically side by side to the beat of the nineteenth-century march of progress. A rational, quantitative approach executed with "scientific precision," explained the *Providence American*, would allow the construction of a canal "without affecting injuriously the rights or property of the ... manufactories."[63]

This "scientific" approach required a radical restructuring of the entire Blackstone River watershed. Along the length of the canal route, every cubic foot of fluid had to be apportioned so that the canal would flow throughout the year. One of the first priorities of the canal commissioners was to begin the process of damming feeder ponds. The dams allowed canal engineers to store and then release impounded pond water when it was needed. In Worcester they purchased land from Samuel Stowell to construct two dams at the outlet of North Pond. In Shrewsbury, they started construction on a dam at the junction of Long Pond and Round Pond. They also planned a dam for Flint Pond and one for others farther downstream. In Douglas, Massachusetts, they built one at Badluck Pond and one across Beaver Brook in Uxbridge. In Millbury, they built dams at the outlets of Dorety and Ramshorn Ponds. In the towns of Douglas and Sutton, they built one at Manchaug Pond. At Upton, another dam went up at Pratt's Pond. Two more dams were erected at the outlets of Mendon Pond. At Mendon, the Canal Company also raised a dam across the Blackstone River by five feet.[64] The *Massachusetts Yeoman* estimated that "when the necessary dams are completed, more than five thousand acres [of water], varying in the increased depth from five to ten feet" would be made available.[65] In Rhode Island, numerous other dams and diversion were either under contract or construction.

Within a few months, the hydrology of the Blackstone River watershed was reconfigured. When the ponds and their dams were completed, water—nearly eight square miles of it—would flow, the logic held, surely and steadily throughout the year regardless of the weather. When a barge operator opened a lock's upper gates and thousands of gallons of water buoyed his boat to the next level, that water, upon clearing the lower gate, washed downstream to be used by the next person. Apart from these temporary, intentional fluxes, canal waters—future Bay waters— lay still and controlled. Of course, water seeped through the lock gates and into the surrounding mud banks. It also evaporated. But these losses had all been taken into consideration. Guided, corralled, and meted out as needed, the waters of the Blackstone Valley—the headwaters of Narragansett Bay—were domesticated in ways that would bolster trade while supporting and perhaps even improving water-powered industry.

But if mills were spared injury by the Canal Company's "scientific approach" to water management, farmers were forced to endure the blunt trauma of the spade. On June 13, 1825, the Rhode Island Court of Common Pleas assigned three independent representatives to appraise lands through which the Blackstone Canal would pass. The charter of 1823 had granted the Canal Company the power to lay the canal in "such place or places as may be deemed most convenient for said company."[66] Although an 1826 amendment to the charter limited the company's ability to impinge upon mills, farmers were left largely unprotected.[67] They appeared in court to voice their opposition to the canal with what the *National Aegis* described as "forced gravity." Anticipating encroachment on their lands, farmers demanded damages. One "respectable elderly gentleman" who approached the courthouse with an "indescribably self complacent smile" told reporters that he "should certainly demand *damages*, though he shouldn't be very hard with the canal folks as he didn't know after all whether the canal would be much *injury* to his land."[68] Other farmers expected far worse. William Arnold of Uxbridge wrote to John Lincoln in February 1826 concerning a proposed reservoir near Buxton meadow, explaining, "The owners of the land have been cleaning it of timber & wood entirely—expecting it to be laid some 8 to 10 ft under water in this season."[69]

Nevertheless, many accepted payment. Published in the papers, the appraiser reports listed those who received money for damages and where. Samuel Stowell, whose farm was located at the outlets of Long Pond, received $62. Abel and John Wesson, who owned land at Flint Pond, were awarded $11 each. At Dorety Pond in Millbury, Jacob Dodge was awarded $16 and Daniel Rice $4 for a dam that the Canal Company built on their lands.[70] In other cases, however, damages were not forthcoming. David Dunn, owner of more than an acre along the canal route in Uxbridge, received nothing because "the advantages of the Canal . . . [were] a full equivalent." Similarly, John Segreve and James and Paul Aldrich were denied compensation on the grounds that the canal would increase the value of their properties.[71] From Providence to Worcester, the Blackstone Canal Company methodically assessed lands and in many cases arranged for their purchase. The surveyors noted the location and dimensions in chains and links of every parcel and published them widely.[72] So many transactions were completed that they often used a preprinted contract onto which names and fees for damages were handwritten. In some cases they even purchased discharges to future claims. In its willingness to pay in denominations as small as $1, the Blackstone Canal Company ensured that every drop of water in the valley fell under its control.[73]

The Canal Company did everything in its power to determine the way water flowed downhill, but the creep of the sea was beyond its control. Even before construction started, Worcester began to experience the effects of the tidewater. Anticipating the canal's arrival, business boomed. "From a very quiet, well disposed, inoffensive kind of place," observed the *National Aegis* of Worcester, "it has become more than half a sea-port city in bargaining, trading, and trafficking [*sic*]. A most money-giving, land-parting, meadow-buying spirit has sprung up on a sudden."[74] With the coming of the canal, an inland farming community, once parochial and polite, had adopted the mercenary and acquisitive ways of a seaport. With its connection to the coast, Worcester, some pundits believed, was well on its way to becoming a great commercial entrepot. "[A]t the head of navigation or the top of a tunnel . . . ," wrote the *Providence Patriot*, "all the surrounding country, as far north perhaps as the southern counties of New-Hampshire and Vermont, will pour in the surplus produce and their manufactures." So powerful was a canal connecting Worcester with Narragansett Bay that it brought to the town the promise of posterity. "[I]t's marriage with the ocean, by means of the Blackstone canal," the *Patriot* continued, "will be an important epoch in its history; its marriage portion will be a monopoly of an inland trade."[75]

The marshy margins of Providence in 1819 before construction began on the Blackstone Canal. Alvan Fisher, *Providence from across the Cove*. Smith Hill, Providence, RI: 1819, Oil on Canvas, RHi X5 11. Courtesy Rhode Island Historical Society.

When construction began in late 1825, the sea surged north into apple country. Digging commenced at the tidewater in Providence, following the logic that building materials could be floated north in the canal and that every completed portion would have access to the Bay.[76] Where the Moshassuck River flowed into Providence Cove, work crews began to excavate the marsh and flats "with great vigor." Nearby at Scott's Pond in Smithfield, between twenty and thirty men, mostly Irish, excavated embankments using pickaxes and shovels and loaded the debris into oxcarts, which they reportedly filled upwards of fifty times per hour. Workmen pushing wheelbarrows carted sand from the growing ditch. Nearby, they had erected a barracks 30 by 20 feet, with a "capacious oven made of brick and stone." The wives and "ruddy" children of two of the laborers were also present.[77]

Just as a connection to the sea had changed the commercial culture of Worcester, so too did it transform its religious landscape. As the canal pushed north, Irish immigrants flooded the region in search of work.

Throngs of "strollers," as the roving Irish laborers were known, had worked on England's extensive canal system and then came to the United States to work on the Erie Canal.[78] When construction there drew to a close, many moved on to other public works. In July 1826, roughly 500 Irish workers arrived in the Blackstone Valley to dig its growing trench.[79] Also from the tidewater came Worcester's first priest. The Reverend Robert Woodley, who had a parish in Providence, established another in Worcester, although until 1830 he only visited his headwaters flock but twice a year.[80]

By October of 1826, most of the canal route in Rhode Island had been excavated. Although rain had hampered progress that fall, a large portion of the canal had been dug and many of the locks had been constructed. Excepting the "tide lock" in Providence, which was built of wood, the others used granite blocks.[81] Work also began near Worcester, but not fast enough. "[G]reat impatience has prevailed in this vicinity, for some near evidence of the progress of the work of improvement," reported the *National Aegis*.[82] Nevertheless, work was underway. On July 6, 1826, William Farnsworth submitted his estimate to the Massachusetts canal commissioners to produce locks of 130 feet long and 13 feet, 6 inches high, at $4,400.[83] Two days later, workers with shovels and barrows began to excavate the meadow between Main and Back Streets in Worcester.[84] What had been impatience soon turned to assurance, and perhaps even hubris. Of the farms and factories dotting the rolling hills and "luxuriant vegetation" of the Blackstone Valley, one reporter marveled at "the extent to which the natural resources of a rich territory have been made subservient to human enterprise and skill."[85] If the plow and waterwheel had subordinated nature, the canal would conquer it completely.

By early 1827, major changes were well underway. At eighteen feet wide at the bottom, thirty-four feet wide at the water's surface, and four feet deep, the canal required the massive movement of soil, sand, and stone along most of its route. Patrick O'Connor, who led an excavation team, was receiving 8 cents per cubic yard for removal and 10 cents per cubic yard for building embankments.[86] The payment he received in February of 1827 for $1,000 suggests that he and his crew had displaced roughly 10,000 cubic yards of soil, or a continuous mound 10 feet high, 9 feet wide, and 3,000 feet long.[87] Many of the embankments were situated around holding ponds, which were altered as well. Alum Pond in Rhode Island was raised six feet. An embankment at North Pond in Worcester allowed engineers to raise it by eight feet. Nearby Long Pond, which had already been raised five feet, was elevated another two. Finally, seventeen of twenty stone locks in Rhode Island had been completed and twenty-five of twenty-nine in Massachusetts had been contracted. All told, forty-nine locks would allow the canal to climb 451½ feet from the Providence tidewater to Worcester.[88]

As the canal progressed, mill owners began to seriously consider the benefits of a fully engineered hydrologic system because the way water moved through the landscape was changing dramatically. When one reporter traveled the Blackstone Valley in 1826, he described "forest retreats" that were "overspread by the harvest of cultivation where wind stirs up the little ripples of gold to chase each other over the surface, and break on the margins of verdure," which "stretch on either side of the pleasant stream."[89] His candied prose aside, the writer described a largely deforested landscape. Stands of trees existed only as "retreats" and "margins." The cleared strips of cultivation were likely covered with goldenrod, the ubiquitous colonizers of New England's un-mowed pastures. By the turn of the nineteenth century, New England forests had been largely cleared. Of coastal southern Maine, Yale President Timothy Dwight observed in 1807, "The forests are not only cut down, but there appears little reason to hope that they will ever grow again."[90] In his turn-of-the century travels through the mid-Atlantic, Isaac Weld observed of Americans, "They have an unconquerable aversion to trees."[91] Traveling at mid-century, the English poet Emmeline Stuart-Wortley lamented of coastal Massachusetts that the few scraggly trees remaining were "almost as sad to look at" as the "*girdled* trees, which look like skeletons of malefactors bleaching in the wind."[92] On John Sanderson's farm in the central-Massachusetts town of Petersham, only 100 of his 850 acres were wooded during the 1830s and 1840s.[93]

Rhode Island experienced similar levels of deforestation. Along the Blackstone River and adjacent to the canal route, giant swaths of forest were cleared for lime burning and potash and charcoal production. Of primary importance to the newly established Hope Furnace in 1765 was firewood. After establishing the Furnace's Cranston dam site, Stephen Hopkins signed agreements with thirty-seven local workers for "Cuting Carting or Coling the Said Wood" from the surrounding forests.[94] By 1769 Rufus Hopkins was managing the Furnace's production of charcoal, a necessary ingredient for smelting iron ore. In May of that year, he explained, they had "five pitts afire . . . & shall have two if not Three more afire this Evening and Shall Continuous Firing as fast as Possible." The speed with which he and his coalers were clearing Rhode Island's forests was accelerated by accidents. Hopkins explained of their coaling operation that Joseph Briggs had accidently set the woods on fire by "Burning Bushes Last Saturday Evening or Sunday morning." He had burnt "60 or 70 acres . . . and Distroyed all the wood and Timber Left on the Land" as well as between 750 and 1,000 split rails that had been piled nearby.[95] That the exact day of the fire's start was blurry suggests the forest had been smoldering for some time. As stands of trees were razed and converted into charcoal, Hope Furnace agents scoured the countryside farther afield. In 1784 Rufus Hopkins negotiated with Christopher Lippit for woodlots west of Cranston in Scituate. Well aware that nothing would be spared once the deal was sealed, Lippit was careful to stipulate "There are 8 or 10 Pine Trees on the Different Lots that I do Reserve as [I] shall want them for my own use."[96]

By the early nineteenth century, Rhode Island, like much of the Atlantic seaboard, was being cleared of trees at a rapid pace. According to the agricultural census in 1850 and again in 1860, just over 64 percent of Rhode Island had been denuded.[97] The expanding population, which had grown to 108,830 people by 1840, required fuel, as did Rhode Island's growing textile industry, which, per capita and square mile, was by far the largest in the United States. Rhode Island's rivers were driving almost 5 spindles per person and 427 spindles per square mile across the state. Although the overall textile output of Massachusetts had surpassed that of Rhode Island by 1840, its rivers drove fewer than 1 spindle per person and only 63 spindles per square mile.[98] Such high concentrations of mills in Rhode Island required a lot from its forests. The buildings, including the mills themselves, worker housing, and machines shops all required thousands of board feet of lumber. Water wheels, raceways, and flumes were made of wood, as were most of the machines. Milldams, too, were made of large wood timbers.[99] And finally, these mills were connected by an extensive network of turnpikes, during the construction of which hundreds, if not thousands, of acres of forest were razed.[100]

As trees were cleared, particularly those lining stream and riverbeds, the movement of water through Rhode Island's rivers changed alongside the growing demand for power. Mills that had largely confined their operations to periods of adequate flow now required a constant supply of water, regardless of the season or weather.[101] As modern water scientists have shown, deforestation dramatically changes the quantity, quality, and movement of water. When trees and their ability to retain moisture are removed, water flows faster through the landscape. Riverbeds that have lost their buffer of trees are prone to flooding.[102] Winter snows, exposed to the sun, melt faster, and, without a forest buffer, inundate rivers at a faster rate.[103] For the same reason that rivers swell immediately after trees drop their fall foliage—their leaves no longer transpiring—rivers filled their banks when trees were chopped for charcoal. During wet periods the rivers roared, and during dry periods they slowed to a crawl. Although these processes are complicated by soil type, hill slope, climate, and the amount of impervious surface nearby, among numerous other variables, the basic hydrologic patterns hold true: when the land is cleared, river basins become "flashier" and less predictable.[104]

It was exactly this predictability that mill owners wanted but that had grown increasingly difficult to achieve. Tree removal had surely altered the Blackstone River, but the complicated matrix of dams, diversions, and holding ponds combined with the proliferation of impervious surfaces, such as mill roofs, roadways, and the stone walls lining river banks, only added to the inconsistency of its flow. The canal promised to restore order to the river. But so unpredictable had the flow of water become, due in part to the canal's construction, that work on it was often delayed. During the spring and fall of 1827, rains hampered excavation and even led some work crews to abandon sections of it altogether. Although the rain-induced damage to the canal was considerable, the flooding was affirmation that the natural system would no longer suffice. "The experience of the present season," wrote the *Massachusetts Spy* of the heavy rains, "has fully demonstrated the capacity of the reservoirs, to afford an abundant supply of water, even more durable than was anticipated, thus improving instead of injuring the water privileges on the route."[105] The canal and its feeder ponds, reasoned the *Spy*, would form a bulwark against such capricious weather. But that wouldn't help trench workers chest-deep in mud. The rains continued to "deluge the earth" through the winter and following spring.[106] Broader climatic trends were beyond human control, but human meddling along the river and in nearby forests would have surely exacerbated the rain's effects. Although work slowed, the canal was finally completed except for the lock doors, which would be hung during the summer of 1828.[107] When the weather broke in July, the *Lady Carrington* made the first excursion up the lower portions of the canal near Providence.

Despite the canal's purported abilities to control the excesses of the weather, there were cracks in the walls of manipulated nature. And in some cases they were actual cracks. Scott's Pond, which had been raised a whopping fourteen feet from its natural level, had sprouted numerous leaks. The porous soil in the area had made its man-made berms susceptible to breaches. The *Rhode Island American and Providence Gazette* reported that "in many places" there were "little rivulets through the embankments."[108] The problems grew worse as the rains returned later in July. The *National Aegis* explained, "The contractors have been delayed in their operations by the fountains and streams bursting from the earth." Not only were sections of embankment giving way, making work impossible, but alterations to the canal's surroundings once again aggravated the situation. Without forest buffers, water "poured down from every hill-side."[109]

The message was clear: the Canal Company's control over the valley's water was far from absolute. The part of the canal that entered the Blackstone River itself was particularly vulnerable to the pressures of flood and drought.[110] When water grew scarce during dry periods, barges often ran aground on the riverbed. In addition, the enclosed portion of the canal system lost considerable water when the locks opened to let boats pass into the main river. The system began to wilt in August 1828 when after months of rain southern New England experienced four weeks without a drop. "[M]any fields have become so dry that vegetation suffers severely," wrote the *Massachusetts Spy*, "The roads are so dry and dusty as to render travelling rather uncomfortable even were it not for the excessive heat."[111] Human error or malicious intent was also to blame. Massachusetts Canal Commissioner John Lincoln announced a $10 reward for any information that led to the conviction of the person who, using a "false key," raised the gates at the North Pond reservoir in Worcester, which caused the loss of valuable water reserves.[112] Apparently the company suspected unruly youths, for they specifically addressed "Parents" and "Guardians" when they announced the fine of $500.[113] In response, the Canal Company made further attempts to secure its growing hydraulic network from natural and human pressures by raising Manchaug and Badluck Ponds by six feet and a dam across the Blackstone River at Mendon by twenty-seven inches.[114]

The need to further bank the canal raised question about funding it. Building dams and turning ponds into walled pools was expensive. The cost of fixing breaches and repairing towpaths from winter frost heaves was also costly. And the dizzying business of paying damages continued to drain the coffers of the Blackstone Canal Company throughout its operation. All told, the canal had cost roughly $750,000 but had taken in only $400,000 in stock subscriptions from Rhode Island and $100,000 from Massachusetts. Those shareholders received only five dividends between 1832 and 1836 that amounted to a total of $2.75.[115] When on October 7, 1828 the *Lady Carrington* tied its lines to the pilings at Worcester, becoming the first boat to travel the full length of the canal, the celebration that followed—replete with cannon shot and banquet—belied the precarious financial footing on which the entire operation rested.

Tolls, it was hoped, would generate enough revenue to maintain the canal and pay its investors. In its first few years, revenues increased. An abstract prepared by the canal commissioners and submitted to the *Providence Journal* showed that in 1828 canal toll revenue was $1,100. In 1829 it climbed to $8,603. In 1830, the Blackstone Canal Company reaped $12,006.[116] From Providence cargos of, among other items, corn, rye, flour, salt, molasses, whale oil, and lime, as well as cotton, wool, iron, bricks, cheese, fish, and leather were sent north. From Worcester, cloth, cordwood, wine, stone, chairs, staves, lead, boots and shoes, paper, and ship timbers were sent south.[117] By 1832 tolls would reach their all-time high of $18,907.[118]

The near constant movement between Providence and Worcester projected Narragansett Bay north. The *Boston Centinel* marveled at the way Worcester papers now included a weekly "Marine Intelligence" section that listed vessels clearing into and out of its port. So "imposing" was the sheer number of boats in Worcester that the *Centinel* estimated the town had roughly the same traffic as Boston did thirty years before.[119] Embracing its new maritime identity, the *Lady Carrington* had become a canal "Packet," making scheduled runs between Providence and Worcester in just over fourteen hours. References to "shoals" along the Blackstone also suggest maritime language had begun to trickle into the interior.[120] Worcester woodcutters had begun targeting ship's timbers in nearby forests.[121] In April 1829, the salty-named Christopher Columbus Baldwin observed that the canal boat *Washington*, "the first built in Worcester," was wheeled through the streets to the distillery basin where it was launched on the 13th. No simple barge to the people of Worcester, the *Washington*, like any proper vessel, received a speech and song the day it was splashed.[122] When, in July 1829, the canal boat *Independence* delivered the Providence Light Infantry Company to Worcester, a Dr. Fiske commented, "*We have* witnessed fleets of commerce wafted on our waters, we *now* see borne to our port, *first rate men of war.*"[123] For Fiske, Worcester had become a full-fledged seaport. Not only was the town shaped by the bustle of trade, but it was also filled with a maritime military presence. Worcester had become an arm of the sea in every way. With the canal, even wharf rats began to arrive.[124]

But the construction of an inland port had come at the expense of mill owners, who felt cheated. In 1833 and 1834 the Blackstone Canal Company was slapped with 149 lawsuits over water rights. Although the Canal Company had placed markers at various points along the river to show the low-water mark, the Rhode Island Supreme Court found that the Canal Company had failed to measure and record the water's fluctuations. The court cases proved that the Canal Company's reservoirs, which claimed to hold 698 million cubic feet of water, only provided 264 million cubic feet.[125] The great effort that had gone into impounding water had not been enough. Engineering an entire watershed, the centerpiece of which was "the hardest working river in America," as the Blackstone has been long called, had been so complicated, costly, and politically treacherous that the goal of improving nature for everyone could not be achieved.

If the mill owners' lawsuits weakened the Canal Company, problems with water availability and the coming of the railroads dealt deadly blows. During the winter months, the canal was forced to close because of ice. As early as mid-November, the stillwater sections froze hard. By December, the locks were typically drained to prevent damage, and canal operation might not open again until March or April.[126] During the summers, when the water level was low, sand bars in the Blackstone often delayed boats for days. Breaches to canal walls also delayed traffic.[127] Talk of a railroad between Boston and Worcester began in 1828. There were many who believed Providence and its canal would usurp Boston's rightful place as the capital of New England. Giving a toast at the Worcester Cattle show, one Boston merchant raised his glass and said of the canal with trepidation, "May it give activity to the *'Heart,'* [of Massachusetts] without depressing the *Head* [Boston]."[128]

In their push to build support for their cause, railroad boosters were keen to address the shortcomings of canals. Railroads, they explained, were less costly to construct and maintain. Canals needed bridges and were subject to droughts, floods, leaks, and freezing. Canals could not climb mountains and required loading at wharves, whereas rails could be laid to the doors of, if not into, warehouses and factories. And of course, canals interfered with water rights.[129] The advantages of the railroad were legion. When the Boston and Worcester Railroad was completed in 1835, it caused many farmers and manufacturers in central Massachusetts to look east. The Norwich and Worcester Railroad, which was under construction, and another anticipated line connecting Worcester with Albany, likewise led inland merchants to turn their trade elsewhere.[130] The canal simply could not compete. In 1836, toll revenues dropped by 20 percent.[131] By 1844 boats no longer ran clear through to Worcester. In 1846 canal navigation stopped altogether.[132] And with the establishment of the Providence and Worcester Railroad in 1848, the water route between the cities was all but abandoned. Although the tracks running parallel to the canal were no more than a foot high, they cast a long, dark shadow. Forever caught within it, the canal withered.

The canal had earned its shareholders only a pittance, but Worcester had nevertheless profited. Between 1825 and 1836, the population grew from 3,650 to 7,500 people.[133] Between 1821 and 1831 Worcester had built almost 400 new buildings. And between 1825 and 1835, the town tax revenue climbed 50 percent.[134] Worcester had gained enough importance to compel William Lincoln, the compiler of such statistics, to write a 383-page history of the town, which he published in 1837.

The dissolution of the Blackstone Canal Company, however, had a profound environmental impact across the entire Blackstone River watershed and into Narragansett Bay. When the company attempted to disband in the 1840s, many mill owners fought it. At their behest, legislators stalled the process.[135] (That Rhode Island was mired in a constitutional crisis during the 1840s surely caused further delays.)[136] Although mill owners had sued the Blackstone Canal Company to restore the "natural run" of the river, they soon realized that the patterns of natural flow were so far gone and the river so intricately engineered that abandoning the system as it existed would cause hydraulic chaos. The Canal Company's corporate model of water control, one that heaped massive amounts of capital at problem after problem, had, for a time, worked. But when it failed, the effects were far-reaching.

A combination of hubris and blind faith in an inexact science led to the Canal Company's demise. Canal managers had failed to account for environmental variability, including extended periods of rain, snow, and drought that either damaged the canal or made it impassable. They had not considered human error. Misguided water storage estimates and massive water losses that occurred when lock gates and dam doors were left open often delayed operations and undermined confidence. And they had not anticipated the possibility of superior technology in the coming of the railroad. For all these reasons, the Blackstone Canal Company failed.

Dissolving the Blackstone Canal Company's assets fragmented the hydraulic system it had taken so long to establish. In May and June 1849, the locks between Providence and Woonsocket were sold for building material. Samuel Saunders bought the Mineral Spring Lock for $217.50.

The Horton Lock was sold to William Randall for $250. The Lonsdale Company, a sprawling textile manufacturer owned by, among others, Nicholas Brown, Thomas P. Ives, and Edward Carrington, a canal commissioner, purchased three locks and the surrounding land at Scott's Pond for $335.[137] Various firms and individuals purchased the remaining locks. The Canal Company was legally dissolved and its charter withdrawn by the end of the year. No longer would one company manage, or attempt to manage, the entire Blackstone River watershed. Individual firms took control of their own dams, holding ponds, and head- and tailraces in a watershed that had, over the course of twenty years, been pushed, pulled, and prodded far from its natural inclination. In some places, such as at Lonsdale, the canal continued to work hard. In other places, the waters lay stagnant. Ignored, the alders, swamp apple bushes, and laurels that had been trimmed periodically by the Canal Company choked the canal's banks.[138] As mills transitioned from water to steam after mid-century, the canal turned from an energy source to a sewer.

By 1880, conditions in the canal had become cause for alarm. Responding to a query by the Rhode Island Superintendent of Health, Professor John H. Appleton at Brown University explained, "The water of the river is polluted and rendered exceedingly foul and offensive before it reaches the limits of the city, and gives off an offensive odor of sulphureted hydrogen gas, like very foul sewers." The pollution the professor described was not limited to the canal. The entire length of the Blackstone River had become an industrial dump and cesspool. Worcester was pumping sewage into the Blackstone, as were other towns along its length. Industrial waste flowed downriver as well. By the time the Blackstone flowed into Providence, it was downright hazardous. The *Providence Evening Press* explained that the river "can be looked upon only as an open sewer, and the time is rapidly approaching when it must be treated as such, and must be covered." This river of sludge coursed into Narragansett Bay. "It is certain that much filth is turned into the river in the city," explained Appleton, "and it must be still more filthy when it reaches the harbor."[139] In 1885, the *Providence Sunday Journal* explained that below Mill Street, where the canal met Narragansett Bay and where the *Lady Carrington* had embarked on its maiden voyage in July 1828, the water was of a "dull greasy brick hue." At Horton's Grove, where the *Lady Carrington*'s

By 1858, the Providence shore had begun to harden into a more clearly defined edge. In the following years, human and industrial waste poured into northern Narragansett Bay. J. P. Newell, *Providence R.I.—View from the West Bank of the River*. Providence, RI 1858 Ink on Paper. Print. RHi X32 30. Courtesy Rhode Island Historical Society.

guests had stopped for refreshments, the canal was altogether gone, "the railroad and freshets," the *Journal* explained, "having perturbed the old artificial conditions and left the Moshassuck [River] to follow its aboriginal inclination in seeking the sea."[140]

The Blackstone Canal marked an important transition for Narragansett Bay. There is no doubt that most of the action between 1825 and 1849 occurred above the tidewater. But in its remarkable ability to transform the natural run of a watershed into a quantified, systematized machine, the Blackstone Canal Company imposed unprecedented order on the freshwater half of an estuary. When that intricate system proved untenable, the canal and the river to which it was connected became a gutter filled with heavy metals, industrial dyes, human waste, and trash. During the second half of the nineteenth century and throughout most of the twentieth, Narragansett Bay bore the brunt of countless environmental transgressions in its northern reaches. Shell and finfish populations declined. Marshes were fouled and filled. And algal blooms became increasingly prevalent.[141] The canal and its many feeders and holdings ponds showcased the brass of human ingenuity and bluster of the new corporate order. But politics, property, the vicissitudes of nature, and the fallibility of human beings tend to complicate things. Alas, absolute control is a chimera. With the ecstasy of improvement comes the agony of decline.

EPILOGUE

Between Progress and the Pull of the Sea

"[W]E ARE IN a sense amphibious," suggested Strabo in his geography of the ancient world, "not exclusively connected with the land, but with the sea as well.... [T]he sea and earth in which we dwell," he continued, "furnish theaters for action; limited for limited actions; vast for grander deeds."[1] The space between land and sea, explained the historian and philosopher, responded to human desires and shaped them in return. For Strabo, humans were inherently intertidal. The nexus of dry land and ocean provided "theaters" for both the steady march of quotidian progress and all things extraordinary and profound. And the push and pull of these two forces shaped not only that space but also history itself.

For Henry David Thoreau, writing nearly two millennia later, the littoral likewise shaped human culture, albeit a distinct one. "Before the land rose out of the ocean, and became dry land," he wrote in his 1865 *Cape Cod,* "chaos reigned; and between high and low water mark, where she is partially disrobed and rising, a sort of chaos reigns still, which only anomalous creatures can inhabit."[2] The powerful tension between land and sea, he believed, created conceptual chaos. Like the Victorian woman "disrobed," dry land exposed at low-water induced cultural vertigo and perhaps even panic. So jarring was this space, he explained, that only "anomalous creatures," whether human or nonhuman, could navigate it. Although thinking and writing in vastly different times and places, Strabo and Thoreau agreed that the peculiarities of littoral space made the muddy interstices between land and sea something special.

For the littoral people of Narragansett Bay, the tension between inland and ocean defined almost every facet of life. Theirs was a watery world of salt creeks and rivers, marshes and beaches, and islands and shoals. Across the Bay, they shuttled themselves and their animals in canoes, lighters, and ferries. Most were just as familiar with anchoring skiffs and hauling seines as they were with haltering horses and hoeing turnips. Native Americans and Europeans alike navigated this physically and imaginatively brackish borderland on a daily basis. And as a result, they, much as Thoreau observed, developed a distinct littoral society. The liminal nature of Rhode Island's coastal world lent itself to cultural and religious diversity, a natural inclination toward adventure and experimentation, perhaps a desire for political autonomy, or even a penchant for mischief. The Bay waterscape shaped relations among people within Rhode Island's borders and with those outside them who clamored for its land and coastal resources. In times of political strife, Narragansett Bay formed a border between Indian tribes, between European and Indians, and between quarreling colonies. At other times, the space between land and sea became a borderland, a porous and permeable geography of diplomacy as well as commercial and cultural exchange.

As the backdrop to littoral life, Narragansett Bay, the "theater" for human action that Strabo described, responded to the pressures of work, war, and an expanding population. Complicated by the amphibious attributes of the estuary, environmental change often came in fits and starts. At times, the sea denied progress, checking human efforts to maintain their harbors and patrol their shores. At others, when rivers were dammed or redirected or when oystermen raked the Bay's beds clean, it bowed under its weight. Humans sliced lines of jurisdiction through this watery frontier, but as evidenced by the contentious squabbling and even outright violence that followed, geographic confusion often held sway. In its ability to internalize the push of a mutable shore and pull of an enduring sea, the estuary became much more than a dent in the coast. It was also a product of instinct and awareness, a place shaped discursively by the people who plied its beaches and worked its waters. In this sense, Narragansett Bay became at once the "theater" *and* an "actor," which in dialogue with its human counterparts changed them and their responses in return. The Bay as theater surely endured the scuff of boots, but as a place molded by historical memory, partitioned by political negotiation, and systematized by scientific debate, it also actively shaped the play and the people who performed it.

Reveling in the promise of progress, the people of Rhode Island remade their Bay to meet the demands of an industrialized world. To the best of their abilities, they extended the Bay geographically north into the heart of New England. In the process, however, they transformed what had been a broad coastal region, an unbounded domain that smelled of seaweed and salt when the wind was just right, into a clearly defined "coastline," or edge. So radical was the transformation of the coastal margin and the environmental fallout that followed that it called into question the aims and achievements of the Enlightenment. By fracturing space and bifurcating time, the alongshore "line" produced by mapmakers, natural philosophers, and engineers, among others, seemed to defy all reason.[3] When progress was pushed to an extreme and the coast was contained, catastrophe followed.[4]

As the clank of industrialization shaped upper Narragansett Bay, the prattle of polite society remade it farther south. Once considered a "gone glory—an ornament of the Past," wrote *Harper's Monthly* in 1854, Newport was staging a recovery by midcentury. Although rotting wooden shacks and decrepit docks still lined the shore, a "new town," the report explained, "is rapidly arising upon the hill." Along the ocean's edge, a "spacious and beautiful avenue has pierced the solitary fields." From the sea, approaching mariners now eyed a "cluster of handsome houses."[5] Hotels began to sprout along Bay shores and tourism expanded, especially after 1847, when a new steamer line began regular service between New York, Newport, and Fall River in the northeast corner of Narragansett Bay.[6] During the early 1850s, one particularly enterprising real estate speculator, Alfred Smith, began purchasing parcels of farmland near Newport, cutting roads through them, and building fashionable neighborhoods.[7] Smith's most ambitious project was a road from downtown Newport that, following the edge of the ocean, ringed the southern tip of Aquidneck Island. Completed in 1868, Ocean Drive added a hard-packed, graded edge to one of the most sumptuous stretches of southern Narragansett Bay.[8]

As developers remade Rhode Island's shores materially to meet the desires of summering elites, painters remade it figuratively. By the 1870s, holidaymakers reimagined the coast of Rhode Island as a place of play. Mirroring trends at other seaside resorts, the beaches of the Bay and Rhode Island Sound were polished to please the bathing bourgeoisie.[9] At Sachuest Beach, artists such as John Frederick Kensett and Thomas Worthington Whittredge, among others, painted romantic visions of the seashore that actively removed weirs and workmen from the waterfront. Their projections of a "pristine" shoreline, in turn, masked the fact that Rhode Island's coastal fisheries were beginning to falter. A romantic representation of coastal nature, one that painted over the blood and guts of coastal work with an image of an eternal sea, in turn masked the grim realities of ecological change alongshore.[10]

While artists softened Rhode Island's coastline for their wealthy clients, fishermen witnessed a much sharper one developing. As sewage and industrial pollutants poured into the rivers at the northern end and tourists flocked to the southern, fishermen noticed that an actual line had developed at the mouth of Narragansett Bay. "[D]own in the neighborhood of Beaver Tail . . . ," explained Portsmouth fisherman Benjamin Tillman in 1870, "it is plainly to be seen where the corrupt and impure waters of the bay, meet the clear, blue waters of the pure ocean." A new edge of filth had developed at the Bay's principal point of entry, one, he attested, that "scores" of other fishermen had witnessed as well. In addition, since fishermen were no longer capable of securing their catch in Bay waters, Tillman explained, they had begun placing their trap seines "upon the margin of the great ocean." A line of nets now separated sea and estuary, which, Tillman explained, roused protest from the "pleasure-seekers" who presumably did not want their views of the ocean obstructed.[11]

If the guests of Newport's grand hotels had commanded clear views of the sea before the Civil War, the construction of opulent "cottages" along Ocean Drive and Easton's Beach increasingly obscured them in the years that followed. As a new ultra-rich set retired to the privacy of their palaces, the once-popular hotels lining Bellevue Avenue closed, the last of which, Ocean House, burned to the ground in 1898. While Newport grew increasingly exclusive, new hotels sprouted elsewhere along Rhode Island's shores, at places like Oakland Beach, Rocky Point, Jamestown, Watch Hill, Narragansett Pier, and Block Island.[12] For the

The growing popularity of seaside resorts encouraged coastal construction projects that transformed the shore at the end of the nineteenth century. "Ocean Road, Narragansett Pier, R.I.," 1880–1899, Detroit Publishing Co., Library of Congress Prints and Photographs Division.

American novelist Henry James, who had lived in Newport as teenager, these changes were distressing. The Aquidneck Island of his childhood, he recalled, consisted of a "thousand delicate secret places," dotted with "mild 'points' and promontories, [and] far away little lonely, sandy coves." He found comfort in the Point District at the northern end of Newport Harbor where he reveled in its "quiet, mild waterside sense . . . in which shores and strands and small coast things" dictated the rhythms of life. At the turn of the twentieth century, James pined for the estuarine Newport of old, one that had yet to be sullied by the garish wall of wealth that lined the "bold bluff outer sea." Newport's beauty, he believed, lay instead in the "wide-curving Bay and dim landward distances that melted into a mysterious . . . Providence." But, he moaned, the "monuments of pecuniary power," had risen "thick and close" alongshore. He wished that Ocean Drive, which was "engineered by landscape artists

The Gilded Age brought new material and symbolic edges to Newport. "Cliff Walk, Newport, R.I.," c. 1880-1899, Library of Congress Prints and Photographs Division.

and literally macadamized all the way, were still in the lap of time." The *"pure"* Newport, he remembered, was "perfectly guarded by a sense of margin and of mystery." Alas, he lamented, the "margin has been consumed and the palaces . . . stare silently seaward."[13]

If for James inspiration flowed from coastal ambiguities, for others it coalesced in the ocean's continuities. In her 1941 *Under the Sea-Wind*, the biologist and naturalist Rachel Carson explained that to "stand at the edge of the sea, to sense the ebb and flow of the tides, to feel the breath of a mist moving over a great salt marsh . . . is to have knowledge of things that are as nearly eternal as any earthly life can be."[14] For Carson, oceans and estuaries subsumed all sensation and redefined the limits of space and time. When she published *The Sea Around Us* ten years later, she avowed that man "has returned to his mother sea only on her own terms. He cannot control or change the ocean as . . . he has subdued and plundered the continents."[15] Her poetic evocation of an eternal sea won the National Book Award, sold more than 200,000 copies within a year of its release, and remained on the *New York Times* bestseller list for eighty-six weeks.[16] So seductive was the notion of an enduring ocean that, when Burt Lancaster and Deborah Kerr embraced in the surf in the 1953 film *From Here to Eternity*, Americans were swept away by the suggestion that a love like theirs awash in the waves would continue without end. But having witnessed the ways nuclear testing had transformed the underwater world of the Pacific, Carson soon realized she had been wrong. In the preface to her 1961 edition of *The Sea Around Us*, she admitted that her earlier belief that the ocean was eternal had been "naïve."[17]

As the ocean environment changed, humans responded by establishing new and even more creative coastlines. When New England's offshore fisheries began to decline in the 1970s, fisheries managers responded with the 1976 Fishery Conservation and Management Act, also known as the Magnuson Act. This established a new 200-mile Exclusive Economic Zone (EEZ) around the United States, which excluded other nations from fishing American waters. While their efforts had intended to protect coastal fisheries by enclosing them, fisheries managers soon realized that the American fishermen inside this new coastline were perfectly capable of catching everything their foreign competitors had left behind. Unprecedented precision in navigation and bottom-scanning technology combined with nets the size of city buildings and hundreds of years of fishing pressure had made recovery all but impossible. By 1992, New England's fisheries had reached an all-time low. Environmental fallout followed the new coastline even when it was pushed well beyond the horizon. In response, regulators added still more lines. In 2010, the New England Fishery Management Council implemented a new "sector-based" management plan, designed to concentrate fishing into specific zones. Their efforts to order the ocean, however, have produced few positive results. And so the fisheries crisis continues.[18]

Cartographers have likewise shaped the ways coasts are imagined and managed. To establish national and state borders, nineteenth-century American mapmakers drew coasts that followed the high-water line. To secure coastal shipping, they penned alternatives that showed low-water lines. As Americans crowded the ocean's edge, new coastlines were

The expansion of transportation infrastructure transformed Providence Harbor during first decade of the twentieth century. The process of "hardening" the harbor's edge has continued to the present day. Archibald Ellis, "View of Providence Harbor & the Four Bridges," c. 1910, Library of Congress Prints and Photographs Division.

The crumbling remnants of the Blackstone Canal. "Moshassuck River/ Blackstone Canal, View Looking North, Canal Street On Left - Blackstone Canal, Charles & Randall Streets, Canal & Haymarket Streets, Providence, Providence County, R.I.," Library of Congress Prints and Photographs Division.

drawn, reflecting a growing recognition that development so close to the sea was fraught with uncertainty. Insurance companies created new inland coastlines that follow the contours of flood plains. And now as sea levels rise in response to global warming, cartographers and politicians are drawing still more coastlines to various degrees of accuracy that show a radically altered world should the polar ice sheets melt into the sea.[19] Although lines on a map can't change the movement of water, they do affect the decisions people and their governments make. Risk assessments, for instance, shape the designs and location of road, rail, and building construction. Maps often determine both the mix of the cement and the duration of the pour. The material world, in sum, often morphs to meet the conjurations of actuaries.

As coasts become increasingly crowded and the margins between land and sea continue to harden, the potential for environmental disaster increases. Fifty-four percent of Americans now live within fifty miles of the ocean.[20] All Rhode Islanders do. A 2013 Federal Emergency Management Agency report, the first of its kind, projected that by 2100 the area at risk of flooding in the United States will increase by 45 percent, and the number of families requiring flood insurance will double. Seventy percent of that increase is attributed to climate change and the other 30 percent to population growth in flood-prone areas.[21] The coasts of the North Atlantic will continue to be battered by storms, but the way humans respond to these changes will determine the outcomes. When Hurricane Sandy hit the coast of New Jersey in October 2012, 94 percent of its beaches were damaged, some losing thousands of cubic yards of sand. The Army Corp of Engineers has promised to rebuild them, but experts fear that the process of restoration will harm coastal ecosystems and exacerbate erosion. Although New Jersey and neighboring New York are actively purchasing homes from people within the coastal flood plain, it will be impossible to buy them all.[22] And so a new, more permanent edge, one of cement and rebar and miles of sand, will doubtless be built along the edge of the mid-Atlantic. And that sand will move, and in all likelihood so will the houses.

Narragansett Bay was born of and shaped by the interplay between impermanence and stability. But the search for order continues to hold sway. By 1995, nearly 6,000 miles of roads surrounded the Bay.[23] Between 1972 and 1999, the amount of impermeable surface, such as roads, parking lots, and rooftops, increased in Rhode Island by 43 percent, at a rate six times faster than population growth.[24] What was once the Great Salt Cove in Providence is now the cement-lined reflecting pool in Waterplace Park.[25] Providence's working harbor is now ringed with new highways, and the city itself is encased in brick and concrete, off of which contaminants slip quickly and effortlessly into Bay waters with every rain. More than one-third of all coastal wetland "buffers" have been developed, almost 50 percent of estuarine marshes have been ditched and dried, and 52 percent of Narragansett Bay's shore has been transformed into "hardened shoreline." In addition, more than 1,100 working and abandoned dams dot the upper reaches of the Narragansett Bay watershed.[26] As the transformation from margin to edge continues, the Bay becomes less permeable and therefore less resilient. When heavy rains doused Rhode Island in March 2010, the towns of Warwick, West Warwick, and Cranston along the western shores of the Bay were submerged when heavy water runoff clashed with centuries of industrial development. Deforestation combined with miles of impermeable surface flushed the water so quickly into the Pawtuxet River that it had nowhere to go except over its banks.[27] When the littoral and its riparian feeders lost their porous, sponge-like qualities, when the edges of the estuary and the rivers that nourish it became an impenetrable line, they could no longer sustain the people who lived there.

But all is not lost. As history has shown, humans will continue to construct the Bay to meet the desires and demands of the times, and the Bay will respond and transform them in return. Narragansett Bay still forms the belly of Rhode Island, a state whose culture is inextricably tied to the estuary. Despite continued development, the boundary between land and sea, a space that naturalist Rachel Carson characterized as "elusive and indefinable," will remain, at least in part, exquisitely obscure.[28] If coasts evoke a commitment to planning and perfection, they also arouse the need for risk and adventure. Alongshore, the security of place is met with the freedom of space and these emotions are made manifest in the margins.[29] Ultimately, the postmodern condition has cast doubt upon and opened deep fissures within the linear legacy of modernity.[30] And this opens the door to incredible planning opportunities. Among the "broads" of East Anglia, the region from which many early New Englanders hailed, communities have begun to let the ocean in, buffering themselves from tidal surges by nurturing their saltmarshes.[31] Similarly, Dutch engineers have begun to redesign their cities in ways that anticipate and accommodate rising sea levels.[32] So perhaps it is time to forge a new definition of progress, one that accepts the necessity of hybridity and impermanence by embracing the sea. For if the push of progress is to define Rhode Island's future relationship with the Bay, it will no doubt be met with the pull of the sea's powerful tides. And somewhere in between, along the slipper-shelled shores of its islands, among the tidal rips at Nayatt Point and Narrow River, and among the mud-lined creeks of Hundred-Acre Cove, there exists something that is perhaps "chaotic" or even "anomalous," as Thoreau once mused, but is also capable of continual renewal.

Notes

Prologue: From Sweetwater to Seawater

1. Roger Williams, *A Key into the Language of America: Or, An help to the Language of the Natives in that part of America, called* New-England (London: Gregory Dexter, 1643), in *The Complete Writings of Roger Williams,* vol. 1, ed. James Hammond Trumbull (New York: Russell & Russell, 1963), 23.
2. *Collins Latin Dictionary,* s.v. "boil," accessed July 5, 2013, <http://www.credoreference.com/entry/hcdlat/boil>.
3. Deposition of Roger Williams, n.d., sworn by his son Joseph Williams, September 28, 1704, in Elisha R. Potter, *The Early History of Narragansett* in *Collections of the Rhode Island Historical Society,* vol. 3 (Providence, RI: Marshall, Brown, 1835), 3–5.
4. Sir John Denham, *Coopers-Hill* (London: H. Hills, 1709), 10–11.
5. Steven Mentz discussed the corrosive effects of the sea in his introductory remarks to a roundtable discussion titled "Saltwater in the Archives: Toward a New Oceanic Studies," John Carter Brown Library 50th Anniversary Fellows Conference, Brown University, Providence, RI, June 7–9, 2012.
6. Greg Dening, *Islands and Beaches* (Chicago: University of Chicago Press, 1988), 34.
7. J. C. Heesterman, "Littora et Intérieur de l'Inde," *Itinerario* 1 (1980): 89.
8. John Gillis, *Islands of the Mind: How the Human Imagination Created the Atlantic World* (New York: Palgrave MacMillan, 2004), 97.
9. John R. Gillis, *The Human Shore: Seacoasts in History* (Chicago: Chicago University Press, 2012), 2.

10. Michael N. Pearson, "Littoral Society: The Concept and Problems," *Journal of World History* 17, no. 4 (2006): 353.
11. Petra Van Dam, *De Amfibische Cultuur: Een Visie op Watersnoodrampen, Oratie, Vrije Universiteit, Amsterdam,* October 29, 2010, accessed June 20, 2013, <http://hdl.handle.net/1871/18457>.
12. See Lance Van Sittert, "The Other Seven Tenths," *Environmental History* 10, no. 1 (January 2005): 106-109; Kären Wigen, "Oceans of History," [*AHR* Forum] *American Historical Review* 111, no. 3 (June 2006): 717-721; W. Jeffrey Bolster, "Opportunities in Marine Environmental History," *Environmental History* 11 (July 2006): 567-597; Helen M. Rozwadowski, "Forum: Ocean's Depths," *Environmental History* 15, no. 3 (July 2010): 1-6; Helen M. Rozwadowski, "The Promise of Ocean History for Environmental History," *Journal of American History* 100, no. 1 (June 2013): 136-139.
13. John F. Richards, *The Unending Frontier: An Environmental History of the Early Modern World* (Berkeley: University of California Press, 2003), 572.
14. W. Jeffrey Bolster, *The Mortal Sea: Fishing the Atlantic in the Age of Sail* (Cambridge, MA: Harvard University Press, 2012); W. Jeffrey Bolster, "Putting the Ocean in Atlantic History," *American Historical Review* 113, no. 1 (February 2008): 19-47.
15. John R. Wennersten, *The Chesapeake: An Environmental Biography* (Baltimore: Maryland Historical Society, 2001), 55.
16. Steven G. Davidson, et al., *Chesapeake Waters: Four Centuries of Controversy, Concern, and Legislation* (Centreville, MD: Tidewater Publishers, 1997), 6, 50-53.
17. Philip D. Curtin, Grace S. Brush, and George W. Fisher, eds., *Discovering the Chesapeake: The History of an Ecosystem* (Baltimore, MD: The Johns Hopkins University Press, 2001).
18. William Cronon, *Changes in the Land: Indians, Colonists, and the Ecology of New England* (New York: Hill and Wang, 1983), 12-15; Brian Donahue, *The Great Meadow: Farmers and the Land in Colonial Concord* (New Haven, CT: Yale University Press, 2004), 21-23.
19. Arthur F. McEvoy, *The Fisherman's Problem: Ecology and Law in the California Fisheries, 1850-1980* (New York: Cambridge University Press, 1990), 14.
20. Alain Corbin, *The Lure of the Sea: The Discovery of the Seaside in the Western World, 1750-1840,* trans. Jocelyn Phelps (Berkeley: University of California Press, 1994), 29.
21. Yi-Fu Tuan, *Space and Place: The Perspective of Experience* (1977; Minneapolis: University of Minnesota Press, 2008), 3-4.
22. Eliga H. Gould, "Entangled Histories, Entangled Worlds: The English-Speaking Atlantic as a Spanish Periphery," *American Historical Review* 112, no. 3 (June 2007): 764-786.
23. Paul W. Mapp, "Atlantic History from Imperial, Continental, and Pacific Perspectives," *The William and Mary Quarterly* 63, no. 4 (October 2006): 713-724; Peter H. Wood, "From Atlantic History to Continental Approach," in *Atlantic History: A Critical Appraisal,* eds. Jack P. Greene and Philip D. Morgan (Oxford: Oxford University Press, 2009), 299-316; Michael Witgen, "Rethinking Colonial History as Continental History," *The William and Mary Quarterly* 69, no. 3 (July 2012): 527-530.

24. Richard White, *The Middle Ground: Indians, Empires, and Republics in the Great Lakes Region, 1650-1815* (Cambridge: Cambridge University Press, 1991).
25. Adam Smith, *An Inquiry into the Nature and Causes of the Wealth of Nations*, ed. Andrew Skinner (1776; New York: Penguin, 1999), 124.
26. Matthew Klingle, *Emerald City: An Environmental History of Seattle* (New Haven, CT: Yale University Press, 2007), 271; Carl Bridenbaugh, *Cities in the Wilderness: The First Century of Urban Life in America, 1625-1742* (New York: The Ronald Press, 1938), 5.
27. Ari Kelman, *A River and Its City: The Nature of Landscape in New Orleans* (Berkeley: University of California Press, 2003), 8; Matthew Booker, *Down by the Bay: San Francisco's History Between the Tides* (Berkeley: University of California Press, 2013), 4; Anthony N. Penna and Conrad Edick Wright, *Remaking Boston: An Environmental History of the City and Its Surroundings* (Pittsburgh, PA: University of Pittsburgh Press, 2009), 16.
28. Michael Rawson, *Eden on the Charles: The Making of Boston* (Cambridge, MA: Harvard University Press, 2012), viii.
29. Eric W. Sanderson, *Mannahatta: A Natural History of New York City* (New York: Abrams, 2009). Sanderson's work formed the basis for a museum exhibit, *Mannahatta/Manhattan: A Natural History of New York City*, Museum of the City of New York and the Wildlife Conservation Society, May 20–October 12, 2009. The Welikia Project continues the Mannahatta work by examining the environmental transformation of the outer boroughs.
30. Paul Carter, *Road to Botany Bay: An Exploration of Landscape and History* (Minneapolis: University of Minnesota Press, 2010), xxii.
31. Mark Monmonier, *Coastlines: How Mapmakers Frame the World and Chart Environmental Change* (Chicago: University of Chicago Press, 2008), ix; John R. Gillis, *The Human Shore*, 49–50; John Gillis, "From Ecotone to Edge: Atlantic Coasts, 1450–1850," unpublished essay e-mailed to author September 6, 2012.
32. Carl Bridenbaugh, *Fat Mutton and Liberty of Conscience: Rhode Island Society, 1636-1690* (Providence, RI: Brown University Press, 1974), 130.

1. Clams, Dams, and the Desiccation of New England

   1. John Winthrop, *Winthrop's Journal: History of New England, 1630–1649*, vol. 1, ed. James Kendall Hosmer (New York: Charles Scribner's Sons, 1908), 183.
   2. William Bradford, *Of Plymouth Plantation, 1620–1647*, ed. Samuel Eliot Morrison (New York: Knopf, 1952), 151, 157.
   3. Winthrop, *Winthrop's Journal*, 1: 83 and n. 1.
   4. Ibid., 90.
   5. Thomas Morton, *The New English Canaan*, ed. Charles Francis Adams, Jr. (1637; Boston: The Prince Society, printed by John Wilson and Son, 1883), 40.
   6. Winthrop, *Winthrop's Journal*, 1: 140.
   7. Bradford, *Of Plymouth Plantation*, 291.
   8. A fathom is six feet, so 400 fathoms equals 2,400 feet. There were roughly 300 wampum beads per fathom, which amounted to a total of 144,000 individual beads. A

single person could produce between thirty-six and forty-eight white wampum per day. Purple wampum took twice as long. See Paul A. Robinson, "The Wampum Trade in 17th-Century Narragansett Country," in *What a Difference a Bay Makes* (Providence, RI: Rhode Island Historical Society, 1993), 27.
9. Winthrop, *Winthrop's Journal*, 1: 184.
10. Ibid.
11. Lauren Benton, *A Search for Sovereignty: Law and Geography in European Empires, 1400–1900* (New York: Cambridge University Press), 45–49.
12. Daniel K. Richter, *Facing East from Indian Country: A Native History of Early America* (Cambridge, MA: Harvard University Press, 2001), 26–27, 18–19.
13. [Edward Winslow], [*Mourt's Relation:*] *The Journal of the Pilgrims at Plymouth in New England, in 1620*, ed. George B. Cheever (1622; New York: John Wiley, 1848), 45.
14. James Rice, *Nature & History in the Potomac Country: From Hunter-Gatherers to the Age of Jefferson* (Baltimore, MD: The Johns Hopkins University Press, 2009), 89.
15. Mark G. Soulsby, *Native American Trade and Exchange Systems in Southern New England* (Ledyard, CT: Mashantucket Pequot Tribal Council, 1994), 50.
16. Daniel Gookin, *Historical collections of the Indians in New England. Of their several nations, numbers, customs, manners, religion and government, before the English planted there. Also a true and faithful account of the present state and condition of the praying Indians . . . Together with a brief mention of the instruments and means, that God hath been pleased to use for their civilizing and conversion . . . Also suggesting some expedients for their further civilizing and propagating the Christian faith among them* (1674; Boston: Apollo Press by Belknap & Hall, 1792), 12.
17. Bradford, *Of Plymouth Plantation*, 203.
18. William Wood, *New England's Prospect: A True, Lively, and Experimentall description of that part of* America, *commonly called New England: discovering the state of that Countrie, both as it stands to our new-come* English *Planters; and to the old Native Inhabitants,* ed. Alden T. Vaughan (1634; Amherst: University of Massachusetts Press, 1977), 81.
19. On wampum as money see Marc Shell, *Wampum and the Origins of American Money* (Urbana, IL: University of Illinois Press, 2013).
20. William B. Weeden, "Indian Money as a Factor in New England Civilization," in *Institutions and Economics*, vol. 2, ed. Herbert B. Adams, Johns Hopkins University Studies in Historical and Political Science (Baltimore, MD: The Johns Hopkins University, 1884), 395. "Wampum," Weeden observed, "was the magnet which drew the beaver out of the interior forests."
21. Neal Salisbury, *Manitou and Providence: Indians, Europeans, and the Making of New England, 1500–1643* (1982; New York: Oxford University Press, 1984), 147–152.
22. William Cronon, *Changes in the Land: Indians, Colonists, and the Ecology of New England* (New York: Hill and Wang, 1983), 95. Cronon explained that "wampum was ideally suited to become the medium for a wider, more commercial exchange."
23. Van Cleaf Bachman, *Peltries or Plantations: The Economic Policies of the Dutch West India Company in New Netherland, 1623–1639* (Baltimore, MD: The Johns

Hopkins University Press, 1969), 21–22, 93; Allen W. Trelease, *Indian Affairs in Colonial New York: The Seventeenth Century* (1960; Lincoln: University of Nebraska Press, 1997), 49.

24. Lynn Ceci, "Native Wampum as a Peripheral Resource in the Seventeenth-Century World-System," in *The Pequots in Southern New England: The Fall and Rise of an American Indian Nation*, eds. Laurence M. Hauptman and James D. Wherry (Norman: University of Oklahoma Press, 1990), 48–63.
25. François Simiand, "La Monnaie, Réalité Sociale," *Annales Sociologiques* D, fasc. 1 (1934): 58. Bronislaw Malinowski, "The Primitive Economics of the Trobriand Islanders," *Economic Journal* 31, no. 121 (March 1921): 15, showed that economics were inextricably tied to "social, customary, legal and magico-religious" traditions. See also Pierre Vilar, *A History of Gold and Money, 1450–1920,* trans. Judith White (1969; London: New Left Books, 1976), 24, which provides a summary of the historiography of money and includes a discussion of Simiand's work.
26. Arjun Appadurai, "Introduction: Commodities and the Politics of Value," in *The Social Life of Things: Commodities in Cultural Perspective*, ed. Arjun Appadurai (New York: Cambridge University Press, 1986), 11–15.
27. Gaston Bachelard, *The Poetics of Space: The Classic Look at How We Experience Intimate Places,* trans. Maria Jolas (Beacon Press: Boston, 1994), 106, 112.
28. Giovanni Da Verrazano [to King François I, July 8, 1524], *Sailors Narratives of Voyages along the New England Coast, 1524–1624,* ed. George Parker Winship (New York: Burt Franklin: 1905), 4.
29. Ibid., 14.
30. Ibid., 14–15.
31. Ibid., 20.
32. Morton, *The New English Canaan*, 173.
33. Winslow, *Mourt's Relation*, 70.
34. Verrazano [to King François I, July 8, 1524], 18.
35. Ibid., 19.
36. Richard A. Chinman and Scott W. Nixon, "Depth-Area-Volume Relationships in Narragansett Bay," *University of Rhode Island Marine Technical Report* 87, no. 67 (1985).
37. Jon C. Boothroyd and Peter V. August, "Geologic and Contemporary Landscapes of the Narragansett Bay Ecosystem," in *Science for Ecosystem-based Management: Narragansett Bay in the 21st Century,* eds. Alan Desbonnet and Barry A. Costa-Pierce (New York: Springer Science, 2008), 26. The total Narragansett Bay watershed including estuarine waters is 4,766.2 square kilometers or 1840.2 square miles. The Wood-Pawcatuck watershed is 300 square miles. Combined, they equal more than 2,000 square miles.
38. Michael E. Q. Pilson, "Narragansett Bay amidst a Globally Changing Climate," in *Science for Ecosystem-based Management: Narragansett Bay in the 21st Century,* eds. Alan Desbonnet and Barry A. Costa-Pierce (New York: Springer Science, 2008), 39.
39. Michael E. Q. Pilson, "On the Residence Time of Water in Narragansett Bay," *Estuaries* 8 (1984): 2–14.
40. Wood, *New England's Prospect*, 55–56.

41. Morton, *The New English Canaan*, 226.
42. Wood, *New England's Prospect*, 56.
43. Morton, *The New English Canaan*, 222.
44. Wood, *New England's Prospect*, 55.
45. H. Bruce Franklin, *The Most Important Fish in the Sea: Menhaden and America* (Washington, DC: Shearwater, 2008), 27.
46. Wood, *New England's Prospect*, 56.
47. Ibid., 57.
48. David J. Bernstein, *Prehistoric Subsistence on the Southern New England Coast: The Record from Narragansett Bay* (New York: Harcourt Brace Jovanovich, 1993), 5. The following section follows Bernstein's thorough survey of archaeological sources for the Northeast coastal region.
49. In *Prehistoric Subsistence,* Bernstein showed that shellfish use began 2,700 years ago; Also see A. Leveillee and P. F. Thorbahn, "An Archaeological Assessment of the Sakonnet River," report 46-1 (Pawtucket, RI: Public Archaeology Laboratory, 1984). Leveillee and Thorbahn date quahog use along the Sakonnet River to 4,000 years ago, plus or minus 110 years. Wilbur Smith and Associates found similar results on Conanicut Island, with shells dating to 3,850 years ago, plus or minus 120 years.
50. D. R. Snow *The Archaeology of New England* (New York: Academic Press, 1980), 49–50.
51. Bernstein, *Prehistoric Subsistence,* 57.
52. Ibid., 52.
53. R. L. Greenspan, "Determination of Seasonality on *Mercenaria mercenaria* Shells from Archaeological Sites on Narragansett Bay, Rhode Island," The Rhode Island Sea Grant College Program Completion Report in *Archaeological Investigations at the Lambert Farm Site, Warwick, Rhode Island: An Integrated Program of Research and Education by the Public Archaeology Laboratory,* vol. 1, ed. J. Kerber (Pawtucket, RI: The Public Archaeology Laboratory, 1990), 184–193.
54. Wood, *New England's Prospect*, 114.
55. Bernstein, *Prehistoric Subsistence,* 149.
56. D. R. Yesner, "Resource Diversity and Population Stability Among Hunter-Gatherers," *Western Canadian Journal of Anthropology* 7, no. 2 (1977): 18–57.
57. Bernstein, *Prehistoric Subsistence,* 81.
58. Dena Dincauze, "A Capsule Prehistory of Southern New England," in *The Pequots in Southern New England,* eds. Laurence Hauptman and James Wherry (Norman: University of Oklahoma Press, 1990), 25–28.
59. Charles Whilloughby, *Antiquities of the New England Indians* (Cambridge, MA: Peabody Museum of American Archaeology and Ethnology, Harvard University, 1935), 86–87, 92–112; James Tuck, "Regional Cultural Development, 3000 to 300 B.C.," in *Handbook of North American Indians, Northeast,* vol. 15, ed. Bruce Trigger (Washington, DC: Smithsonian Institution, 1978), 28–43; James Fitting, "Regional Cultural Development, 300 B.C. to A.D. 1000," in *Handbook of North American Indians, Northeast,* vol. 15, ed. Bruce Trigger (Washington, DC: Smithsonian Institution, 1978), 45.

60. William Turnbaugh, "Post Contact Smoking Pipe Development: The Narragansett Example," in *Proceedings of the 1989 Smoking Pipe Conference, Selected Papers*, Research Record 22, ed. Charles Hayes III (Rochester, NY: Research Division of the Rochester Museum and Science Center, 1992), 113–124.
61. Roger Williams, *A Key into the Language of America: Or, An help to the Language of the Natives in that part of America, called New-England* (London: Gregory Dextor, 1643), in vol. 1 of *The Complete Writings of Roger Williams*, ed. James Hammond Trumbull (New York: Russell & Russell, 1963), 72–73.
62. Wood, *New England's Prospect*, 81.
63. Roger Williams, *A Key into the Language of America*, 173.
64. Daniel Gookin, *Historical Collections of the Indians in New England*, 12.
65. Wood, *New England's Prospect*, 81.
66. Salisbury, *Manitou and Providence*, 25–30. Salisbury estimated that there were between 37,000 and 38,000 Narragansett Indians.
67. Ibid., 147–148.
68. William Bradford, *History of Plymouth Plantation, 1620–1647*, vol. 2 (1912, 1940; New York: Russell & Russell, 1968), 43.
69. Williams, *A Key into the Language of America*, 173.
70. Williams, *A Key into the Language of America*, 175–176, 177–178; Wood, *New England's Prospect*, 81, 104, 111.
71. Williams, *A Key into the Language of America*, 175–176, 177–178.
72. Barbara A. Mann, "The Fire at Onondaga: Wampum as Proto-Writing," *Akwesasne Notes* 1, no. 1, (Spring 1995): 43. Also see Christopher L. Miller and George R. Hamell, "A New Perspective on Indian-White Contact: Cultural Symbols and Colonial Trade," *Journal of American History* 73, no. 2 (1986): 315, 325.
73. Miller and Hamell, "A New Perspective on Indian-White Contact," 316.
74. George R. Hamell, "Strawberries, Floating Islands, and Rabbit Captains: Mythical Realities and European Contact in the Northeast during the Sixteenth and Seventeenth Centuries," *Journal of Canadian Studies* 21, no. 4 (Winter 1986–1987): 78.
75. Williams, *A Key into the Language of America*, 24.
76. William Scranton Simmons, *Cauntantowwit's House: An Indian Burial Ground on the Island of Conanicut in Narragansett Bay* (Providence, RI: Brown University Press, 1970), 46, 138; Wood, *New England's Prospect*, 111.
77. Hamell, "Strawberries, Floating Islands, and Rabbit Captains," 78.
78. Williams, *A Key into the Language of America*, 150, 24.
79. Ezra Stiles, *Extracts from the Itineraries and Other Miscellanies of Ezra Stiles, D.D., LL.D., 1755–1794*, ed. Franklin B. Dexter (New Haven, CT: Yale University Press, 1916), 157.
80. Helen Manning, *Moshup's Footsteps: The Wampanoag Nation Gay Head/Aquinnah, The People of First Light*. (Aquinnah, MA: Blue Cloud Across the Moon Publishing, 2001), 22–25.
81. William S. Simmons, *Spirit of the New England Tribes: Indian History and Folklore, 1620–1984* (Hanover, NH: University Press of New England, 1986), 172–234.
82. Williams, *A Key into the Language of America*, 178.

83. Williams, *A Key into the Language of America*, 135.
84. Psalms, 107:24, NRSV; Wood, *New England's Prospect*, 70; Gen. 7-8; Acts 27.
85. Williams, *A Key into the Language of America*, 142.
86. Hamell, "Strawberries, Floating Islands, and Rabbit Captains," 84. Hamell explains that the "degree to which a native community's traditional mythical reality was synthesized or syncretized was inversely proportional to that community's distance from the Atlantic Coast and the nodes of face-to-face European contact experience."
87. Ibid., 64.
88. David Thompson, *The "Travels," 1850 Version* in vol. 1 of *The Writing of David Thompson*, ed. William E. Moreau (Montreal and Kingston: McGill-Queen's University Press; Seattle: University of Washington Press; Toronto: The Champlain Society, 2009), 194-195; Susan M. Preston, introduction to "A Pair of Hero Stories," in *Algonquian Spirit: Contemporary Translations of the Algonquian Literatures of North America*, ed. Brian Swann (Lincoln: University of Nebraska Press, 2005), 231. Within the Cree mythic tradition, Preston shows that water often had "conjuring power" and was often the "locus of species transformation, both literal and symbolic."
89. Jaap Jacobs, *The Colony of New Netherland: A Dutch Settlement in Seventeenth-Century America* (Ithaca, NY: Cornell University Press, 2009), 28.
90. Ibid., 30.
91. Samuel Greene Arnold, *History of the State of Rhode Island and Providence Plantations*, vol. 1 (New York: D. Appleton, 1859), 155 n. 1.
92. J. Franklin Jameson, introduction to *Historisch Verhael*, in *Narratives of New Netherland: 1609-1664*, ed. J. Franklin Jameson (New York: Charles Scribner's Sons, 1909), 63-64.
93. Nicolaes Van Wassenaer, *Historisch Verhael*, in *Narratives of New Netherland: 1609-1664*, ed. J. Franklin Jameson (New York: Charles Scribner's Sons, 1909), 86.
94. Salisbury, *Manitou and Providence*, 148-149.
95. Van Wassenaer, *Historisch Verhael*, 68.
96. Adriaen van der Donck, *A Description of New Netherlands*, ed. Thomas F. O'Donnell (1655; Syracuse, NY: Syracuse University Press, 1968), 15-16.
97. Ibid., 71-72.
98. Bradford, *History of Plymouth Plantation*, 42-44.
99. Isaack de Rasieres to Samuel Blommaert, 1628, *Narratives of New Netherland: 1609-1664*, ed. J. Franklin Jameson (New York: Charles Scribner's Sons, 1909), 110.
100. Bradford, *History of Plymouth Plantation*, 42-43.
101. Roger Williams, "Testimony of Roger Williams relative to the first settlement of the Narragansett Country by Richard Smith, July 21, 1679," in *The Letters of Roger Williams*, vol. 6 of *The Complete Writings of Roger Williams* ed. John Russell Bartlett (New York: Russell & Russell, 1963), 399.
102. Francis X. Moloney, *The Fur Trade in New England, 1620-1676* (Cambridge, MA: Harvard University Press, 1931), 43; also see Howard Millar Chapin, *The Trading*

Post of Roger Williams with Those of John Wilcox and Richard Smith (Providence, RI: E. L. Freeman, 1933).

103. Stephen Innes, *Labor in a New Land: Economy and Society in Seventeenth-Century Springfield* (Princeton, NJ: Princeton University Press, 1983), 30–33.
104. Moloney, *The Fur Trade in New England*, 110–112.
105. Sylvester Judd, "The Fur Trade on the Connecticut River in the Seventeenth Century," in *The New England Historical and Genealogical Register*, vol. 11, ed. Samuel G. Drake (Boston: C. Benjamin Richardson, 1857), 218–219.
106. Williams, *A Key into the Language of America*, 127.
107. Robert J. Naiman, Jerry M. Melillo, and John E. Hobbie, "Ecosystem Alteration of Boreal Forest Streams by Beaver (*Castor canadensis*)," *Ecology* 67, no. 5 (October 1986): 1254–1269.
108. Robert T. Paine, "A Note on Trophic Complexity and Community Stability," *The American Naturalist* 103, no. 929 (1969): 91–93.
109. Ernest Thompson Seton, *Lives of Game Animals: An Account of those Land Animals in America, north of the Mexican Border, Which Are Considered "Game," either because they have held the Attention of Sportsmen, or received the Protection of Law*, vol. 4, *Rodents, Etc.* (Garden City, NY: Doubleday, Doran, 1929), 447–448.
110. R. Rudemann and W. J. Schoonmaker, "Beaver Dams as Geological Agents," *Science* 88 (1938): 523–525.
111. Present-day scientists examining eastern gray squirrel populations typically focus their studies on discrete locations. In 1920, however, Ernest Thompson Seton estimated—a shot in the dark, really—that there were 1 billion gray squirrels in the United States. See Ernest Thompson Seton, "Migrations of the Graysquirrel (*Sciurus carolinensis*)," *Journal of Mammalogy* 1, no. 2 (February 1920): 57; Seton's estimate was based on New York City's Central Park, where the population density of squirrels was likely higher than that in natural forests. On the urbanization of gray squirrels, see Etienne Benson, "The Urbanization of the Eastern Gray Squirrel in the United States," *Journal of American History* 100, no. 3 (December 2013): 691–710.
112. Thompson, *The "Travels," 1850 Version*, 191.
113. Jeremy Belknap, *The History of New Hampshire*, vol. 3 (Boston: Belknap and Young, 1792), 160.
114. Ronald L. Ives, "The Beaver-Meadow Complex," *Journal of Geomorphology* 5, no. 3 (October 1942): 194–197.
115. M. Novak, "Beaver," in *Wild Furbearer Management and Conservation in North America*, eds. M. Novak, J. A. Baker, M. E. Obbard, and B. Malloch (Toronto: Ontario Ministry of Natural Resources, 1987), 282–312.
116. David R. Butler and George P. Malanson, "The Geomorphic Influences of Beaver Dams and Failures of Beaver Dams," *Geomorphology* 71 (2005): 55–56.
117. Van Wassenaer, *Historisch Verhael*, 70.
118. Van der Donck, *A Description of New Netherlands*, 114.
119. Ibid., 16.
120. Van Wassenaer, *Historisch Verhael*, 83.
121. Van der Donck, *A Description of New Netherlands*, 111.

122. Ibid.
123. Belknap, *The History of New Hampshire*, 3: 160–161.
124. Van der Donck, *A Description of New Netherlands*, 18.
125. Morton, *The New English Canaan*, 240.
126. R. A. Marston, "River Entrenchment in Small Mountain Valleys of the Western USA: Influence of Beaver, Grazing and Clearcut Logging," *Revue de Geographie de Lyons* 69 (1994): 11.
127. R. S. Rupp, "Beaver-Trout Relationships in the Headwaters of the Sunkhaze Stream, Maine," American Fisheries Society *Transactions* 84 (1955): 75–85. J. D. Stock and I. J. Schlosser, "Short-Term Effects of a Catastrophic Beaver Dam Collapse on a Stream Fish Community," *Environmental Biology of Fishes* 31 (1991): 123–129.
128. Johannes Megaolensis, Jr., "A Short Account of the Mohawk Indians," in *Narratives of New Netherland: 1609–1664*, ed. J. Franklin Jameson (New York: Charles Scribner's Sons, 1909), 171.
129. Van der Donck, *A Description of New Netherlands*, 17.
130. David R. Butler and George P. Malanson, "The Geomorphic Influences of Beaver Dams and Failures of Beaver Dams," *Geomorphology* 71 (2005): 55–56.
131. Salisbury, *Manitou and Providence*, 147–152.
132. Nixon based his calculations on Naiman, Melillo, and Hobbie, "Ecosystem Alteration."
133. Scott W. Nixon, "Prehistoric Nutrient Inputs and Productivity in Narragansett Bay," *Estuaries* 2, no. 2 (June 1997): 253–261.
134. J. E. Ridley and J. A. Steel, "Ecological Aspects of River Impoundments," in *River Ecology*, ed. B. A. Whitton (Berkeley: University of California Press, 1975): 565–587; R. M. Baxter, "Environmental Effects of Dams and Impoundments," *Annual Review of Ecology and Systematics* 8 (1977): 255–283. See also J. C. Varekamp, "The Historic Fur Trade and Climate Change," *Eos* 87, no. 52 (December 26, 2006): 593, 596–597.
135. Nixon, "Prehistoric Nutrient Inputs and Productivity in Narragansett Bay," 258.
136. Charles S. Hopkinson, Jr., and Joseph J. Vallino, "The Relationships among Man's Activities in Watersheds and Estuaries: A Model of Runoff Effects on Patterns of Estuarine Community Metabolism," *Estuaries* 18, no. 4 (December 1995): 598–621.
137. F. Gross, "An Experiment in Marine Fish Cultivation: Introduction," *Proceedings of the Royal Society of Edinburgh* 63B (1947): 1–2. In 1947, Gross experimented with fertilizing Loch Craiglin in Scotland with the hope of increasing British food production during World War II. As Gross and his associates had hoped, dumping inorganic fertilizer into the loch at regular intervals produced more fish. See also F. Thurow, "Estimation of Total Fish Biomass in the Baltic Sea during the 20th Century," *ICES Journal of Marine Sciences* 54 (1997): 444–461. An examination of twentieth-century Baltic fish stocks revealed increases after 1950, which Thurow attributed to eutrophication. Also see J. D. H. Cushing, *Marine Ecology and Fisheries* (Cambridge: Cambridge University Press, 1975). Cushing outlined

the mechanics of what he called a nutrient-induced "agricultural model," of fish production.

138. Scott W. Nixon, "Quantifying the Relationship Between Nitrogen Input and the Productivity of Marine Ecosystems," *Proceedings of Advanced Marine Technology Conference*, eds. M. Takahashi, K. Nakata, and T. R. Parsons (Tokyo: Advanced Marine Technology Conference, 1992): 57–83; Scott W. Nixon, "Nutrient Dynamics, Primary Production and Fishery Yields of Lagoons," *Oceanologica Acta* 4 (1982): 357–381; S. Nixon and B. Buckley, "'A Strikingly Rich Zone'—Nutrient Enrichment and Secondary Production in Coastal Marine Ecosystems," *Estuaries* 25, no. 4b (August 2002): 782–796.

139. See Charles C. Mann, *1491: New Revelations of the Americas before Columbus* (New York: Alfred A. Knopf, 2008). Mann posits that the extraordinary abundance witnessed by European observers could have been human-induced. Disease and land clearance could have altered environmental conditions in ways that caused population explosions. Noting that many experts contest this thesis, he cites the example of the passenger pigeon, explaining that some scientists believe that changing ecological conditions caused "outbreak populations," pp. 315–318. Similarly, referencing the work of William S. Preston, Mann explained that as coastal California Indian populations crashed due to disease, clams and mussels grew larger and their populations "exploded," p. 321.

140. William Hubbard, *The History of the Indian Wars in New England from the First Settlement to the Termination of the War with King Philip, in 1677*, ed. Samuel G. Drake (1677; Roxbury, MA: Printed for W. Elliot Woodward, 1865), 31.

2. Shoveling Dung Against the Tide

1. James M[a]cSparran, D.D., *America dissected, being a full and true account of all the American Colonies, shewing the Intemperance of the Climates, excessive Heat and Cold, and sudden violent Changes of Weather, terrible and Mischievous Thunder and Lightning, bad and unwholesome air, destructive to Human Bodies.–Badness of Money, Danger from Enemies, but above all, the Danger to the souls of the Poor People that remove thither from the multifarious wicket and pestilent Heresies that prevail in those parts. In several letters from a Reverend Divine of the Church of England, Missionary to America and Doctor of Divinity, Published as a Caution to Unsteady People who may be tempted to leave their Native Country* (Dublin: S. Powell, Dame Street, 1753) in *Collections of the Rhode Island Historical Society*, vol. 3 (Providence: Marshall Brown and Company, 1835), 133–134.

2. Ibid., 133.

3. Sydney V. James, *Colonial Rhode Island: A History* (New York: Charles Scribner's Sons, 1975), xvi.

4. Carl Bridenbaugh, *Fat Mutton and Liberty of Conscience: Society in Rhode Island, 1636–1690* (Providence, RI: Brown University Press, 1974), 130.

5. Howard M. Chapin, *Illustrations of the Seals, Arms and Flags of Rhode Island* (Providence: Rhode Island Historical Society, 1930), 1. After the four original towns of

Rhode Island (Providence, Newport, Portsmouth, and Warwick) united under the charter of 1643, they adopted the anchor as their symbol in 1647.
6. John Winthrop, *History of New England from 1630–1649*, vol. 1, ed. James Savage (1853; Baltimore: Clearfield, 2003), 209.
7. Ibid., 198.
8. Roger Williams to Major Mason, Providence, June 22, 1670, *Letters of Roger Williams, 1632–1682*, vol. 6 of *The Complete Writings of Roger Williams*, ed. John Russell Bartlett (New York: Russell & Russell, 1963), 335.
9. Edwin S. Gaustad, *Roger Williams* (New York: Oxford University Press, 2005), 2–5; for an engaging and meticulous biography of Williams that places him firmly in the context of early modern English law, see John M. Barry, *Roger Williams and the Creation of the American Soul: Church, State, and the Birth of Liberty* (New York: Viking, 2012).
10. William Bradford, *Of Plymouth Plantation: 1620–1647*, ed. Samuel Elliot Morison (New York: Alfred A. Knopf, 1952), 257.
11. Roger Williams, "Mr. Cotton's Letter Examined and Answered," in *The Complete Writings of Roger Williams*, vol. 1, ed. Reuben Aldridge Guild (New York: Russell & Russell, 1963), 324–325.
12. Roger Williams, *Bloody Tenet yet More Bloody*, vol. 4 of *The Complete Writings of Roger Williams*, ed. Perry Miller (1652; New York: Russell & Russell, 1963), 461.
13. Winthrop, *History of New England*, 1: 189.
14. Glenn W. LaFantasie, ed., editorial note to *The Correspondence of Roger Williams*, vol. 1 (Providence, RI and Hanover, NH: Brown University Press and University Press of New England, 1988), 17–18.
15. Ibid., 18–19.
16. Winthrop, *History of New England*, 1: 204.
17. Roger Williams to John Cotton, March 25, 1671, *The Complete Writings of Roger Williams*, 1: 355.
18. Winthrop, *History of New England*, 1: 204.
19. Nathaniel B. Shurtleff, ed., *Records of the Governor and Company of the Massachusetts Bay in New England*, vol. 1 (Boston: W. White, 1853–1854), 160.
20. Winthrop, *History of New England*, 1: 209.
21. Roger Williams to Major Mason, Providence, June 22, 1670, *Letters of Roger Williams*, 335–336.
22. The compass is held in the Rhode Island Historical Society collections.
23. Roger Williams to Major Mason, Providence, June 22, 1670, *Letters of Roger Williams*, 335–336.
24. Glenn W. LaFantasie, "Introduction," in *The Correspondence of Roger Williams*, 1: xxxiv; Winthrop, *History of New England*, 1: 209.
25. Roger Williams to Major Mason, 335.
26. John Russell Bartlett, ed., *Records of the Colony of Rhode Island and Providence Plantation in New England*, vol. 1 (Providence, RI: A. Crawford Greene and Brother, State Printers, 1856), 22.
27. Bartlett, *Records of the Colony of Rhode Island*, 1: 18–20.
28. James, *Colonial Rhode Island*, 20.

29. Bartlett, *Records of the Colony of Rhode Island*, 1: 16.
30. Roger Williams to John Winthrop, [Providence], October 28, 1637, *Letters of Roger Williams*, 70-71.
31. Virginia DeJohn Anderson, "Animals into the Wilderness: The Development of Livestock Husbandry in the Seventeenth-Century Chesapeake," in *Environmental History and the American South: A Reader*, eds. Paul S. Sutter and Christopher J. Manganiello (Athens: University of Georgia Press, 2009), 30. Anderson explained that on seventeenth-century Chesapeake farms sheep "all but disappeared."
32. Roger Williams to John Winthrop, Providence, January 10, 1637-1638, *Letters of Roger Williams*, 85.
33. Williams overestimated the distance. Providence is only about fifteen miles to the northern end of Prudence Island.
34. James, *Colonial Rhode Island*, 30.
35. Ibid., 28.
36. Daniel Berkeley Updike, *Richard Smith: First English Settler of the Narragansett Country, Rhode Island* (Boston: The Merrymount Press, 1937), 13, 16.
37. [Edward Winslow], [*Mourt's Relation:*] *The Journal of the Pilgrims at Plymouth in New England, in 1620*, ed. George B. Cheever (1622; New York: John Wiley, 1848), 69.
38. "Roger Williams' Testimony in Favor of Richard Smith's Title to the Wickford Land," in *Collections of the Rhode Island Historical Society*, vol. 3 (Providence, RI: Marshall Brown & Co., 1835), 166.
39. Ibid.
40. Elisha R. Potter Jr., *The Early History of Narragansett: With an Appendix of Original Documents, Many of Which are Now for the First Time Published* in *Collections of the Rhode Island Historical Society*, vol. 3 (Providence, RI: Marshall, Brown and Company, 1835), 32.
41. Bartlett, *Records of the Colony of Rhode Island*, 3: 57.
42. Updike, *Richard Smith*, 15.
43. Ibid., 14-15.
44. Wilkins Updike, *A History of the Episcopal Church in Narragansett, Rhode Island Including a History of the Other Episcopal Churches in the State* (Boston: D. B. Updike, The Merrymount Press, 1907), 13-14.
45. Potter, *The Early History of Narragansett*, 32.
46. "Records of King's Province, nos. 56-59," in Potter, *The Early History of Narragansett*, 33.
47. Irving Berdine Richman, *Rhode Island: Its Making and Its Meaning* . . . (New York: G. P. Putnam's Sons, 1908), 498. See also Edgar Mayhew Bacon, *Narragansett Bay: Its Historic and Romantic Associations and Picturesque Setting* (New York: G. P. Putnam's Sons, 1904), 218-223.
48. James Kendall Hosmer, ed., *Winthrop's Journal: History of New England, 1630-1649*, vol. 1 (New York: Charles Scribner's Sons, 1908), 138.
49. Edward Channing, *The Narragansett Planters: A Study of Causes*, Johns Hopkins University Studies in Historical and Political Science, vol. 4, ed. Herbert B. Adams (Baltimore, MD: The Johns Hopkins University, 1886), 117.

50. *Collections of the Rhode Island Historical Society*, 3: 275.
51. Roger Williams to Major Mason, Providence, June 22, 1670, *Letters of Roger Williams*, 342.
52. Simon Ryan, *The Cartographic Eye: How Explorers Saw Australia* (Cambridge: Cambridge University Press, 1996), 105. See also Philip E. Steinberg, *The Social Construction of the Ocean* (Cambridge: Cambridge University Press, 2001), 37–38. Steinberg discusses Ryan's thesis in the context of ocean space.
53. Channing, *The Narragansett Planters*, 17. For a detailed discussion of the ways Connecticut and Massachusetts maneuvered for Narragansett Country lands, see Richard S. Dunn, "John Winthrop Jr., and the Narragansett Country," *The William and Mary Quarterly* 13, no. 1 (January 1956): 68–86.
54. Roger Williams to Major Mason, Providence, June 22, 1670, *Letters of Roger Williams*, 343.
55. Potter, *The Early History of Narragansett*, 290.
56. "Mr. Samuell Sewall's Deed," in Appendix to Caroline Hazard, *Thomas Hazard son of Robt call'd College Tom: A Study of Life in Narragansett in the XVIIIth Century* (Boston: Houghton Mifflin, 1893), 217–220.
57. Caroline Hazard, *Thomas Hazard son of Robt*, 16.
58. Ibid., 30–31.
59. W. Noel Sainsbury, ed., *Calendar of State Papers, Colonial Series: American and West Indies, 1661–1668* (London: Longman & Co., 1880), 25.
60. Ibid., 343; Carl Bridenbaugh, *Fat Mutton and Liberty of Conscience: Society in Rhode Island, 1636–1690* (Providence, RI: Brown University Press, 1974), 53 n. 38. Bridenbaugh's analysis of seventeenth-century Rhode Island husbandry was invaluable to this chapter. His synoptic survey of primary sources made it necessary, in many cases, to follow his lead. This study, however, has drawn from the original sources.
61. Edward Randolph, *Edward Randolph: Including His Letters and Officials Papers from the New England, Middle, and Southern Colonies in America, with Other Documents Relating Chiefly to the Vacating of the Royal Charter of the Colony of Massachusetts Bay, 1676–1703*, vol. 2, ed. Robert Noxon Toppan (Boston: The Prince Society, 1898), 246.
62. Clarence S. Brigham, ed., *The Early Records of the Town of Portsmouth*, [vol. 1] (Providence, RI: E. L. Freeman & Sons, State Printers, 1901), 119.
63. Bacon, *Narragansett Bay*, 249, 283; Howard S. Russell, *A Long Deep Furrow: Three Centuries of Farming in New England* (Hanover, NH: University Press of New England, 1976), 156.
64. John Osborn Austin, *Genealogical Dictionary of Rhode Island, Comprising Three Generations of Settlers Arriving Before 1690* (1887; Baltimore, MD: Clearfield Publishing, 2008), 254.
65. "Harris Papers," in *Collections of the Rhode Island Historical Society*, vol. 10 (Providence: Rhode Island Historical Society, 1902), 144–145.
66. William Coddington to John Winthrop, Jr., Newport, April 20, 1647, *Winthrop Papers, 1645–1649*, vol. 5, ed. Allyn Bailey Forbes (Boston: Massachusetts

Historical Society, 1929–1947), 149–150; also see William Coddington to John Winthrop Jr., [Newport], October 14, 1648, *Winthrop Papers*, 269–270.
67. M. De La Mothe Cadillac, "Memoire of M. De La Mothe Cadillac," *Collections of the Maine Historical Society*, vol. 6 (Portland, ME: Published for the Society, 1859), 288.
68. Howard M. Chapin, *Illustrations of the Seals, Arms, and Flags of Rhode Island* (Providence: Rhode Island Historical Society, 1930), 6, 57. Also see Bridenbaugh, *Fat Mutton and Liberty of Conscience*, 57.
69. Bridenbaugh, *Fat Mutton and Liberty of Conscience*, 49.
70. *South Kingstown Town Records*, 1: 101, 197. Cited in Hazard, *Thomas Hazard son of Robt*, 101–102.
71. *Calendar of State Papers Colonial Series, America and West Indies 1675–1676*, ed. W. Noel Sainsbury (1893; Vaduz, Liechtenstein: Kraus Reprint, 1964), 213.
72. Clarence S. Brigham, ed., *The Early Records of the Town of Portsmouth* (Providence, RI: E. L. Freeman & Sons, State Printers, 1901), 189.
73. John Russell Bartlett, ed., *Records of the Colony of Rhode Island and Providence Plantations in New England, 1678–1706*, vol. 3 (Providence, RI: Knowles, Anthony, & Co., 1858), 202.
74. John Josselyn, *Account of Two Voyages to New-England*, in *John Josselyn, Colonial Traveler: A Critical Edition of Two Voyages to New-England*, ed. Paul J. Lindholdt (1674; Hanover, NH: University Press of New England, 1988), 132.
75. John Hull to Mr. Arnold, April 16, 1677, John Hull Letter Book, vol. 2, typescript, 335, American Antiquarian Society.
76. *Proceedings of the Massachusetts Historical Society*, 2d. ser., vol. 1 (Boston: Massachusetts Historical Society, 1885), 247.
77. Joseph Dudley to William Blathwayt, New England, July 31, 1686, in *Edward Randolph; Including His Letters and Official Papers from the New England, Middle, and Southern Colonies in America and the West Indies, 1678–1700*, vol. 6, ed. Alfred Thomas Scrope Goodrick (Boston: The Prince Society, 1909), 196–197.
78. Thomas M. Truxes, *The Irish-American Trade* (Cambridge: Cambridge University Press, 1998), 15.
79. Ibid., 9–14.
80. William Douglass, M.D., *A Summary, Historical and Political, of the First Planting, Progressive Improvements, and Present State of the British Settlements in North-America*, vol. 2 (London: Printed for R. and J. Dodsley, in Pall-mall, 1760), 99.
81. John Russell Bartlett, ed., *Records of the Colony of Rhode Island and Providence Plantations in New England*, vol. 5, 1741–1756 (Providence, RI: Knowles, Anthony & Co., 1860), 242.
82. Howard M. Chapin, *A Documentary History of Rhode Island*, vol. 2, (Providence, RI: Preston and Rounds, Co., 1919), 34.
83. Brigham, *The Early Records of the Town of Portsmouth*, 389.
84. Ibid., 72–73.
85. Ibid., 262–265.
86. Russell, *A Long, Deep Furrow*, 158.

87. Howard M. Chapin, ed., *The Early Records of the Town of Warwick* (Providence: E. A. Johnson, 1926), 52–53. Also in Bridenbaugh, *Fat Mutton and Liberty of Conscience*, 41–42; Virginia DeJohn Anderson, *Creatures of Empire: How Domestic Animals Transformed Early America* (New York: Oxford, 2004), 149.
88. William Wood, *New England's Prospect*, ed. Alden T. Vaughan (Amherst: University of Massachusetts Press, 1977), 57.
89. Charles J. Hoadly, ed., *Records of the Colony and Plantation of New Haven, from 1638–1649*, vol. 1 (Hartford, CT: Case, Tiffany, and Company, 1857), 312. Also in Anderson, *Creatures of Empire*, 160.
90. Russell, *A Long, Deep Furrow*, 153.
91. Ibid., 152.
92. Thomas Williams Bicknell, *The History of the State of Rhode Island and Providence Plantations*, vol. 3 (New York: The American Historical Society, 1920), 1192.
93. R. H. Platt, "Overview of Developed Coastal Barriers," in *Cities on the Beach: Management Issues of Developed Coastal Barriers*, Research Paper No. 224, eds. Rutherford H. Plat, Sheila G. Pelczarski, and Barbara K. R. Burbank (Chicago: University of Chicago Department of Geography, 1987), 4–6.
94. Thomas B. Hazard, *Nailer Tom's Diary: Otherwise the Journal of Thomas B. Hazard of Kingstown Rhode Island, 1778 to 1840 Which includes Observations on the Weather Records of Births Marriages and Deaths Transactions by Barter and Money of Varying Value Preaching Friends and Neighborhood Gossip*, (Boston: The Merrymount Press, 1930), 173.
95. *Collections of the Rhode Island Historical Society*, 7: 334.
96. Robert V. Wells, *The Population of the British Colonies in America Before 1776: A Survey of Census Data* (Princeton, NJ: Princeton University Press, 1975), 98.
97. US Environmental Protection Agency, "Municipally Owned Wastewater Treatment Facilities in New England," (Boston: US EPA New England Region, Office of Ecosystem Protection, 2002).
98. S. P. Hamburg, Donald Pryor, and Matthew A. Vadeboncoeur, "Nitrogen Inputs to Narragansett Bay: An Historic Perspective," in *Science for Ecosystem-based Management: Narragansett Bay in the 21st Century*, eds. A. Desbonnet and B. A. Costa-Pierce (New York: Springer Science, 2008), 196.
99. R. B. Alexander, et al., "Atmospheric Nitrogen Flux from the Watersheds of Major Estuaries of the United States: An Application of the SPARROW Watershed Model," in *Nitrogen Loading in Coastal Water Bodies: An Atmospheric Perspective*, eds. R. Valigura, et al., American Geophysical Union Monograph 57 (Washington, DC: American Geophysical Union, 2001), 119–170.
100. Hamburg et al., "Nitrogen Inputs to Narragansett Bay," 194, 196; Horsely, Witten, Hegemann, Inc. showed that in Buttermilk Bay, where most residences were located close to the water, upwards of 3.1 kg of nitrogen per year were introduced. See "Quantification and Control of Nitrogen Inputs to Buttermilk Bay," Report prepared for the US Environmental Protection Agency, Massachusetts Executive Office of Environmental Affairs, and New England Interstate Water Pollution Control Commission, (Barnstable, MA: Horsley, Witten, Hegemann, Inc., 1991).

101. E. M. Snow, "Report upon the Census of Rhode Island, 1865," (Providence, RI: Providence Press, 1867); C. D. Wright, "The Census of Massachusetts 1885: Agricultural Products and Property," vol. 3 (Boston: Wright & Potter Printing Company, 1887).
102. Russell, *A Long Deep Furrow*, 158, 153, 165.
103. Bridenbaugh contended that this estimate was probably not exaggerated because it represented only a doubling of the Rhode Island flock since 1661. Bridenbaugh, *Fat Mutton and Liberty of Conscience*, 57.
104. US Department of Agriculture, "Costs Associated with Development and Implementation of Comprehensive Nutrient Management Plans. Part 1: Nutrient Management, Land Treatment, Manure and Wastewater Handling and Storage, and Recordkeeping (Washington DC: US Department of Agriculture, 1997).
105. P. Johnes, "Evaluation and Management of the Impact of Land Use Change on the Nitrogen and Phosphorus Load Delivered to Surface Waters: The Export Coefficient Modeling Approach," *Journal of Hydrology* 183 (1996): 323-349.
106. S. P. Hamburg, et al., "Nitrogen Inputs to Narragansett Bay: An Historic Perspective" in A. Desbonnet and B. A. Costa-Pierce, eds., *Science for Ecosystem-based Management: Narragansett Bay in the 21st Century* (New York: Springer, 2008), 181.
107. Scott W. Nixon, "An Extraordinary Red Tide and Fish Kill in Narragansett Bay," in *Novel Phytoplankton Blooms, Causes and Impacts or Recurrent Brown Tides and Other Unusual Blooms*, eds. E. M. Cosper, V. M. Bricelj, and E. J. Carpenter (Berlin: Springer-Verlag, 1989), 429-447. Nixon discusses a massive phytoplankton bloom and fish die-off that occurred in the Bay in 1898.
108. Hamburg, et al., "Nitrogen Inputs to Narragansett Bay," 182, 184.
109. *Rhode Island State Census, 1885*, Amos Perry, Superintendent (Providence, RI: E. L. Freeman & Son, 1887), 95.
110. Rhode Island State Board of Health, *Ninth Annual Report of the State Board of Health of the State of Rhode Island, For the Year Ending December 31, 1886 and Including the Report Upon the Registration of Births, Marriages, and Deaths in 1885* (Providence, RI: E. L. Freeman & Son, 1887), 102.
111. Rhode Island State Board of Health, *Twenty-Fifth Annual Report of the State Board of Health of the State of Rhode Island for the Year Ending December 31, 1902, and Including the Report Upon the Registration of Births, Marriages, and Deaths in 1901* (Providence, RI: E. L. Freeman Company, 1910), 260.
112. John H. Ryther, "The Ecology of Phytoplankton Blooms in Moriches Bay and Great South Bay, Long Island, New York," *Biological Bulletin* 106, no. 2 (April 1954): 198-209.
113. James Helme, Draft of Capt. Henry Bull's Lots, S. Kingston 1729, Map 0150, 1729 Cartography Collection, Rhode Island Historical Society.
114. Hamburg, et al., "Nitrogen Inputs to Narragansett Bay," 204. Also see Sindya N. Bhanoo, "Amish Farming Draws Rare Government Scrutiny," *New York Times*, June 8, 2010, A12. Bhanoo explained that EPA officials imposed environmental mandates on Amish farmers, whose "traditional" husbandry practices were

polluting Chesapeake Bay. Home to a large Amish community, Lancaster Country, the article showed, produced 61 million pounds of manure per year, by far the highest in Pennsylvania. Six of nineteen Lancaster County wells surveyed contained E. coli bacteria and sixteen had nitrate levels exceeding EPA limits.

115. Bridenbaugh, *Fat Mutton and Liberty of Conscience*, 31–33; Jane Fletcher Fiske, *Thomas Cook of Rhode Island: A Genealogy of Thomas Cook, alias Butcher of Netherbury, Dorsetshire, England, who came to Taunton, Massachusetts in 1637 and settled in Portsmouth, Rhode Island in 1643* ([Boxford, MA]: J. F. Fiske, 1987), 27, 164, 254.

116. Reverend Mr. Mellen, "A Topographical Description of Barnstable," *Collections of the Massachusetts Historical Society*, vol. 3 (1794, Boston: Munroe & Francis: 1810), 14; See also, William R. Putnam, "Bone Disorder in Cows," in *The New England Farmer; A Semi-Monthly Journal, Devoted to Agriculture, Horticulture, and their Kindred Arts and Sciences*, vol. 2, ed. S. W. Cole (Boston: J. Nourse, 1850), 44.

117. Bridenbaugh, *Fat Mutton and Liberty of Conscience*, 33.

118. W. Noel Sainsbury, ed., *Calendar of State Papers, Colonial Series: American and West Indies, 1661–1668* (London: Longman & Co., 1880), 343.

119. Roger Williams to John Winthrop, Jr., May 28, 1647, *Winthrop Papers*, 5: 168.

120. Russell, *A Long Deep Furrow*, 175.

121. J. Bradford Hubeny, John W. King, and Mark Cantwell, "Anthropogenic Influences on Estuarine Sedimentation and Ecology: Examples from the Varved Sediments of the Pettaquamscutt River Estuary, Rhode Island," *Journal of Paleolimnology* 41, no. 2 (February 2009): 297–314.

122. Bartlett, *Records of the Colony of Rhode Island*, 4: 510.

123. Ibid., 5: 304.

124. Ibid., 4: 510.

125. Samuel Truesdale Livermore, *A History of Block Island from Its Discovery in 1514 to the Present Time 1876* (Hartford, CT: The Case, Lockwood, and Brainard Co., 1877), 16.

126. Bartlett, *Records of the Colony of Rhode Island*, 2: 125, 305.

127. Bartlett, *Records of the Colony of Rhode Island*, 6: 308.

128. Chapin, *Illustrations Of The Seals, Arms And Flags Of Rhode Island*, 1–2.

129. Bartlett, *Records of the Colony of Rhode Island*, 4: 24.

130. Ibid., 263.

131. Bartlett, *Records of the Colony of Rhode Island*, 5: 462–463.

3. The Geographic Quicksilver of Narragansett Bay

1. Ezra Stiles, *A History of the Three of the Judges of King Charles I, Major-General Whalley, Major-General Coffe, and Colonel Dixwell: Who, at the Restoration, 1660, Fled to America; and were Secreted and Concealed, in Massachusetts and Connecticut, for Near Thirty Years. With an Account of Mr. Theophilus Whale, of Narragansett, Supposed to have been also one of the Judges* (Hartford, CT: Printed by Elisha Babcock, 1794), 342.

2. Ibid., 342, 348.
3. Ibid., 343.
4. Patricia Seed, *Ceremonies of Possession in Europe's Conquest of the New World, 1492-1640* (New York: Cambridge University Press, 1995), 16-40. Seed showed that houses, gardens, and fences were important symbols of ownership among early English settlers. Where they could not be built—namely, among the arms of the sea—the establishment of possession and jurisdiction was considerably more difficult.
5. Jeremy Adelman and Stephen Aron, "From Borderland to Borders: Empires, Nation-States, and the Peoples in between in North American History," *The American Historical Review* 104, no. 3 (June 1999): 816-817. On oceanic borders and borderlands, see Lissa K. Wadewitz, *The Nature of Borders: Salmon, Boundaries, and Bandits on the Salish Sea* (Seattle: University of Washington Press, 2012).
6. *Report of the Commission on State Boundary, made to the General Assembly at its January Session, A.D. 1887* (Providence, RI: E. L. Freeman & Son, State Printers, 1887), 16.
7. Governor Samuel Cranston to Richard Partridge, Newport, November 26, 1723, *Correspondence of the Colonial Governors of Rhode Island, 1723-1775*, vol. 1, ed. Gertrude Selwyn Kimball (Boston: Houghton, Mifflin & Co., 1902), 8-12; Richard Partridge to the Lords of Trade, Providence Plantations, February 10, 1723-1724, *Correspondence of the Colonial Governors of Rhode Island*, 1: 12-15.
8. *Report of the Commission on State Boundary*, 9-10, 6.
9. *A Copy of the Record of the Proceedings of the Commissioners for settling adjusting and determining the Boundary of the colony of Rhode Island and Providence Plantations Eastward toward the Province of the Massachusetts Bay*, America No. 378, Transcribed by Henry Stevens (London: State Paper Office, 1741), English Codex 111, John Carter Brown Library, 2-3. Hereafter cited as *Record of the Boundary Proceedings*.
10. Ibid., 10.
11. Ibid., 10-14.
12. Martin W. Lewis, "Dividing the Ocean Sea," *Geographical Review* 89 (April 1999): 188.
13. E. Bowen, *World Atlas* (London, 1744), cited in Lewis, "Dividing the Ocean Sea," 196, 203, 207.
14. *Oxford English Dictionary* Online, s.v. "bay" (New York: Oxford University Press, 2000) http://www.oed.com/view/Entry/16384 (accessed March 29, 2014). I have followed the first set of examples listed in the *OED* for bay, n.2, but in each case I have drawn from the original source.
15. Ranulph Higden, *Polychronicon*, ed. Churchill Babington, trans. John Travisa and unknown fifteenth-century writer, vol. 1 (London: Longman, Green, Longman, Roberts, and Green, 1865), 57.
16. "The Libel of English Policy," in *Political Poems and Songs Related to English History*, vol. 2, ed. Thomas Wright (1437; London: Longman, Green, Longman, and Roberts, 1861), 187.

17. William Shakespeare, *The Merchant of Venice,* ed. Roma Gill (New York: Oxford University Press, 2001), 34.
18. William Shakespeare, *As You Like It,* ed. William Aldis Wright (Oxford: The Clarendon Press, 1877), 60.
19. Daniel Defoe, *The Wonderful Life and Surprising Adventures of that Renown Hero Robinson Crusoe: Who Lived Twenty Years on an Uninhabited Island, which he afterwords Colonized* (New York: Hugh Gaine, 1774), 64.
20. Gayl S. Westerman, *The Juridical Bay* (New York: Oxford University Press, 1987), 25. For jurisdictional purposes, legal debate has simmered into the present over what degree of shoreline indentation qualifies as a bay.
21. *Record of the Boundary Proceedings,* 17, 20.
22. Commissions were convened on March 11, 1663 and October 26, 1664, *Record of the Boundary Proceedings,* 270, 367–368.
23. "A true copy as appears recorded Anno 1666 in the Old Leather Book folio 228 in the Secretary's Office for the Colony of Rhode Island," *Record of the Boundary Proceedings,* 354.
24. "Some Reasons humbly presented to the Right Honorable Edward Earl of Clarendon Lord High Chancellor of England by the Governor & Company of his Majesty's Colony of Rhode Island and Providence Plantations for settling the Eastern Line . . . A True Copy as appears recorded Anno 1666 in the Old Leather Book folio 229 & 230," in *A Copy of the Record of the Proceedings of the Commissioners,* 357–359.
25. Martin Brückner, *The Geographic Revolution in Early America: Maps, Literacy, and National Identity* (Chapel Hill: University of North Carolina Press, 2006), 16–19.
26. William Chandler, *Journal of the Survey of Narragansett Bay 1741* (Newport, RI: Franklin Ann Publisher, 1741), G1157 Broadsides 1741, no. 1, Rhode Island Historical Society Library.
27. *Record of the Boundary Proceedings,* 25.
28. James Helme and William Chandler, *An exact Plan of the Sea coast of the Continent from Paucautuck River Eastwards* (1741), map no. 1780, RI Map, vol. 17, 1–4, Rhode Island Historical Society Library.
29. Brückner, *The Geographic Revolution in Early America,* 19.
30. Chandler, *Journal of the Survey of Narragansett Bay 1741,* broadside.
31. Ibid.
32. Ibid.
33. Ibid.
34. "Charter of Rhode Island and Providence Plantations," July 8, 1663, in *Calendar of State Papers, Colonial, America and the West Indies, 1661–1668,* vol. 5, ed. W. Noel Sainsbury (London: Public Record Office, 1880), 148–150.
35. Chandler, *Journal of the Survey of Narragansett Bay 1741,* broadside.
36. Ibid.
37. William Zartman, ed., Introduction to *Understanding Life in the Borderlands: Boundaries in Depth and in Motion* (Athens: University of Georgia Press, 2010), 5–6.
38. John White, *Planters Plea or the Grovnds of Plantations Examined and Vsual Objections Answered Together with a manifestation of the causes moving such as have lately*

undertaken a Plantation in New-England (1630; Rockport, MA: Sandy Bay Historical Society and Museum, 1930), 31.
39. William Wood, *New England's Prospect*, ed. Alden T. Vaughan (Amherst: University of Massachusetts Press, 1977), 64.
40. John Russell Bartlett, ed., *Records of the Colony of Rhode Island and Providence Plantations in New England*, vol. 1 (Providence, RI: A. Crawford Greene and Brothers, 1856–1865), 151.
41. Ibid., 45.
42. Roger Williams, *A Key into the Language of America: Or, An help to the* Languages *of the* Natives *in that part of America, called New-England*, in vol. 1 of *The Complete Writings of Roger Williams*, ed. James Hammond Trumbull (1643; New York: Russell & Russell, 1963), 134.
43. John Josselyn, *John Josselyn, Colonial Traveler: A Critical Edition of Two Voyages to New England*, ed. Paul J. Lindholdt (Hanover, NH: University Press of New England, 1988), 101.
44. Roger Williams to Governor John Leverett, Providence, October 11, 1675, *The Correspondence of Roger Williams*, vol. 2, ed. Glenn W. LaFantasie (Hanover, NH: Brown University Press and University Press of New England, 1988), 705.
45. Williams, *A Key into the Language of America*, 99. Also, Williams noted, "If an enemie approach they remove into a Thicket, or Swampe, unless they have some Fort to remove unto," 74.
46. William Hubbard, *The History of the Indian Wars in New England From the First Settlement to the Termination of the War with King Philip, in 1677*, ed. Samuel G. Drake (2nd ed. 1677; Roxbury, MA: For W. Elliot Woodward, 1865), 87–88.
47. Benjamin Thompson, *New England's Crisis or a Brief Narrative of New Englands Lamentable Estate at present, compar'd with the former (but few) years of Prosperity* (1676; Boston: The Club of Odd Volumes, 1894), 18.
48. Roger Williams to Governor John Leverett, Providence, October 11, 1675, *The Correspondence of Roger Williams*, 2: 705.
49. Jill Lepore, *The Name of War: King Philip's War and the Origins of American Identity* (New York: Alfred A. Knopf, 1998), 85–86. See n. 61; Lepore cites Noah Newman to Lieutenant Thomas, September 30, 1675, transcribed in Bowen, *Early Rehoboth*, 3: 89–90.
50. *A farther Brief and True Narration of the Great Swamp Fight in the Narragansett Country December 19, 1675. Written a few days later and first printed at London in February, 1675* (Providence: Printed by S. P. C. for the Society for Colonial Wars of Rhode Island and Providence Plantations, 1912), 10.
51. Alexander Boyd Hawes, *Off Soundings: Aspects of the Maritime History of Rhode Island* (Chevy Chase, MD: Posterity Press, 1999). See Part 1, "Pirates and Piracies."
52. "Letter from the Earl of Bellomont, 1699," in William Eaton Foster, *Stephen Hopkins: A Rhode Island Statesman. A Study in the Political History of the Eighteenth Century*, vol. 1 (Providence, RI: Sidney S. Rider, 1884), 2 n. 3.
53. Samuel Greene Arnold, *History of the State of Rhode Island and Providence Plantation*, vol. 1, 1636–1700 (New York: D. Appleton, 1859), 547.

54. J. W. Fortescue, ed., *Calendar of State Papers, Colonial, America and the West Indies, 1681-1685*, vol. 11 (London: Her Majesty's Stationery Office, 1898), 443.
55. Thomas Paine, "Deposition of Thomas Paine, Sept. 26, 1699," *Rhode Island Historical Magazine* 6 (October 1885): 156; Hawes, *Off Soundings*, 9-16.
56. Governor Lord Carlisle to Lords of Trade and Plantations, St. Jago de la Vega, November 23, 1679, *Calendar of State Papers, Colonial, America and the West Indies, 1677-1680*, vol. 10, ed. W. Noel Sainsbury (London: Public Record Office, 1880), 443.
57. Gertrude Selwyn Kimball, "Introduction," *The Correspondence of the Colonial Governors of Rhode Island*, 1: xxxvii. For a list of Rhode Island privateers, see William P. Sheffield, *Privateersmen of Newport: An Address Delivered by William P. Sheffield before the Rhode Island Historical Society, in Providence* (Newport, RI: John P. Sanborn, 1883), 52-55.
58. Richard Partridge, Agent for Rhode Island and Providence Plantations, to William Sharpe, May 22, 1745, in *Correspondence of the Colonial Governors of Rhode Island, 1723-1775*, vol. 1, ed. Gertrude Selwyn Kimball (Boston: Houghton, Mifflin, 1902), 360.
59. Samuel Cranston to the Board of Trade, December 5, 1708, *Calendar of State Papers, Colonial, America and the West Indies, 1708-1709*, ed. Cecil Headlam (London: His Majesty's Stationery Office, 1922), 172-173; editor's brackets and italics.
60. Samuel Cranston to the Board of Trade, December 5, 1708, in *Records of the Colony of Rhode Island*, 4: 55.
61. Bartlett, *Records of the Colony of Rhode Island*, 4: 34, 471.
62. Robert K. Fitts, *Inventing New England's Slave Paradise: Master/Slave Relations in Eighteenth-Century Narragansett, Rhode Island* (New York: Garland, 1998), 82.
63. *Newport Mercury*, June 12, 1769; Bartlett, *Records of the Colony of Rhode Island*, 4: 59, cited in Hawes, *Off Soundings*, 116.
64. Fitts, *Inventing New England's Slave Paradise*, 27.
65. Rhys Isaac, "Imagination and Material Culture: The Enlightenment on a Mid-Eighteenth-Century Virginia Plantation," in *The Art and Mystery of Historical Archaeology: Essays in Honor of James Deetz*, eds. Anne Yentsch and Mary Beaudry (Boca Raton, FL: CRC Press, 1992), 401-423, in Fitts, *Inventing New England's Slave Paradise*, 137.
66. Susan Allport, *Sermons in Stone* (New York: W. W. Norton, 1990), 89, 109.
67. Robert M. Thorson, *Stone by Stone: The Magnificent History in New England's Stone Walls* (New York: Walker, 2002), 153.
68. Fitts, *Inventing New England's Slave Paradise*, 139.
69. Dr. James MacSparran, *Letterbook and Abstract of Out Services, 1743-1751*, ed. Daniel Goodwin (Boston: The Merrymount Press, 1899), 54.
70. Ibid., 12, 14.
71. Eugene Genovese, *Roll Jordan Roll: The World the Slaves Made* (New York: Pantheon, 1974), 648-657; Richard Price, ed., *Maroon Societies: Rebel Slave Communities in the Americas* (Baltimore, MD: The Johns Hopkins University Press, 1979).

72. C. Edwin Barrows, ed., *The Diary of John Comer*, in *Collections of the Rhode Island Historical Society*, vol. 8 (Providence: Rhode Island Historical Society, 1893), 57.
73. *Newport Mercury*, November 21, 1774, in Maureen Alice Taylor and John Wood Sweet, eds., *Runaways, Deserters, and Notorious Villains From Rhode Island Newspapers*, vol. 2, *Additional Notices from the* Providence Gazette, *1762–1800 as well as Advertisements from All Other Rhode Island Newspapers from 1732–1800* (Rockport, ME: Picton Press, 2001), 77. Also see Appendix A, Fitts, *Inventing New England's Slave Paradise*, 228.
74. MacSparran, *Letterbook and Abstract of Out Services*, 52.
75. Bartlett, *Records of the Colony of Rhode Island*, 4: 179.
76. Ibid., 6: 65.
77. W. Jeffrey Bolster, *Black Jacks: African American Seamen in the Age of Sail* (Cambridge, MA: Harvard University Press, 1998), 135.
78. Genovese, *Roll Jordan Roll*, 639.
79. Bartlett, *Records of the Colony of Rhode Island*, 4: 27.
80. *Record of the Boundary Proceedings*, 43–47.
81. Ibid., 29.
82. Ibid., 32.
83. Ibid., 45.
84. William Cronon, *Changes in the Land: Indians, Colonists, and the Ecology of New England* (1983; New York: Hill & Wang, 1997), 65.
85. *Record of the Boundary Proceedings*, 46.
86. Ibid., 37–38.
87. Edward Winslow, *Good Newes from New England: A True Relation of Things Very Remarkable at the Plantation of Plimoth in New England* (1624; Bedford, MA: Applewood Books, 1996).
88. Ibid., 25.
89. *Record of the Boundary Proceedings*, 63–64. See William Bradford, *Of Plymouth Plantation, 1620–1647*, ed. Samuel Eliot Morrison (New York: Knopf, 1952), 192–193.
90. *Record of the Boundary Proceedings*, 65.
91. See also Thomas Prince, *A Chronological History of New England in the Form of Annals: Being a Summary and exact Account of the most material Transactions and Occurrences relating to this Country, in the order of Time wherein they happened, from the discovery of Capt. Gosnold, in 1602, to the Arrival of Governor Belcher, in 1730* (1736; Boston: Cummings, Hilliard, and Company, 1826), 208.
92. *Record of the Boundary Proceedings*, 65.
93. Ibid., 65.
94. Ibid., 71.
95. Ibid., 71–72.
96. Ibid., 78–79.
97. Ibid., 79–80.
98. Thomas Prince, *A Chronological History of New-England* (Boston: Kneeland & Green, 1736).

99. *Record of the Boundary Proceedings*, 77–78.
100. Prince, *A Chronological History of New-England*, 239.
101. *Record of the Boundary Proceedings*, 104–105.
102. Ibid., 116–117.
103. Ibid., 121–122.
104. Ibid., 127.
105. Ibid., 221–222. See also William Brigham, ed., *The Compact with the Charter and Laws of the Colony of New Plymouth: Together with the Charter of the Council at Plymouth and an Appendix, Containing the Articles of Confederation of the United Colonies of New England and Other Valuable Documents* (Boston: Dutton and Wentworth, 1836), 22–23.
106. *Record of the Boundary Proceedings*, 130.
107. Ibid., 133.
108. Ibid., 136.
109. Ibid., 137.
110. Ibid.
111. "Report of the King's Commission," March 11, 1663, *Record of the Boundary Proceedings*, 270.
112. Order of Thomas Prence, Massachusetts Bay Governor, October 1670, in *Record of the Boundary Proceedings*, 276–277.
113. King Charles to the Governor of Massachusetts, April 23, 1664, in *Record of the Boundary Proceedings*, 234.
114. Ibid.
115. Ibid., 237.
116. King Charles to Colonel Richard Nicholls, May 2, 1665, in *Record of the Boundary Proceedings*, 278.
117. Lepore, *The Name of War*, 39.
118. "Oath taken by King Philip and later by Takamunna," September 28, 1671, *Record of the Boundary Proceedings*, 302–304.
119. Lepore, *The Name of War*, xi. Also see Douglas Leach, *Flintlock and Tomahawk: New England in King Philip's War* (New York: Macmillan, 1958).
120. Lepore, *The Name of War*, 21–22, 97–111, 94–96.
121. Virginia DeJohn Anderson, "King Philip's Herds: Indians, Colonists, and the Problem of Livestock in Early New England," *The William and Mary Quarterly* 51, no. 4 (October 1994): 602.
122. Lepore, *The Name of War*, 112. "The historian who pores over the records of King Philip's War," Lepore explained, "will search in vain for a coherent political ideology or a single legal, moral, or religious justification of the war."
123. Most notably, Douglas Edward Leach, *Flintlock and Tomahawk: New England in King Philip's War* (New York: Macmillan, 1958).
124. Roger Williams to John Winthrop, Jr., Providence, December 18, 1675, *The Correspondence of Roger Williams*, 2: 708.
125. Richard R. Johnson, "The Search for a Usable Indian: An Aspect of the Defense of Colonial New England," *Journal of American History* 64, no. 3 (December 1977):

623-651; James David Drake, *King Philip's War: Civil War in New England, 1675-1676* (Amherst, MA: The University of Massachusetts Press, 1999). Both works show that alliances were never so clearly defined during King Philip's War.
126. *Record of the Boundary Proceedings*, 305-306.
127. Jeremy Adelman and Stephen Aron, "From Borderland to Borders: Empires, Nation-States, and the Peoples in between in North American History," *The American Historical Review* 104, no. 3 (June 1999): 816.
128. *Record of the Boundary Proceedings*, 10, 403.
129. Ibid., 404-405.
130. Ibid., 405.
131. Ibid.
132. It wasn't until 1862 that Rhode Island annexed East Providence from Massachusetts.
133. John Hutchins Cady, *Rhode Island Boundaries, 1636-1936* (Providence: Rhode Island Tercentenary Commission, 1936), 14-16.
134. Massachusetts Bay to the Commissioners for Setting the Eastern Boundary, September 4, 1741, *Record of the Boundary Proceedings*, 412.
135. Rhode Island to the Commissioners for Setting the Eastern Boundary, September 4, 1741, *Record of the Boundary Proceedings*, 416.
136. Paul Carter, *The Road to Botany Bay: An Exploration of Landscape and History* (1987; Minneapolis: University of Minnesota Press, 2010), 81.

4. Natrual Knowledge and a Bay in Transition

1. William Greene, "A Memorandum of the Winter of 1740," copy by Henry Bull, New York, January 15, 1840, Vault A, Box 47, Folder 2, Newport Historical Society, Newport, Rhode Island. See also Willis P. Hazard, *Recollections of Olden Times: Genealogies of the Robinson, Hazard, and Sweet Families of Rhode Island, Also Genealogical Sketch of the Hazards of the Middle States* (Bowie, MD: Heritage Books, 1998), 87.
2. Samuel Tillinghast, Diary Containing "Remarkable Events, Wind & Weather" Warwick, January 1761-1787, Codex Eng. 43, John Carter Brown Library. See also *The Diary of Capt. Samuel Tillinghast of Warwick, Rhode Island, 1757-1766*, ed. Cherry Fletcher Bamberg (Greenville: Rhode Island Genealogical Society, 2000).
3. Tillinghast, *The Diary of Capt. Samuel Tillinghast*, xxii-xxviii.
4. Tillinghast, Diary Containing "Remarkable Events, Wind & Weather."
5. Jan Golinski, *British Weather and the Climate of Enlightenment* (Chicago: University of Chicago Press, 2007).
6. Immanuel Kant, "An Answer to the Question: What is Enlightenment," in *Kant: The Political Writings*, ed. H. S. Reiss (New York: Cambridge University Press, 2003), 54.
7. Roy Porter and Mikulás Teich, eds. *The Enlightenment in National Context* (New York: Cambridge University Press, 1981), vii.
8. Golinski, *British Weather and the Climate of Enlightenment*, xii.
9. Tillinghast, *The Diary of Captain Samuel Tillinghast*, 88.

10. Ibid., 118–119. He does not describe sailing here but describes the shifting wind and "Hard Squals" on the evening of August 12 into the morning of August 13, 1759. This is a common weather pattern over Narragansett Bay during the summers.
11. See "Oyster Act," November 7, 1734, Petitions to the Rhode Island General Assembly, 1728–1734, 191; "Act to Prevent Destroying Oysters," October 31, 1766, Acts and Resolutions of the Rhode Island General Assembly, 1766–1769; Petitions to the Rhode Island General Assembly1755–1757, vol. 9, 1st Monday February 1755, 84–86, 88, 95. Rhode Island State Archives.
12. "The Journal of a Captive, 1745–1748," in Isabel M. Calder, ed., *Colonial Captivities, Marches and Journeys* (New York: Macmillan, 1935), 316.
13. Gen. 1: 6–8. New Revised Standard Version.
14. Gen. 1: 9.
15. Alain Corbin, *The Lure of the Sea: The Discovery of the Seaside, 1750–1840*, trans. Jocelyn Phelps (1988; New York: Penguin, 1995), 2.
16. Jonah 1: 1–4, 17; 2:10, NRSV.
17. Georges Louis Leclerc, Comte de Buffon, *Natural History General and Particular*, vol. 6, trans. William Smellie (1749–1804; London: Printed for A. Strahan and T. Cadell, 1785), 255–257.
18. Henry Steele Commager, *The Empire of Reason: How Europe Imagined and America Realized the Enlightenment* (Garden City, NY: Anchor Press/Doubleday, 1977), 76.
19. Sarah Rivett, *The Science of the Soul in Colonial New England* (Chapel Hill: University of North Carolina Press, 2011), 6.
20. Golinski, *British Weather and the Climate of Enlightenment*, 8.
21. Susan Scott Parish, *American Curiosity: Cultures of Natural History in the Colonial British Atlantic World* (Chapel Hill: University of North Carolina Press, 2006), 57.
22. Steve Mentz, "God's Storms: Shipwreck and the Meaning of Ocean in Early Modern England and America," in *Shipwreck in Art and Literature: Images and Interpretations from Antiquity to the Present Day*, ed. Carl Thompson (New York: Routledge, 2014), 80–82.
23. Georges Louis Leclerc, Comte de Buffon, *Buffon's Natural History Containing a Theory of the Earth, A General History of Man, of the Brute Creation, and of Vegetables, Minerals, &c. &c.*, vol. 7 (1749–1804; London: Printed for the Proprietor and Sold by H. D. Symonds, 1797), 45.
24. Ibid., 46–48.
25. William Robertson, D.D., *The History of America*, vol. 2 (London: A. Strahan, 1803), 15–16.
26. Ibid., 4–5.
27. John Callender, *An historical discourse on the civil and religious affairs of the colony of Rhode-Island and Providence plantations in New-England in America: from the first settlement 1638, to the end of the first century* (Boston: S. Kneeland and T. Green, 1739), 98–99.
28. Ibid.
29. Andrew Buraby, A.M., *Travels Through the Middle Settlements in North-America in the Year 1759 and 1760 with Observations Upon the State of the Colonies* (London: T. Payne, 1775), 69–70.

30. Robert Rogers, *A Concise Account of North America: Containing a Description of the Several British Colonies on that Continent, including the Islands of Newfoundland, Cape Breton, &c.* . . . (London: Printed for the Author, 1765), 36.
31. Daniel Neal, A.M., *The History of New-England Containing an Impartial Account of the Civil and Ecclesiastical Affairs Of the Country, To the Year of our Lord, 1700. To which is added, The Present State of New-England. With a New and Accurate* Map of the *Country. And an Appendix Containing their Present Charter, their Ecclesiastical Discipline, and their Municipal-Laws,* vol. 2 (London: Printed for J. Clark, 1720), 595.
32. Jean Palairet, *A Concise Description of the English and French POSSESSIONS in North-America: For the better explaining of the MAP published with that Title* (London: J. Haberkorn, 1755), 30.
33. Golinski, *British Weather and the Climate of Enlightenment*, 63.
34. Thomas Sprat, *The History of the Royal-Society of London, for the Improving of Natural Knowledge* (London: Printed by T. R. for J. Martyn, 1667), 114–115.
35. Richard Grove, *Green Imperialism: Colonial Expansion, Tropical Island Edens and the Origins of Environmentalism, 1600–1860* (Cambridge: Cambridge University Press, 1995), 6–15.
36. Karen Ordahl Kupperman, "The Puzzle of the American Climate in the Early Colonial Period," *American Historical Review* 87, no. 5 (December 1982): 1265.
37. Ibid., 1280–1282, 1287.
38. Hugh Williamson, "An Attempt to Account for the Change of Climate, Which Has Been Observed in the Middle Colonies in North America," *Transactions of the American Philosophical Society*, vol. 1 (January 1, 1769–January 1, 1771): 277–280.
39. Thomas Jefferson, *Notes on the State of Virginia* (London: Printed for John Stockdale, 1787), 134.
40. In 1799, Noah Webster rejected the idea that human modifications to the landscape had broader climatic effects, although he acknowledged that deforestation modified climates at the local level. Noah Webster, *A Collection of Papers on Political, Literary, and Moral Subjects* (New York: Webster and Clark, 1843), 134. Also see Clarence J. Glacken, *Traces on the Rhodian Shore: Nature and Culture in Western Thought from Ancient Times to the End of the Eighteenth Century* (Berkeley: University of California Press, 1967), 661–663.
41. Jan Golinski, "American Climate and the Civilization of Nature," in *Science and Empire in the Atlantic World,* eds. James Delbourgo and Nicholas Dew (New York: Routledge 2008), 154.
42. Callender, *An Historical Discourse on the Civil and Religious Affairs of the Colony of Rhode-Island*, 99.
43. Golinski, *British Weather and the Climate of Enlightenment*, 78–79.
44. C. Edwin Barrows, ed., *The Diary of John Comer,* in *Collections of the Rhode Island Historical Society,* vol. 8 (Providence: Rhode Island Historical Society, 1893), 32.
45. Ibid., 63–64.
46. Ibid., 112.
47. Vladimir Janković, *Reading the Skies: A Cultural History of English Weather, 1650–1820* (Chicago: University of Chicago Press, 2000), 3.

48. Ezra Stiles, *The Literary Diary of Ezra Stile,* vol. 1, ed. Franklin Bowditch Dexter (New York: Charles Scribner's Sons, 1901), 19.
49. Rev. Stephen Hosmore to Rev. Thomas Prince, August 13, 1729, in *Collections of the Connecticut Historical Society,* vol. 3 (Hartford: Connecticut Historical Society, 1895), 280-281.
50. Tillinghast, *The Diary of Capt. Samuel Tillinghast,* 53-54.
51. Ibid., 102-110.
52. Ibid., 217-220.
53. Robert Hook, "A Method for Making a History of the Weather," in *The History of the Royal-Society of London, for the Improving of Natural Knowledge,* Thomas Sprat (London: Printed by T. R. for J. Martyn, 1667), 173-179.
54. An excellent corresponding example is the diary of the Boston merchant, farmer, and patriot William Powell. See "William Powell Diary, 1777," MSS Dept. Octavo vol. "P," American Antiquarian Society.
55. Peter Eisenstadt, "Almanacs and the Disenchantment of Early America," *Pennsylvania History* 65, no. 2 (Spring 1998): 143.
56. Tillinghast, *The Diary of Capt. Samuel Tillinghast,* 60 and n. 6.
57. Golinski, *British Weather and the Climate of Enlightenment,* 103-104.
58. Tillinghast, *The Diary of Captain Samuel Tillinghast,* 102-103.
59. Ibid., 128-129, 133, 139, 165, 171, 174.
60. Ibid., 130; Isaac Watts, *The Knowledge of the Heavens and the Earth Made Easy: Or the First Principles of Astronomy and Geography Explain'd by the Use of Globes and Maps . . .* (London: T. Longman, 1760), 134-137.
61. Tillinghast, *The Diary of Capt. Samuel Tillinghast,* 71, 105.
62. Golinski, *British Weather and the Climate of Enlightenment,* 194.
63. James Delbourgo, *A Most Amazing Scene of Wonders: Electricity and the Enlightenment in Early America* (Cambridge, MA: Harvard University Press, 2006), 52.
64. Matthew Mulcahy, *Hurricanes and Society in the British Greater Caribbean, 1624-1783* (Baltimore, MD: The Johns Hopkins University Press, 2006), 34.
65. Grove, *Green Imperialism,* 401.
66. Tillinghast, *The Diary of Capt. Samuel Tillinghast,* 22, 62-63, 66.
67. "The Journal of a Captive, 1745-1748," in *Colonial Captivities, Marches and Journeys,* ed. Isabel M. Calder (New York: The Macmillan Company, 1935), 3, 132-134.
68. Ibid., 134-135; Also see Daniel Vickers, *Farmers and Fishermen: Two Centuries of Work in Essex County Massachusetts, 1630-1850* (Chapel Hill: University of North Carolina Press, 1994). Vickers showed that among coastal people in Essex County, Massachusetts, farming and fishing went hand-in-hand. Specialization in one or the other was rare.
69. "The Journal of a Captive, 1745-1748," 136.
70. "Preservation of Oysters," November 7, 1734, *Acts & Resolutions of the Rhode Island General Assembly: 1728-1734,* 195. Rhode Island State Archives, Providence, Rhode Island.
71. Benjamin Franklin, *Journal of Occurrences in My Voyage to Philadelphia on Board the Berkshire, Henry Clark, Master,* in *The Works of Benjamin Franklin,* vol. 1, ed.

Jared Sparks (Boston: Charles Tappan, 1844), 564. On Franklin's scientific investigations, particularly those at sea, see Joyce E. Chaplin, *The First Scientific American: Benjamin Franklin and the Pursuit of Genius* (New York: Basic Books, 2006).
72. Clyde L. McKenzie, Jr., "History of Oystering in the United States and Canada, Featuring the Eight Greatest Oyster Estuaries," *Marine Fisheries Review* 58, no. 4 (1996): 3.
73. James Thacher, *History of the Town of Plymouth from its First Settlement in 1620, to the Year 1832* (Boston: Marsh, Capen & Lyon, 1832), 172.
74. Benjamin Lincoln to Jeremy Belknap, December 12, 1791, in Ernest Ingersoll *The Oyster Industry*, in G. Brown Goode, *The History and Present Condition of the Fishery Industries*, (Washington DC: Government Printing Office, 1881), 21.
75. Ingersoll, *The Oyster Industry*, 51.
76. Ibid., 55; Rhode Island's penchant for planting oysters contrasted with Maryland, whose watermen resisted the attempts of state managers and scientists to partition and "plant" portions of Chesapeake Bay during the late nineteenth and early twentieth centuries. Instead, oyster fishermen held fast to traditional, "natural" gathering methods. See Kristine Keiner, *The Oyster Question: Scientists, Watermen, and the Maryland Chesapeake Bay since 1880* (Athens: University of Georgia Press, 2009).
77. Ingersoll, *The Oyster Industry*, 57, 55, 53.
78. "July 14, 1729 Providence Town Council Meeting," in *The Early Records of the Town of Providence*, vol. 12, comps. Horatio Rogers and Edward Field (Providence, RI: Remington Printing Company, 1897), 88–89.
79. *Acts and laws, of His Majesty's colony of Rhode-Island, and Providence-Plantations, in New-England, in America* (Newport, RI: Printed by the Widow Franklin, 1745), 184.
80. November 7, 1734 Oyster Act, Acts and Resolutions of the Rhode Island General Assembly, 1728–1734, bound MSS, Rhode Island State Archives, 191.
81. 1st Monday, 1755, Petitions to the Rhode Island General Assembly, 1755–1757, bound MSS, Rhode Island State Archives, vol. 9, 84–86, 88, 95. Five original copies of the petition exist in the Rhode Island State Archives, but it is conceivable that other additional copies were lost.
82. October 29, 1756, "Petition for Shutting up Hoggs in Sd. NW part Town of Providence," Petitions to the Rhode Island General Assembly, 1755–1757, vol. 9, bound MSS, Rhode Island State Archives, 141.
83. October 31, 1766, "An Act to prevent dragging and destroying of Oysters in any of the Bays, Coves, Rivers or Harbours within this Colony," Acts and Resolutions of the Rhode Island General Assembly, 1766–1769, vol. 10, bound MSS, Rhode Island State Archives, 38.
84. Carl Linnaeus, *Reflections on the Study of Nature* (1745; London: Printed for George Nicol, 1785), 17.
85. Lisbet Koerner, *Linnaeus: Nature and Nation* (Cambridge, MA: Harvard University Press, 1999), 7.
86. Emma Spary, "Political, Natural and Bodily Economies," in *Cultures of Natural History*, eds. N. Jardine, J. A. Secord, and E. C. Spary, (Cambridge: Cambridge

University Press, 1996), 179. On Smith's efforts toward agricultural improvement and the connections between ecology and his economic theories, see Fredrik Albritton Jonsson, "Rival Ecologies of Global Commerce: Adam Smith and the Natural Historians," *American Historical Review* 115, no. 5 (December 2010): 1342–1363.
87. Lisbet Koerner, "Carl Linnaeus in His Time and Place," in *Cultures of Natural History*, eds. N. Jardine, J. A. Secord, and E. C. Spary, (Cambridge: Cambridge University Press, 1996), 153.
88. Carl Hårleman to Linnaeus, Stockholm, July 21, 1748, cited in Koerner, "Carl Linnaeus in His Time and Place," 153.
89. On the history of the polyp's discovery, see Virginia P. Dawson, *Nature's Enigma: The Problem of the Polyp in the Letters of Bonnet, Trembley, and Réaumure* (Philadelphia: American Philosophical Society, 1987); Sylvia G. Lenhoff and Howard M. Lenhoff, *Hydra: And the Birth of Experimental Biology—1744* (Pacific Grove, CA: The Boxwood Press, 1986).
90. Abraham Trembley, *Memoirs Concerning the Natural History of the Polyps: First Memoir* in *Hydra: And the Birth of Experimental Biology—1744*, eds. Sylvia G. Lenhoff and Howard M. Lenhoff (Pacific Grove, CA: The Boxwood Press, 1986), 6.
91. Aram Vartanian, "Trembley's Polyp, La Mettrie, and Eighteenth-Century French Materialism," *Journal of the History of Ideas* 11, no. 3 (June 1950): 259.
92. Abraham Trembley, *Memoirs Concerning the Natural History of the Type of Freshwater Polyp with Arms Shaped Like Horns* (Leiden: Jean and Herman Verbeek, 1744).
93. Vartanian, "Trembley's Polyp . . . ," 263.
94. Spary, "Political, Natural and Bodily Economies," 183.
95. See Mary Douglas, *Purity and Danger: An Analysis of Concept of Pollution and Taboo* (1966; New York: Routledge, 2002), 69. Douglas explains, "Those species are unclean which are imperfect members of their class, or whose class itself confounds the general scheme of the world." In this context, shellfish, which lacked the scales and fins that defined true fish according to Leviticus, were impure.
96. "The Fate of the Mouse," *New-England Weekly Journal*, March 1, 1737, 1.
97. Count de Buffon, *Natural History General and Particular*, vol. 1, trans. William Smellie (London: Printed for W. Strahan and T. Cadell, 1785), 6.
98. Ibid., 190.
99. Ibid., 502.
100. Ibid., 484, 487–489.
101. Ibid., 490–491, 493, 495.
102. Georges Louis Leclerc, Comte de Buffon, *The natural history of animals, vegetables, and minerals; with the theory of the earth in general*, vol. 4, trans. W. Kenrick and J. Murdock (London: Printed for, and sold by T. Bell, (no. 26.) Bell-Yard, Temple-Bar, 1776), 109–110.
103. Georges Louis Leclerc Comte de Buffon, *Barr's Buffon. Buffon's Natural History: Containing a theory of the earth, a general history of man, of the brute creation, and of vegetables, minerals, &c.*, vol. 3 (London: Printed for the proprietor [J. S. Barr] and sold by H. D. Symonds, 1797), 124–125.

104. D. Graham Burnett, *Trying Leviathan: The Nineteenth-Century New York Case that Put the Whale on Trial and Challenged the Order of Nature* (Princeton, NJ: Princeton University Press, 2007), 90; W. Jeffrey Bolster, *The Mortal Sea: Fishing the Atlantic in the Age of Sail* (Cambridge, MA: Harvard University Press, 2012), 89–92.
105. Franklin Stuart Coyle, "Welcome Arnold, 1745–1798, Providence Merchant: The Founding of an Enterprise" (Ph.D. dissertation, Brown University, 1978), 26–27.
106. Buffon noted, "Lime made from oysters or other shells, is weaker than that made with hard stone." In Buffon, *Barr's Buffon*, 92.
107. Eugene W. Banks, "Lime and Lime Kilns," *Journal of Chemical Education* 17, no. 11 (1940): 506.
108. Jacques Savary Brûlons, *The Universal Dictionary of Trade and Commerce*, vol. 2, trans. Malachy Postlethwayt (London, 1774), in Coyle, "Welcome Arnold," 22.
109. Coyle, "Welcome Arnold," 29.
110. "Preservation of Oysters," November 7, 1734, *Acts & Resolutions of the Rhode Island General Assembly: 1728–1734*, 195. Rhode Island State Archives, Providence, Rhode Island.
111. Coyle, "Welcome Arnold," 32.
112. Peter Kalm, *Travels into North America: containing its natural history, and a circumstantial account of its plantations and agriculture in general. With The Civil, Ecclesiastical And Commercial State Of The Country, The Manners of the Inhabitants, and several curious and Important Remarks on various Subjects*, vol. 1 (Warrington, England: William Eyres, 1770), 293.
113. Bolster, *The Mortal Sea*, 110.
114. Coyle, "Welcome Arnold," 38.
115. Richard L. Bushman, *The Refinement of America: Persons, Houses, Cities* (New York: Knopf, 1992), 5.

5. Improving Coastal Space During a Century of War

1. Robert V. Wells, *The Population of the British Colonies in America before 1776: A Survey of Census Data* (Princeton, NJ: Princeton University Press, 1975), 98.
2. Carl Bridenbaugh, *Cities in the Wilderness: The First Century of Urban Life in America, 1625–1742* (New York: Ronald Press, 1938), 362.
3. Petitions to the Rhode Island General Assembly, vol. 2, 1728–1733, Bound MSS, Rhode Island State Archives, 6.
4. Bridenbaugh, *Cities in the Wilderness*, 172.
5. Petitions to the Rhode Island General Assembly, vol. 2, 1728–1733, bound MSS, Rhode Island State Archives, 6.
6. John Russell Bartlett, ed., *Records of the Colony of Rhode Island and Providence Plantations in New England*, vol. 4 (Providence, RI: Knowles, Anthony & Co., 1859), 568.
7. "Upon the Petition for a Lighthouse," March 3, 1748, Journal of the House of Deputies, 1747–1748, bound MSS, Rhode Island State Archives, 312.
8. D. Alan Stevenson, *The World's Lighthouses Before 1820* (New York: Oxford University Press, 1959), 173–177.

9. In 1753 this wooden lighthouse burned down, but the General Assembly hired Peter Harrison to construct a new three-story stone tower and a keeper house, which were completed in 1755 and 1756, respectively. In 1761 Peter Harrison also served on a committee that equipped the new lighthouse with a brighter lantern. Carl Bridenbaugh, *Peter Harrison: First American Architect* (Chapel Hill: University of North Carolina Press, 1949), 89–90.
10. John Stilgoe, *Common Landscape of America, 1580–1845* (New Haven, CT: Yale University Press, 1982), 109.
11. Bartlett, *Records of the Colony of Rhode Island*, 331 n.
12. Ibid., 329.
13. *Boston News-letter,* July 18, 1723, 2.
14. Alexander Boyd Hawes, *Off Soundings: Aspects of the Maritime History of Rhode Island* (Chevy Chase, MD: Posterity Press, 1999), 61.
15. *Boston News-Letter,* July 25, 1723, 2.
16. Bartlett, *Records of the Colony of Rhode Island,* 4: 331 n; *Boston News-Letter,* July 25, 1723, 2.
17. The *Boston News-Letter* reported it was a "deep Blew Flagg."
18. "Postscript," *New-England Courant,* July 15–22, 1723, 2.
19. Bartlett, *Records of the Colony of Rhode Island,* 4: 426.
20. Samuel Greene Arnold, *History of the State of Rhode Island and Providence Plantations,* vol. 2 (New York: D. Appleton, 1860), 119.
21. Ibid., 225.
22. Lauren Benton, "Oceans of Law: The Legal Geography of the Seventeenth-Century Seas." Paper presented at Seascapes, Littoral Cultures, and Trans-Oceanic Exchanges, Library of Congress, Washington DC, February 12–15, 2003, accessed June 21, 2011, <http://www.historycooperative.org/proceedings/seascapes/benton.html>.
23. Hawes, *Off Soundings,* 60.
24. John Stilgoe, *Shallow Water Dictionary: A Grounding in Estuary English* (New York: Princeton Architectural Press, 2004), 14–15.
25. Julie Sanders, *The Cultural Geography of Early Modern Drama, 1620–1650* (New York: Cambridge University Press, 2011), 63–64; Claire Jowitt, "Scaffold Performances: The Politics of Pirate Execution," in *Pirates?: The Politics of Plunder, 1550–1650,* ed. Claire Jowitt (New York: Palgrave Macmillan, 2007), 151–68.
26. Stephen Saunders Webb, *Governors-General: The English Army and the Definition of the Empire, 1569–1681* (Chapel Hill: University of North Carolina Press, 1979), xv.
27. Ian Steele, "Governor's or Generals?: A Note on Martial Law and the Revolution of 1689 in English America," *William and Mary Quarterly* 46, no. 2 (April 1989): 313.
28. For a detailed discussion of Webb's arguments, see Richard R. Johnson, "The Imperial Webb: The Thesis of Garrison Government in Early America Considered," *William and Mary Quarterly* 43, no. 3 (July 1986): 408–430.
29. Quoted in George W. Cullum, *Historical Sketches of the Fortification Defenses of Narragansett Bay since the Founding in 1638 of the Colony of Rhode Island* (Washington, DC: , 1884), 6.

30. Bartlett, *Records of the Colony of Rhode Island*, 3: 445.
31. Bartlett, *Records of the Colony of Rhode Island*, 4: 190–191.
32. Ibid., 241, 248, 271.
33. Cullum, *Historical Sketches of the Fortification Defenses of Narragansett Bay*, 6.
34. Bartlett, *Records of the Colony of Rhode Island*, 4: 473.
35. Ibid., 475–476.
36. Ibid., 57.
37. Hugo Grotius, *Mare Liberum: The Freedom of the Seas or The Right Which Belongs to the Dutch to Take Part in the East India Trade*, ed. James Brown Scott, trans. Ralph Van Deman Magoffin (1608; New York: Oxford, 1916).
38. John Selden, *Of the dominion, or ownership, of the sea* (London: William Da-Gard, 1652), e2–f.
39. Gayl S. Westerman, *The Juridical Bay* (New York: Oxford University Press, 1987), 43 n. 43.
40. Selden, *Of the dominion, or ownership, of the sea*, 141.
41. Ibid., 143.
42. R. R. Churchill and A. V. Lowe, *The Law of the Sea* (Manchester: Manchester University Press, 1824) 72, 77–78.
43. Quoted in Sayre A. Swarztrauber, *The Three-Mile Limit of Territorial Seas* (Annapolis, MD: Naval Institute Press, 1972), 54–56.
44. Bartlett, *Records of the Colony of Rhode Island*, 5: 498.
45. "Report of Governor Ward, to the Board of Trade, on paper money," January 9, 1740, in Bartlett, *Records of the Colony of Rhode Island*, 5: 11, 8–10.
46. Ibid., 10–11, 54, 74, 121, 124.
47. Ibid., 11, 58.
48. The Rhode Island General Assembly allocated funds for a "highway from ferry to ferry, across the Island" in its meeting on February 28, 1709–1710, in Bartlett, *Records of the Colony of Rhode Island*, 4: 85.
49. "Report of Governor Ward, to the Board of Trade, on paper money," January 9, 1740, in Bartlett, *Records of the Colony of Rhode Island*, 5: 11, 8–10.
50. On the declining value of money, see Lynne Withey, *Urban Growth in Colonial Rhode Island: Newport and Providence in the Eighteenth Century* (Albany: State University of New York Press, 1984), 25. On counterfeiting in Rhode Island, see Kenneth Scott, *Counterfeiting in Colonial America* (1957; Philadelphia: University of Pennsylvania Press, 2000), 40, 57, 111, 196. For examples of Rhode Island paper currency, see "Rhode Island Colonial Currency Collection, 1715–1786," vol. 1, American Antiquarian Society. For counterfeit bills, see "Continental Currency Collection, 1777–1792," vol. 2, American Antiquarian Society.
51. Phillip E. Steinberg, *The Social Construction of the Ocean* (Cambridge: Cambridge University Press, 2001), 21. "The political-economic logic and structures of a given society," explained Steinberg, "lead social actors to implement a series of uses, regulations, and representations in specific spaces, including ocean-space. Once implemented in a particular space, each aspect of the social construction (each use, regulation, and representation) impacts the others, effectively creating a new 'nature' of that space."

52. Ibid. Steinberg explained, "This 'second nature' is constructed both material and discursively, and it is maintained through regulatory institutions."
53. Sydney V. James, *The Colonial Metamorphosis in Rhode Island: A Study of Institutions in Change*, eds. Sheila L. Skemp and Bruce C. Daniels (Hanover, NH: University Press of New England, 2000), 132.
54. Bridenbaugh, *Peter Harrison*, 21, 92.
55. Mary Sponberg Pedley, *The Commerce of Cartography: Making and Marketing Maps in Eighteenth-Century France and England* (Chicago: University of Chicago Press, 2005), 122.
56. Ibid., 125.
57. "A British navy yard contemplated in Newport, RI, in 1764," in *Rhode Island Historical Magazine* 6, no. 1 (1885): 44. An introduction to the document explains that the report's author was likely Robert Melville, Governor of Grenada, but Pedley disagreed (p. 124 n. 18, 274), arguing that the dates of the report don't correspond with Melville's time in Rhode Island. That said, in the letter he specifically thanks the official to whom the letter is written for his appointment as "Governor and Commander in Chief of Grenada."
58. Ibid., 43-44.
59. US House, "Letter from the Secretary of the Navy: Transmitting a Report and Survey of Narragansett Bay, & c.," 22nd Cong., 2d Sess., December 20, 1832, Doc. No. 19, 2; US House, "Fortifying Narragansett Bay: Resolutions from the State of Rhode Island and Providence Plantations," 24th Cong., 1st Sess., June 6, 1836, Doc. No. 277; also see US House, "Resolution of the Legislature of Rhode Island: Recommending Provision for the Defense of Narragansett Bay, in that State," 24th Cong., 1st Sess., June 17, 1836, 408.
60. "A British navy yard contemplated in Newport, RI, in 1764," 45-46.
61. Bridenbaugh, *Peter Harrison*, 45, 47-51, 98-102.
62. *A Geographic Description of the Coasts, Harbours, and Sea Ports of the Spanish West-Indies; Particularly of Porto Bello, Cartagena, and the Island of Cuba with Observations of the CURRENTS, and the Variations of the COMPASS in the Bay of Mexico, and the North Sea of America. Translated from A Curious and Authentic Manuscript, written in Spanish by Domingo Gonzales Carranza, his Catholick Majesty's Principal Pilot of the Flota in New Spain, Anno 1718. To which is added, An Appendix, containing Capt. Parker's own Account of his Taking the Town of Porto Bello, in the Year 1601. WITH An Index, and a New and Correct CHART of the Whole; AS ALSO Plans of the HAVANNAH, PORTO-BELLO, CARTEGENA, and LA VERA CRUZ* (London: Printed for the Editor Caleb Smith, 1740), appendix. John Carter Brown Library, Providence, RI.
63. Bartlett, *Records of the Colony of Rhode Island*, 5: 497.
64. Silvio A. Bedini, "Benjamin King of Newport, RI—Part II" *Professional Surveyor* (September 1997): 41-42.
65. Bartlett, *Records of the Colony of Rhode Island*, 5: 512.
66. "An Abstract of Mr. Benjamin West's Account of the transit of Venus, as observed at Providence, in New England, June 3d, 1769," *Transactions of the American Philosophical Society*, vol. 1 (Philadelphia: Aitken and Son at Pope's Head on Market Street, 1789), 91.

67. "Extract of a Letter from a Peer of the First Blood in Britain, to His Friend in N. America, Dated April 13," *Essex Gazette*, July 1769, 209; Also see *Connecticut Journal*, July 28, 1769, 4.
68. "Majesty's Ship; Bonnetta; Admiral Montague; Gaspee; Lieut. Dudington; Providence; Rhode-Island," *The New-York Gazette, and the Weekly Mercury*, September 28, 1772, 2. Also see Rockwell Stensrud, *Newport: A Lively Experiment, 1639–1969* (Newport, RI: Redwood Library and Athenaeum and Rockwell Stensrud, 2006), 187–188.
69. Pedley, *The Commerce of Cartography*, 128–134.
70. Charles Blaskowitz, *A Topographical Chart of the Bay of Narragansett in the Province of New England* (London: Engraved & Printed for Wm. Faden, Charing Cross, 1777).
71. Ibid.
72. Steinberg, *The Social Construction of the Ocean*, 33–34.
73. Frederic C. Lane, "The Economic Meaning of the Invention of the Compass," *American Historical Review* 68, no. 3 (April 1963): 605–617.
74. Francis Bacon, *The New Organon*, eds. Lisa Jardine and Michael Silverthorne (1620; Cambridge: Cambridge University Press, 2000), 100.
75. John Gillis, *Islands of the Mind: How the Human Imagination Created the Atlantic World* (New York: Palgrave Macmillan, 2009), 47–51.
76. Mark Monmonier, *Rhumb Lines and Map Wars: A Social History of the Mercator Projection* (Chicago: The University of Chicago Press, 2004), 3–6.
77. Dava Sobel, *Longitude: The Story of a Lone Genius Who Solved the Greatest Scientific Problem of His Time* (New York: Walker and Company, 1995), 108.
78. N. A. M. Rodger, *The Command of the Ocean: A Naval History of Britain, 1649–1815* (New York: W. W. Norton, 2004), 383.
79. Michael S. Reidy, *Tides of History: Ocean Science and Her Majesty's Navy* (Chicago: University of Chicago Press, 2008), 42–56.
80. Ezra Stiles, *The Literary Diary of Ezra Stiles*, vol. 1, ed. Franklin Bowditch Dexter (New York: Scribner, 1901), 500; "Extract of a letter from Captain Wallace, to Vice Admiral Graves, dated on board His Majesty's ship Rose, at Newport, Rhode Island, December 12, 1774," Bartlett, *Records of the Colony of Rhode Island*, 7: 306. The cannon mounted in the fort at Goat Island was removed by Thomas Freebody, per the direction of the General Assembly on June 28, 1775, p. 357.
81. "The Deputy Governor of Rhode Island, East Greenwich, RI, to Capt. James Wallace, Commander of His Majesty's Ship Rose," June 17, 1775, Bartlett, *Records of the Colony of Rhode Island*, 7: 338.
82. "Capt. James Wallace, of His Majesty's Ship Rose, to the Deputy Governor of Rhode Island," June 15, 1775, Bartlett, *Records of the Colony of Rhode Island*, 7: 338.
83. Bartlett, *Records of the Colony of Rhode Island*, 7: 341.
84. Ibid., 310–311, 309. These 1,500 soldiers formed one brigade, under the command of a brigadier general. The brigade was divided into three regiments, each of which was commanded by a colonel, lieutenant colonel, and one major. Each regiment consisted of eight companies, each of which was commanded by a captain, one lieutenant, and one ensign. Ibid., 317–318.

85. Ibid., 311; also see "Governor Wanton to the General Assembly of Rhode Island," June 13, 1775, ibid., 336-337.
86. Ibid., 346, 383.
87. Ibid., 367.
88. Ibid., 364.
89. Ibid., 358.
90. Ibid., 362, 372.
91. Ibid., 373-374, 379-380, 432.
92. Stiles, *The Literary Diary of Ezra Stiles*, 1: 625.
93. Bartlett, *Records of the Colony of Rhode Island*, 7: 381-382.
94. Stiles, *The Literary Diary of Ezra Stiles*, 1: 625.
95. Ibid., 624-625.
96. Ibid., 623-624.
97. Bartlett, *Records of the Colony of Rhode Island*, 7: 439.
98. *Diary of Frederick Mackenzie: Giving a Daily Narrative of his Military Service as an Officer in the Regiment of Royal Welch Fusiliers During the Years 1775-1781 in Massachusetts Rhode Island and New York*, vol. 1 (Cambridge, MA: Harvard University Press, 1930), 73. Excerpts from the Mackenzie diary also appear in Elizabeth Covell, "Newport Harbor and the Lower Narragansett Bay Rhode Island during the American Revolution," *Bulletin of the Newport Historical Society* 68 (January 1933): 3-37.
99. *Diary of Frederick Mackenzie*, 1: 121-124, 126.
100. Ibid., 135, 140-141.
101. Ibid., 138.
102. Samuel Greene Arnold, *The History of the State of Rhode Island and Providence Plantations from the Settlement of the State, 1636, to the Adoption of the Federal Constitution, 1790*, vol. 2 (New York: D. Appleton, 1860), 402-403; *Diary of Frederick Mackenzie*, 1: 148-150, 153-154, 142.
103. Fleet S. Greene, *Diary 1777-1789*, "Newport in the Hands of the British," in *The Historical Magazine* 4 (January 1860): 1, 70.
104. *Diary of Frederick Mackenzie*, 1: 288, 289, 304.
105. "Mrs. Almy's Journal; Siege of Newport, August 1778," *Newport Historical Magazine* 1 (July 1880): 18-19.
106. Greene, *Diary 1777-1789*, 72.
107. *Diary of Frederick Mackenzie*, 2: 321, 330.
108. Ibid., 340-341; Greene, *Diary 1777-1789*, 72.
109. *Diary of Frederick Mackenzie*, 2: 330.
110. Ibid., 1: 332-333, 341.
111. "Mrs. Almy's Journal; Siege of Newport, August 1778," 23.
112. *Diary of Frederick Mackenzie*, 2: 347, 351, 359, 358.
113. Greene, *Diary 1777-1789*, 105-106.
114. Ibid., 105.
115. *Diary of Frederick Mackenzie*, 1: 353.
116. "Mrs. Almy's Journal; Siege of Newport, August 1778," 34.

117. Greene, *Diary 1777–1789*, 106; Arnold, *The History of the State of Rhode Island and Providence Plantations*, 428; Arnold states that 211 Americans died in the battle, while 1,023 British and Hessian soldiers died. See also Paul F. Dearden, *The Rhode Island Campaign of 1778* (Providence: The Rhode Island Publications Society, 1980).
118. Greene, *Diary 1777–1789*, 106.
119. Sydney V. James, *Colonial Rhode Island: A History* (New York: Charles Scribner's Sons, 1975), 356.
120. Paul A. Gilje, *Liberty on the Waterfront: American Maritime Culture in the Age of Revolution* (Philadelphia: University of Pennsylvania Press, 2004), xii. See also Gary Nash, *The Urban Crucible: Social Changes, Political Consciousness, and the Origins of the American Revolution* (Cambridge, MA: Harvard University Press, 1979).
121. David Hackett Fischer, *Washington's Crossing* (New York: Oxford University Press, 2004), 33.
122. Alex Roland, W. Jeffrey Bolster, and Alexander Keyssar, *The Way of the Ship: America's Maritime History Reenvisioned, 1600–2000* (New York: John Wiley, 2007), 1–5. Roland, Bolster, and Keyssar argued that America was a brown-water maritime nation, one whose connections with the continental interior have been buried beneath more romantic blue-water narratives. In its orientation inland, America's maritime history was more similar to that of Germany than Britain.
123. David Hackett Fischer, *Paul Revere's Ride* (New York: Oxford University Press, 1994), 113–123.
124. Robert Middlekauff, *The Glorious Cause: The American Revolution, 1763–1789* (New York: Oxford University Press, 1982), 293; Howard H. Peckham, *The War for Independence: A Military History* (Chicago: University of Chicago Press, 1958), 17.
125. W. J. Wood, *The Battles of the Revolutionary War, 1775–1781* (Chapel Hill, NC: Algonquin Books, 1990), 55–58, 92–95.
126. Edward H. Tatum, Jr., ed., *The American Journal of Ambrose Serle, Secretary to Lord Howe, 1776–1778* (San Marino, CA: The Huntington Library, 1940), 41.
127. J. R. McNeill, *Mosquito Empires: Ecology and War in the Greater Caribbean, 1620–1914* (New York: Cambridge University Press, 2010), 232–234.
128. John Ferling, *Almost A Miracle: The American Victory in the War of Independence* (New York: Oxford University Press, 2007), 360–361.
129. Ibid., 362, 367, 372–374. See also Joseph Callo, *John Paul Jones: America's First Sea Warrior* (Annapolis, MD: Naval Institute Press, 2006).
130. Andrew Jackson O'Shaughnessy, *An Empire Divided: The American Revolution and the British Caribbean* (Philadelphia: University of Pennsylvania Press, 2000) 169–170.
131. Selwyn H. H. Carrington, "The American Revolution and the Sugar Colonies, 1775–1783," in *A Companion to the American Revolution*, eds. Jack P. Greene and J. R. Pole (Malden, MA: Blackwell, 2000), 517–518.
132. *Diary of Frederick Mackenzie*, 1: 127.
133. Stiles, *The Literary Diary of Ezra Stiles*, 2: 427.

134. Jedidiah Morse, *The American Geography; or a View of the Present Situation of the United States of America* (Elizabethtown, NJ: Shepard Kollock, 1789), 202-203.
135. Greene, *Diary 1777-1789*, 136; *Diary of Frederick Mackenzie*, 2: 431.
136. Francisco Miranda, *Diary of Francisco Miranda: Tour of the United States, 1783-1784*, ed. William Spence Robertson (New York: The Hispanic Society of America, 1928), 99. I have used the translation by Don Juan de Riano, "A Spaniard's Visit to Newport in 1784," *Bulletin of the Newport Historical Society* no. 85 (October 1932): 8. For a survey of post-Revolution sources pertaining to Newport, see Stensrud, *Newport: A Lively Experiment, 1639-1969*, 236-237.
137. Miranda, *Diary of Francisco Miranda*, 106-107; Riano, "A Spaniard's Visit to Newport in 1784," 15.
138. Jacques-Pierre Brissot de Warville, *New Travels in the United States of America: Performed in 1788* (London: J. S. Jordan, 1792), 145.
139. Benjamin Waterhouse, Newport, RI, to Thomas Jefferson, September 14, 1822, Thomas Jefferson Papers Series 1, General Correspondence, 1651-1827, Library of Congress, accessed March 30, 2014, <http://hdl.loc.gov/loc.mss/mtj.mtjbib024450>.
140. François-Alexandre-Frédéric, duc de La Rochefoucauld-Liancourt, *Travels through the United States of North America, the Country of the Iroquois, and Upper Canada, in the Years 1795, 1796, 1797*, 2nd ed., vol. 2 (London: R. Phillips, 1800), 280, 277.
141. Ibid., 287-291.
142. Pedley, *The Commerce of Cartography*, 145-153.
143. James, *Colonial Rhode Island*, 371-372.
144. Martin Brückner, *The Geographic Revolution in Early America: Maps, Literacy, and National Identity* (Chapel Hill: University of North Carolina Press, 2006), 119.
145. James D. Drake, *The Nation's Nature: How Continental Presumptions Gave Rise to the United States of America* (Charlottesville, VA: University of Virginia Press, 2011), 268-270.

6. Carving the Industrial Coastline

1. "First Trip on the Blackstone," *Rhode Island American and Providence Gazette*, July 4, 1828.
2. Ibid.
3. Samuel Eliot Morrison, *The Maritime History of Massachusetts, 1783-1860* (Boston: Houghton Mifflin, 1921), 124.
4. Lynne Withey, *Urban Growth in Colonial Rhode Island: Newport and Providence in the Eighteenth Century* (Albany: State University of New York Press, 1984), 91-94.
5. Edward Peterson, *History of Rhode Island and Newport in the Past* (New York: John S. Taylor, 1853), 301-302.
6. Withey, *Urban Growth in Colonial Rhode Island*, 97-98.
7. Ibid., Appendix A-2, 115.
8. John Hutchins Cady, *The Civil & Architectural Development of Providence, 1636-1950* (Providence, RI: The Book Shop, 1957), 81-83.

9. William R. Staples, *Annals of the Town of Providence from Its First Settlement to the Organization of the City Government, in June, 1832* (Providence, RI: Knowles and Vose, 1843), 353, 386.
10. "Population of the 33 Urban Places: 1800," U.S. Bureau of the Census, internet release date June 15, 1998, accessed July 15, 2011 <http://www.census.gov/population/www/documentation/twps0027/tab03.txt>; "Population of the 61 Urban Places: 1820," U.S. Bureau of the Census, internet release date June 15, 1998, accessed July 15, 2011, <http://www.census.gov/population/www/documentation/twps0027/tab05.txt>.
11. Nancy Seasholes, *Gaining Ground: A History of Landmaking in Boston* (Cambridge, MA: MIT Press, 2003); also see Peter K. Weiskel, Lora K. Barlow, and Tomas W. Smieszek, "Water Resources and the Urban Environment, Lower Charles River Watershed, Massachusetts, 1630–2005," US Geological Survey Circular 1280 (Reston, Virginia: US Geological Survey, 2005).
12. Carolyn Merchant, "The Theoretical Structure of Ecological Revolutions," *Environmental Review* 11, no. 4, Special Issue: *Theories of Environmental History* (Winter 1987): 273. See also Carolyn Merchant, *Ecological Revolutions: Nature, Gender, and Science in New England* (Chapel Hill: University of North Carolina Press, 1989).
13. Theodore Steinberg, *Nature Incorporated: Industrialization and the Waters of New England* (1991; Amherst: University of Massachusetts Press, 1994), 12.
14. See C. J. Vörösmarty, et al., "Global Threats to Human Water Security and River Biodiversity," *Nature* 467 (September 30, 2010): 555–561. The authors show that, during the twentieth century, industrialized nations created complex and costly systems for mitigating threats to water security. Massive investment in hydraulic engineering and pollution controls allowed industrialized nations to "offset high stressor levels without remedying their underlying causes." This chapter examines what happens when massive investment in hydraulic engineering is placed in the hands of a private corporation that fails.
15. James B. Hedges, *The Browns of Providence Plantations: The Nineteenth Century* (Providence, RI: Brown University Press, 1968), 209.
16. Col. Israel Plummer, "History of the Blackstone Canal" (Worcester, MA: Printed by O. B. Wood, 1887), 3. This was originally published as "History of the Blackstone Canal," *Proceedings of the Worcester Society of Antiquity* 1, no. 3 (1878): 41–50.
17. Brenton H. Dickson, "Comparison of the Blackstone and Middlesex Canals," *Old-Time New England,* Bulletin for the Society for the Preservation of New England Antiquities 58, no. 4 (April–June 1968): 89–92.
18. Carol Sheriff, *The Artificial River: The Erie Canal and the Paradox of Progress, 1817–1862* (New York: Hill and Wang, 1996), xvi.
19. Plummer, "History of the Blackstone Canal," 4.
20. "Petition to the Massachusetts General Court Concerning the Blackstone Canal," May 31, 1822, Worcester Public Library, Canal Folder, 1; also printed in *The Massachusetts Spy,* September 4, 1822; *National Aegis,* September 11, 1822.
21. Ibid., 2–4.
22. "Canals," *The Massachusetts Spy,* April 17, 1822.

23. "Public Meeting," *National Aegis,* May 15, 1822. See also "Canal," *National Aegis,* May 15, 1822; "Canal," *Rhode Island American,* May 18, 1822.
24. "The Proposed Canal," *Providence Gazette,* May 18, 1822.
25. Reported in the *Providence Gazette,* April 27, 1822.
26. Reported in *National Aegis,* May 1, 1822.
27. "Providence Canal," *Boston Centinel,* printed in *The Massachusetts Spy,* May 22, 1822.
28. "Account of the Proposed Canal from Worcester to Providence Containing the Report of the Engineer, Together with some Remarks upon Inland Navigation" (Providence, RI: John Miller, Journal Office, 1825), 4.
29. Ibid., 5.
30. "Charter Granted by the Legislature of Massachusetts to the Blackstone Canal Company," in *The Charter, By-laws, &c., of the Blackstone Canal Corporation* (Providence, RI: Printed by Knowles, Vose & Co., 1836), 3–11.
31. *National Aegis,* January 15, 1823.
32. "Blackstone Canal," *Massachusetts Spy and Worcester County Advertiser,* June 25, 1823.
33. "Charter Granted by the Legislature of Rhode Island. An Act to Incorporate the Blackstone Canal Company, and Other Purposes," in *The Charter, By-laws, &c., of the Blackstone Canal Corporation,* 14–15.
34. Name illegible to John W. Lincoln, April 1823, Manuscript Collection, Box 1, Folder 1, Blackstone Canal Folder, American Antiquarian Society.
35. Stephen Smith to John W. Lincoln, June 10, 1823, Manuscripts, Blackstone Canal Folder, Box 1, Folder 1, American Antiquarian Society.
36. Stephen Smith to John W. Lincoln, March 16, 1824, Manuscripts, Blackstone Canal Folder, Box 1, Folder 1, American Antiquarian Society.
37. "Canal to Boston," *The Massachusetts Spy [and] Worcester Advertiser,* October 6, 1824.
38. Gary Kulik, "Dams, Fish, and Farmers: Defense of Public Rights in Eighteenth-Century Rhode Island," in *The Countryside in the Age of Capitalist Transformation,* eds. Steven Hahn and Jonathan Prude (Chapel Hill: University of North Carolina Press, 1985), 28–31.
39. "Joseph Bucklin, Jeremiah Angell, and Isaac Madbery arbitration of dispute between James Matheson and owners of Hope Furnace," August 8, 1765, Hope Furnace Papers, Box 177, Folder 5, John Carter Brown Library.
40. Daniel Vickers, "Those Damned Shad: Would the River Fisheries of New England Have Survived in the Absence of Industrialization?" *William and Mary Quarterly* 61 (October 2004): 685–712. Vickers argued that if industrialization had not occurred, overfishing would have nevertheless decimated New England river fisheries.
41. Kulik, "Dams, Fish, and Farmers," 40–42.
42. John Russell Bartlett, ed., *Records of the Colony of Rhode Island and Providence Plantations, in New England,* vol. 5 (Providence, RI: Knowles, Anthony & Co., 1860), 57, 73, 279.
43. Ibid., 463, 479.

44. Robert Grieve, *An Illustrated History of Pawtucket, Central Falls, and Vicinity: A Narrative of the Growth and Evolution of the Community*, (Pawtucket, RI: Pawtucket Gazette and Chronicle, 1897), 20.
45. Joseph Story, *Opinion Pronounced by the Hon. Judge Story in the case of Ebenezer Tyler & Others, vs. Abraham Wilkinson & Others: At the Last Full Term at the Circuit Court, for Rhode-Island District* (Pawtucket, RI: Randall Meacham, 1827).
46. Ibid., 4.
47. Betty Buckley and Scott W. Nixon, *An Historical Assessment of Anadromous Fish in the Blackstone River*, Final Report to the Narragansett Bay Estuary Program, the Blackstone River Valley National Heritage Corridor, and Trout Unlimited (Narragansett, RI: University of Rhode Island Graduate School of Oceanography, 2001), 16-18.
48. Bartlett, *Records of the Colony of Rhode Island*, 6: 574.
49. Ibid., 7: 7-8.
50. Mack Thompson, *Moses Brown: Reluctant Reformer* (Chapel Hill: University of North Carolina Press, 1962), 231.
51. Kulik, "Dams, Fish, and Farmers," 42.
52. Ibid., 44.
53. Story, *Opinion Pronounced by the Hon. Judge Story in the case of Ebenezer Tyler & Others, vs. Abraham Wilkinson & Others*; Morton J. Horwitz, *The Transformation of American Law, 1780-1860* (Cambridge, MA: Harvard University Press, 1977), 39.
54. Joseph K. Angell, *A Treatise on the Common Law in relation to Water-Courses* (Boston: Wells and Lilly, 1824), 5.
55. Ibid., 39-40.
56. *The Massachusetts Spy [and] Worcester Advertiser*, October 6, 1824. Newspaper's emphasis.
57. US House. "Message of the President of the United States, Transmitting a Report of the Examination Which Has Been Made By the Board of Engineers with a View to Internal Improvement, &c," 18th Cong., 2nd Sess., February 14, 1825, Doc. 82 (Washington, DC: Gales & Slator, 1825), 68-69, 78; *Massachusetts Spy [and] Worcester Advertiser*, November 17, 1824.
58. "Canal to Worcester," *Providence Gazette*, November 17, 1824.
59. Stephen Smith to John Lincoln, November 15, 1824, Manuscript Files, Blackstone Canal Folder. Box 1, Folder 1, American Antiquarian Society.
60. T. Beckwith to William Lincoln, Manuscript Files, Blackstone Canal Folder. Box 1, Folder 1, American Antiquarian Society.
61. "Blackstone Canal," *Providence Gazette*, reported in *Massachusetts Spy [and] Worcester Advertiser*, April 13, 1825.
62. "Blackstone Canal," *Massachusetts Spy [and] Worcester Advertiser*, April 20, 1825; "Blackstone Canal," *National Aegis*, May 4, 1825.
63. Reported in "Blackstone Canal," *National Aegis*, May 4, 1825.
64. "Notice," *The Massachusetts Spy and Worcester County Advertiser*, July 13, 1825; "Canal Notice," *The Massachusetts Spy and Worcester County Advertiser*, October 12, 1825; "Blackstone Canal," *National Aegis*, December 7, 1825.
65. "Blackstone Canal," *Massachusetts Yeoman*, January 21, 1826, 87.

66. "Charter Granted by the Legislature of Rhode Island. An Act to Incorporate the Blackstone Canal Company, and Other Purposes," in *The Charter, By-laws, &c., of the Blackstone Canal Corporation*, 13.
67. Ibid., 26-30.
68. "Blackstone Canal," *National Aegis,* June 22, 1825, 110.
69. William Arnold to John Lincoln, February 18, 1826, Blackstone Canal Company Records, Box 1, Folder 2, American Antiquarian Society.
70. *The Massachusetts Spy and Worcester County Advertiser,* January 18, 1826.
71. "Commonwealth of Massachusetts: Court of Sessions, at Worcester, September Term, 1826," *The Massachusetts Spy and Worcester County Advertiser,* October 4, 1826.
72. "Canal Notice," *National Aegis,* October 4, 1826, 172. These assessments frequent the pages of the *The Massachusetts Spy and Worcester County Advertiser* and *National Aegis.*
73. Blackstone Canal Company paid $1 to Daniel G. Wheeler, July 17, 1826, to "discharge all claims of damage . . . ," Blackstone Canal Company Records, Box 1, Folder 2, American Antiquarian Society.
74. "Canal! Canal! Canal!!!" *National Aegis,* May 11, 1825.
75. "A Week's Ramble in the Country," *Providence Patriot,* published in *The Massachusetts Spy and Worcester County Advertiser,* August 16, 1826.
76. "Blackstone Canal," *National Aegis,* September 14, 1825, 159.
77. "Blackstone Canal," *Providence Microcosm,* published in *National Aegis,* October 19, 1825, 178.
78. Sheriff, *The Artificial River,* 36-37.
79. Ronald E. Shaw, *Canals for a Nation: The Canal Era in the United States, 1790-1860* (Lexington: University of Kentucky Press, 1990), 51.
80. Timothy J. Meagher, *To Preserve the Flame: St. John's Parish and 150 Years of Catholicism in Worcester* (Worcester, MA: Mercantile Printing, 1984), 15.
81. "Blackstone Canal," *Massachusetts Yeoman,* January 21, 1826; "Blackstone Canal," *Rhode Island American and Providence Gazette,* June 2, 1826.
82. "Blackstone Canal," *National Aegis,* June 7, 1826, 103.
83. William Farnsworth to John Lincoln, July 6, 1826, Blackstone Canal Company Records, Box 1, Folder 2, American Antiquarian Society.
84. *National Aegis,* July 12, 1826, 123.
85. "Blackstone Canal," *National Aegis,* July 19, 1826, 127.
86. Patrick O'Connor to John W. Lincoln, October 31, 1826, Blackstone Canal Company Records, Box 1, Folder 1, American Antiquarian Society.
87. Stephen Salisbury Jr., to John W. Lincoln, February 1, 1827, Salisbury Family Papers, Box 22, Folder 4, American Antiquarian Society.
88. *Pawtucket Chronicle and Manufacturers and Artisans Advocate,* January 13, 1827, 1; "Blackstone Canal," *The Massachusetts Spy and Worcester County Advertiser,* January 24, 1827, ; Also see the *Providence American* excerpt printed in the *Massachusetts Yeoman,* January 27, 1827, 92; A typographical error reported a rise of only "415½" feet in "Providence Canal," *National Aegis,* January 17, 1827, 10.
89. "Blackstone Canal," *National Aegis,* July 19, 1826, 127.

90. Timothy Dwight, "Journey to the White Mountains, Letter XX," *Travels in New England and New York*, vol. 2 (New Haven, CT: Timothy Dwight, 1821), 205.
91. Isaac Weld, Jr., *Travels Through the States of North America, and the Provinces of Upper and Lower Canada During the Years 1795, 1796, and 1739*, vol. 1 (London: John Stockdale, Piccadilly, 1799), 39.
92. Emmeline Stuart-Wortley, *Travels in the United States, etc., During 1849 and 1850* (New York: Harper & Brothers, 1851), 61.
93. Hugh M. Raup, "The View from John Sanderson's Farm: A Perspective for the Use of the Land," *Forest History* 10 (April 1966): 2–11; reprint, *Forest History Today* (1997): 3.
94. "Agreement by sundry persons to cut and cart wood to the Hope Furnace," August 10, 1765, Box 177, Folder 5, Brown Family Papers, Letters and Papers, Nicholas Brown & Co., 1762–1782, ser. 7, Hope Furnace, John Carter Brown Library.
95. Rufus Hopkins to Nicholas Brown & Co., May 17, 1769, Box 3, Folder 16, Personal Correspondence, ser. 3, Nicholas & John Brown, 1746–1762, Brown Family Papers, John Carter Brown Library.
96. Christopher Lippit to Rufus Hopkins, n.d., and Rufus Hopkins to N. Brown, January 22, 1784, Box 28, Folder 7, Brown Family Papers, Letters and Papers, Nicholas Brown & Co., 1762–1782, ser. 7, Hope Furnace, John Carter Brown Library
97. Joseph C. G. Kennedy, superintendent, *Agriculture of the United States in 1860; Compiled from the Original Returns of the Eighth Census, Under the Direction of the Secretary of the Interior* (Washington DC: Government Printing Office, 1864), 126; Samuel B. Ruggles, ed., *Agricultural Products of the United States of America, Classified by Their Proximity to the Oceans and Other Navigable Waters, Natural and Artificial* (New York: Press of the Chamber of Commerce, 1874), 5.
98. US Department of State, *Compendium of the Enumeration of the Inhabitants and Statistics of the United States as Obtained at the Department of State, from the Return of the Sixth Census, by Counties and Principal Towns Exhibiting the Population, Wealth, and Resources of the Country* . . . (Washington DC: Thomas Allen, 1841), 14, 10, 112.
99. Nancy M. Gordon, "The Economic Uses of Massachusetts Farmers," in *Stepping Back to Look Forward: A History of the Massachusetts Forest*, ed. Charles H. W. Foster (Petersham, MA: Harvard Forest, 1998), 82. See also Brook Hindle, ed., *America's Wooden Age: Aspects of its Early Technology* (Tarrytown, NY: Sleepy Hollow Restorations, 1975).
100. Daniel P. Jones, "Commercial Progress versus Local Rights: Turnpike Building in Northwestern Rhode Island in the 1790s," *Rhode Island History* 48 (February 1990): 20–32.
101. Richard Greenwood, "A Landscape of Industry," in *Landscape of Industry: An Industrial History of the Blackstone Valley* (Hanover, NH: University Press of New England, 2009), 17.
102. Jack Rothacher, "Increases in Water Yield Following Clear-Cut Logging in the Pacific Northwest," *Water Resources Research* 6, no. 2 (April 1970): 653–658; J. A. Jones and G. E. Grant, "Peak Flow Responses to Clear-Cutting and Roads in

Small and Large Basins, Western Cascades, Oregon," *Water Resources Research* 32 (1996): 959–974.
103. R. D. Harr, "Effects of Clearcutting on Rain-on-Snow Runoff in Western Oregon: A New Look at Old Studies," *Water Resource Research* 22 (1986): 1095–1100.
104. Günter Blöschl, et al. "At What Scales Do Climate Variability and Land Cover Change Impact on Flooding and Low Flows?" *Hydrological Processes* 21 (2007): 1241–1247.
105. "Blackstone Canal," *The Massachusetts Spy and Worcester County Advertiser*, October 24, 1827.
106. "The Blackstone Canal," *National Aegis*, June 18, 1828.
107. *Massachusetts Spy and Worcester County Advertiser*, April 16, 1828.
108. "Canals," *Rhode Island American and Providence Gazette*, 11 July 1828.
109. "Blackstone Canal," *National Aegis*, July 23, 1828; "Blackstone Canal," *Providence Journal*, July 26, 1828.
110. Plummer, "History of the Blackstone Canal," 8.
111. "The Weather," *The Massachusetts Spy and Worcester County Advertiser*, August 27, 1828.
112. "Ten Dollars Reward," *Massachusetts Yeoman*, September 6, 1828, 11.
113. "Extract from the Charter of the Blackstone Canal Company," *Massachusetts Yeoman*, September 6, 1828, 11.
114. "Canal Notice," *National Aegis*, August 13, 1828.
115. Welcome Arnold Greene, *The Providence Plantations for Two Hundred Fifty Years: An Historical Review of the Foundations, Rise, and Progress of the City of Providence* . . . (Providence, RI: J. A. & R. A. Reid, 1886), 75; "Worcester and Providence Canal," *National Aegis*, February 5, 1840, 3.
116. "Blackstone Canal," *Providence Journal*, published in the *The Massachusetts Spy and Worcester County Advertiser*, January 19, 1831; *Massachusetts Yeoman*, January 29, 1831, 95.
117. Ibid.
118. Richard E. Greenwood, "Natural Run and Artificial Falls: Waterpower and the Blackstone Canal," *Rhode Island History* 49, no. 2 (May 1991): 57.
119. "Internal Improvements," *Boston Centinel*, published in *Pawtucket Chronicle*, May 16, 1829.
120. "Canal Packet," *The Massachusetts Spy and Worcester County Advertiser*, June 24, 1829; *The Massachusetts Spy and Worcester County Advertiser*, July 1, 1829
121. "Ship Timber," *The Massachusetts Spy and Worcester County Advertiser*, December 22, 1830.
122. *Diary of Christopher Columbus Baldwin*, Transactions and Collections of the American Antiquarian Society, vol. 8 (Worcester, MA: American Antiquarian Society, 1901), 16–17.
123. *National Aegis*, July 8, 1829.
124. Edith W. Shute, "Reminiscences of the Blackstone Canal," in *The Blackstone Canal: Papers Read before the Deborah Wheelock Chapter, D.A.R., at a Meeting Held at the Home of Mrs. Shute, November 9, 1907* (1910; Uxbridge, MA: Deborah Wheelock Chapter, D.A.R., 1928), [15]. American Antiquarian Society.

125. Richard E. Greenwood, "Natural Run and Artificial Falls: Waterpower and the Blackstone Canal," *Rhode Island History* 49, no. 2 (May 1991): 57-58.
126. *National Aegis*, November 19, 1828; "Blackstone Canal," *National Aegis*, December 17, 1828.
127. *The Massachusetts Spy and Worcester County Advertiser*, July 1, 1829; *The Massachusetts Spy and Worcester County Advertiser*, November 26, 1828.
128. *National Aegis*, October 15, 1828.
129. "Railways," *National Aegis*, December 3, 1828.
130. William Lincoln, *History of Worcester, Massachusetts from Its Earliest Settlement to September 1836 with Various Notices Relating to the History of Worcester County* (Worcester, MA: Moses D. Phillips and Co., 1837), 340.
131. Greenwood, "Natural Run and Artificial Falls," 58.
132. "The Blackstone Canal: History of Its Construction and Abandonment . . . ," *The Providence Evening Press*, July 3, 1880 and *The Providence Morning Star*, July 6, 1880, reprinted by Robert Bellerose Bookseller, Slatersville, RI (1998), 11-12.
133. Lincoln, *History of Worcester*, 311.
134. Ibid., 313-314. According to Lincoln, tax revenue climbed from $2,437,550 in 1825 to $3,667,250 in 1835.
135. Greenwood, "Natural Run and Artificial Falls," 58.
136. George M. Dennison, *The Dorr War: Republicanism on Trial, 1831-1861* (Lexington, KY: The University Press of Kentucky, 1976).
137. Plummer, *History of the Blackstone Canal*, 9. On the Lonsdale Company, see Greenwood, "Natural Run and Artificial Falls," 59-60. Greenwood proposes that Brown, Ives, and Carrington, working through an intermediary, had, prior to the canal's construction, purchased land near Scott's Pond and then subsequently steered the course of the canal through it for the express purpose of developing the Lonsdale Company, which became one of the largest textile firms in New England.
138. "The Blackstone Canal: History of Its Construction and Abandonment," 10.
139. Ibid., 13.
140. "The Blackstone Canal: Canal Street. Building of the Canal. History and Navigation. Up the Towpath To-Day," *The Providence Sunday Journal*, October 18, 1885, reprinted Robert R. Bellerose Bookseller, Slatersville, RI (1998), 7-8.
141. A. W. Sweet, "A Sanitary Survey of the Seekonk River," (Ph.D. dissertation, Brown University, 1915), 121. Cited in Scott W. Nixon, et al., "Nitrogen and Phosphorus Inputs to Narragansett Bay: Past, Present, Future," in *Science for Ecosystem-Based Management: Narragansett Bay in the 21st Century*, eds. Alan Desbonnet and Barry A. Costa (New York: Springer, 2008), 126.

Epilogue: Between Progress and the Pull of the Sea

1. *The Geography of Strabo*, vol. 1, trans. H. C. Hamilton and W. Falconer (London: Henry G. Bohn, 1854), 15-16.
2. Henry David Thoreau, *Cape Cod* (Boston: Ticknor and Fields, 1866), 64.
3. Paul Carter, "Dark with Excess of Bright: Mapping the Coastlines of Knowledge," in *Mapping*, ed. Denis Cosgrove (London: Reaktion Books, 1999), 146-147.

4. Max Horkeimer and Theodor W. Adorno, *The Dialectic of Enlightenment,* trans. John Cumming (1944; New York: Herder & Herder, 1972), 3–6.
5. "Newport—Historical and Social," *Harper's New Monthly Magazine* 9 (June–November 1854): 315.
6. Jon Sterngass, *First Resorts: Pursuing Pleasure at Saratoga Springs, Newport & Coney Island* (Baltimore, MD: The Johns Hopkins University Press, 2001), 50.
7. Rockwell Stensrud, *Newport: A Lively Experiment, 1639–1969* (Newport, RI: Redwood Library & Athenaeum, 2006), 311–312.
8. Ibid., 315–316.
9. Jean-Didier Urbain, *At the Beach,* trans. Catherine Porter (1994; Minneapolis: University of Minnesota Press, 2003), 134–136.
10. Matthew McKenzie, *Clearing the Coastline: The Nineteenth-Century Ecological and Cultural Transformation of Cape Cod* (Lebanon, NH: University Press of New England, 2010), 138, 161–163, 173–177; Alice Garner, *A Shifting Shore: Locals, Outsiders, and the Transformation of a French Fishing Town, 1823–2000* (Ithaca, NY: Cornell University Press, 2005), 188–189. Garner shows that oyster fishermen of Arcachon, France, were not completely removed from tourist postcards, but were often depicted in supporting roles to the tourist industry.
11. Benjamin Tallman, "Scup! Scup!! Scup!!!" (Portsmouth, RI: s.n., February 21, 1870), 5. American Antiquarian Society Collections.
12. Bryant F. Tolles, Jr., *Summer by the Seaside: The Architecture of New England Coastal Resort Hotels, 1820–1950* (Lebanon, NH: University Press of New England, 2008), 38–62.
13. Henry James, "The Sense of Newport," *Harper's Monthly Magazine* 113, no. 675 (August 1906): 345–346, 350, 353.
14. Rachel Carson, *Under the Sea-Wind* (1941; New York: Penguin, 2007), 3.
15. Rachel Carson, *The Sea Around Us* (1951; New York: Oxford University Press, 1989), 15.
16. Ann H. Zwinger, "Introduction," in Carson, *The Sea Around Us,* xxvi.
17. Gary Kroll, *America's Ocean Wilderness: A Cultural History of Twentieth-Century Exploration* (Lawrence: University Press of Kansas, 2008), 122; Carson, *The Sea Around Us,* 15.
18. W. Jeffrey Bolster, *The Human Shore: Fishing the Atlantic in the Age of Sail* (Cambridge, MA: Harvard University Press, 2012), 273–276.
19. Mark Monmonier, *Coastlines: How Mapmakers Frame the World and Chart Environmental Change* (Chicago: University of Chicago Press, 2008), 163–166.
20. John Gillis, *The Human Shore: Seacoasts in History* (Chicago: University of Chicago Press, 2012), 187. Monmonier, *Coastlines,* 164.
21. "The Impact of Climate Change and Population Growth on the National Flood Insurance Program through 2100," prepared by AECOM for the Federal Insurance Mitigation Administration and Federal Emergency Management Agency, June 2013, ES-7.
22. Jenny Anderson, "Rebuilding the Coastline, but at What Cost," *New York Times,* May 18, 2013, accessed May 25, 2013, <http://www.nytimes.com/2013/05/19/nyregion/rebuilding-the-coastline-but-at-what-cost.html?pagewanted=all&_r=0>.

23. Thomas E. Kutcher, "Human Impacts on Narragansett Bay," in *An Ecological Profile of the Narragansett Bay National Estuarine Research Reserve,* eds. Kenneth B. Raposa and Malia L. Schwartz (Narragansett: Rhode Island Sea Grant, 2009), 151.
24. Yuyu Zhou and Y. Q. Wang, "An Assessment of Impervious Surface Area in Rhode Island," *Northeastern Naturalist* 14, no. 4 (2007): 649.
25. Sandor Bodo, "The Ebb and Flow of the Great Salt Cove," *Providence Journal,* August 11, 2013, G1, G8.
26. Thomas E. Kutcher, "Human Impacts on Narragansett Bay," 151.
27. Tim Roberts, et al., "The Floods of March 2010: What Have We Learned?" *Brown University Center for Environmental Studies* (Spring 2010): 16.
28. Rachel Carson, *The Edge of the Sea* (1955; Boston: Mariner, 1998), 1.
29. Yi-Fu Tuan, *Topophilia: A Study of Environmental Perception, Attitudes, and Values* (Englewood Cliffs, NJ: Prentice-Hall, 1974), 115; Yi-Fu Tuan, *Space and Place: The Perspective of Experience* (Minneapolis: University of Minnesota Press, 1977), 3–4.
30. Tim Ingold, *Lines: A Brief History* (New York: Routledge, 2007), 4.
31. Richard Mabey, *Turned out Nice Again: On Living with the Weather* (London: Profile Books, 2013), 87.
32. Russell Shorto, "How to Think Like the Dutch in a Post-Sandy World," *New York Times Magazine,* April 9, 2014, accessed April 26, 2014, <http://nyti.ms/1qhZCcf>.

Acknowledgments

I owe considerable thanks to numerous people and institutions for making this book possible. First, I would like to thank the University of New Hampshire Department of History and in particular Jeff Bolster, my advisor and mentor, who, in his scholarly and practical experience in all things watery, helped me navigate the shoals of marine environmental history while encouraging me to follow my interests. I would also like to thank Eliga Gould, who, with endless enthusiasm, provided me with valuable ideas and encouragement. Cynthia Van Zandt inspired me to pursue my interests in estuaries, and Kurk Dorsey helped me think about the relationship between humans and the natural world in new ways. Joyce Chaplin gave me her unwavering support as an outside commitee member, and Jan Golinski gave me a deeper understanding of Enlightenment thought and the history of science. Thank you also to David and Ginny Steelman for their generous support of the Steelman Fellowship, which funded a portion of my research travel. And I would like to extend a special thanks to the University of New Hampshire library staff.

This book would not have been possible without the help and support of many colleagues. I would like to extend special thanks to those in the History Department at the University of Montana. I will forever be grateful for my two years in Missoula. I would especially like to thank Kyle Volk and Tobin Shearer for their sage advice, unwavering support, and, of course, for a lot of laughs. I would also like to thank Ken Lockridge for reading an early draft of this book in its entirety and for providing so many rich and insightful suggestions. Finally, I would like to thank my colleagues at the University at Albany, State University of New York, for such a warm welcome to my new academic home.

I am deeply indebted to the Rachel Carson Center for Environment and Society at Ludwig Maximilian University in Munich, where I spent eleven months working on this book as a Carson Fellow. I owe special thanks to Carson Center Directors Christof Mauch and Helmuth Trischler for their unwavering enthusiasm for this project. The Carson Center gave me the financial supported for image licensing and indexing but also the time to think, write, and discuss this project with a community of scholars gathered from around the globe. I would like to extend special thanks to my friends and colleagues there, especially to Thomas Zeller, Karen Oslund, Thomas Lekan, Louis Warren, Lawrence Culver, Franz-Josef Brüggemeier, Bridget Love, Samuel Temple, Kenichi Matsui, Siddhartha Krishnan, Katie Ritson, Sarah Cameron, Ellen Arnold, Melanie Arndt, John Meyer, Frank Zelko, Keiko Matteson, Peter Boomgaard, Amy Hay, Poul Holm, Grace Karskens, Matthew Kelly, Claudia Leal, Shawn Van Ausdal, Uwe Lübken, Michelle Mart, Josh Berson, Wilko Graf von Hardenberg, Kim Coulter, Arielle Helmick, Giacamo Parrinello, Shen Hou, Carmel Finley, Fei Sheng, Luke Keogh, Angela Kreutz, Anka Liepold, Nicole Seymour, Maya Peterson, Melanie Arndt, Andrea Kiss, Maurits Ertsen, and many others. I would like to extend special thanks to Don Worster, who read parts of the manuscript and gave me valuable feedback. Finally, I would like to thank Antonia Mehnert and Rebecca Hofmann for inviting me to work with the Rachel Carson Center doctoral program writing workshop.

I received support in the form of several research fellowships that laid much of the groundwork for this project. The New England Regional Fellowship Consortium funded my work at several archives. Specifically, I would like to thank Conrad Wright of the Massachusetts Historical Society, Laura Linard of the Harvard Baker Library, Paul O'Pecko at the G. W. Blunt Library at Mystic Seaport, and Lee Teverow, Jordan Goffin, Delia Kovak, Karen Eberhart, Natash Brooks-Sperduti, and J. D. Kay of the Rhode Island Historical Society. The John Carter Brown Library funded my research there as the Paul W. McQuillen Memorial Fellow. I would like to thank Ted Widmer, Kim Nusco, Susan Danforth,

Ken Ward, Leslie Tobias-Olsen, Valerie Andrews, John Minichiello, and Margot Nishimura, among many others. During my time there, I also received wonderful recommendations and advice from Jim Muldoon, Ralph Bauer, Charlie Foy, Cynthia Radding, Jonathan Bordo, and Warren Prell. I was also fortunate to conduct research at the American Antiquarian Society as the Kate B. and Hall J. Peterson Fellow. I would like to extend special thanks to Paul Erickson, Thomas Knoles, Elizabeth Watts Pope, and Molly O'Hagan Hardy. I am particularly indebted to Dick Wilson, who not only took me on a tour of the Blackstone River headwaters near Worcester but also let me examine his impressive collection of Blackstone Canal articles and information. I would also like to thank Lisa Wilson, Kyle Volk, Sean Harvey, and Dan Rood for their ideas and our many wonderful conversations. Finally, my research took me to several other important repositories. I would like to thank Kevin Klyberg at the John H. Chafee Blackstone River Valley National Heritage Corridor, Bert Lippincott at the Newport Historical Society, Ken Carlson at the Rhode Island State Archives, as well as the South County Museum in Narragansett and the Gilbert Stuart Museum in Saunderstown for providing information and tours of their grounds.

During the writing process, I was also fortunate to present and receive valuable feedback on parts of this book. I would like to extend special thanks to Louis Warren, who invited me to present a chapter to the Mellon Research Initiative in the Humanities Institute at the University of California, Davis, and to Cori Knudten and Chad Anderson, who provided helpful commentary. I would like to thank Stella Capoccia for having me present at Montana Tech. I owe special thanks to John Gillis for organizing an incredibly fruitful Rachel Carson Center-sponsored workshop at the Island Institute in Rockland, Maine, in 2011. John later read this book in its entirety and provided deeply informed feedback and enthusiastic encouragement. I was also fortunate to present parts of this book at the 2010 American Historical Association annual meeting, the 2010 North American Society for Oceanic History meeting, the American Society for Environmental History annual conference in 2012 and 2014, the 2012 John Carter Brown Library 50th Anniversary Fellows' Conference, the 2013 Omohundro Institute annual conference, the 2014 Organization of American Historians conference, and a Rachel Carson

Center-sponsored workshop hosted by the Nehru Memorial Museum and Library in Delhi, India in 2014. I owe special thanks to, among others, Joyce Chaplin, Matthew Booker, Charles Vörösmarty, Elizabeth Pillsbury, Jeff Bolster, Bill Leavenworth, Karen Alexander, Howard Stewart, John McNeill, Christine DeLucia, Strother Roberts, Karen Halttunen, Steven Mentz, Hester Blum, Cameron Strang, Chris Parsons, Cynthia Radding, Jim Rice, Karen Kupperman, and Derek Nelson, Sam Temple, Siddartha Krishnan, Christof Mauch, and Mahesh Rangarajan.

My thinking about this project was also shaped in many ways by two summer institutes. In 2008 I spent six weeks participating in a program funded by the National Science Foundation, sponsored by the Northeast Consortium for Hydrologic Synthesis, and hosted by MIT. As the lone historian working with a group of physical and biological scientists to recreate America's colonial hydrology, I learned how to work in truly interdisciplinary ways and made some wonderful friends and colleagues in the process. I would like to thank the entire team and particularly Charles Vörösmarty and Mark Green. During summer 2010 I spent six weeks participating in a National Endowment for the Humanities summer institute titled "The American Maritime People," which was hosted by the Frank C. Munson Institute at Mystic Seaport in Connecticut. I am indebted to the smart, fun, interdisciplinary group of faculty and graduate students who attended. I am particularly grateful to Glenn Gordinier and Eric Roorda for organizing such a valuable program.

As this book developed, I benefited from the experience and talents of many. I am deeply indebted to my editor, Joyce Seltzer, whose enthusiasm for and careful reading of this book has made it fundamentally better. I am also grateful to her assistant Brian Distelberg for his help, advice, and encouragement. Philip Schwartzberg of Meridian Mapping in Minneapolis created the wonderful maps. I owe special thanks to Nathalie Wolfram, who, with an eagle eye, read the manuscript in its entirety and gave me detailed feedback throughout. Thank you to my copy editor, Leslie Ellen Jones, who caught and corrected my mistakes. And thank you also to Marianna Vertullo, the editorial production manager, who helped this book finally sail across the finish line. Of course, any factual or interpretive errors that remain are mine and mine alone.

Finally, this book would not have been possible without the help of my family. I would like to thank my parents, Karen and Fred Pastore, who love Narragansett Bay as much as I do. They hosted me while I researched this book in Rhode Island and were always willing to drive to New Hampshire or fly to Montana to watch my kids, make dinner, and generally take care of things while I sat quietly writing and reading for days on end. I would like to thank my sister, Cara Pastore, for always being a source of inspiration and Michael Foster for being a part of our family. Thank you to my grandmother, Sue Pastore, for all her prayers and wonderful cooking. I would like to thank my mother- and father-in-law, Judy and Chip Detwiler, for their generous support. They, too, watched our kids, fed us, and kept the house standing. They flew to visit in Munich and Montana and never flinched while I continued to pound away on the computer. I also owe special thanks to my sister- and brother-in-law, Laura Detwiler and Doug Keene, for their willingness to enmesh their lives with ours in one of the most complicated work/childcare schedules this world has ever known. The two of them helped make this book a reality. I would like to thank Kate Detwiler, James Gray, Anne Detwiler, and Kristian Carson, for always being there. And a special thanks goes to Nate, Sam, and Mary Keene, Celina and June Gray, and Finn Carson for being the best nephews and nieces ever. Thank you also to Jared, Katie, Maeve, Nora, and Charlotte Kelly and Jesse, Danielle, Mari, and Theo Kachapis for opening your homes and providing a lot of laughs whenever I'm in Rhode Island.

But above all, I am forever indebted to my wife, Susan Detwiler, and our kids, Rosie and Abram. With a careful eye and a gift for style, Sue read this book inside and out, providing help every step of the way. She also bore the brunt of my absence while I spent my days scribbling. In short, this book would not have possible without her. Rosie and Abe, well, they kept it loud and they kept it real. While I wrote this, we lived in New Hampshire, moved to Montana, then moved to Munich, then back to Montana, and then finally we landed in Albany, New York. So Susie, Rosie, and Abe, thank you for your patience and your humor. Thank you for your love. And thank you for going on this wild adventure.

Index

Account of North America, 138
Albion, Rhode Island, 197
Alcock, Dr. John, 78
Aldrich, James, 214
Aldrich, Paul, 214
algae, 24, 48, 53, 72-73, 77, 80
Algonquins, 16, 122
Almy, Christopher, 162
Almy, Job, 162
Almy, Mary, 187
Almy, William, 209
American Industrial Revolution, 198. *See also* industrialization
American Revolution, 9, 164, 188-194
Ames, Nathaniel, 144
Anderson, Virginia DeJohn, 122
Angell, Joseph K., 210
animal kingdom, 152-153
Anne, Queen, 82, 103, 166
Antinomian, 59
Appleton, John H., 225
Apponaug, 130-133, 145
aquatic creatures, 8-9, 153-155. *See also* marine animals
Aquidneck Island, 9, 22, 59, 65-68, 87, 89-90, 98-99, 108, 111, 120, 137-142, 146, 174-175, 182-194, 230, 232
Arlington, Lord, 64
Arnold, Josiah, 70, 162
Arnold, Welcome, 160
As You Like It, 89
Asamequin, 110-111
Aspinwall, William, 59
Astronomical Diary, 144
Assawampset, 110, 122. *See also* Sawamsett
Assonet, Town of, 87, 98, 116, 126, 127
Atlantic Neptune, 173

Bachelard, Gaston, 16
Bacon, Sir Francis, 181
baia, 88
Baldwin, Christopher Columbus, 222
Barnard, Mary, 54
barrier beaches, 70, 95, 99, 105
Barrington: River, 123; Town of, 91, 97, 110, 118-120, 193
Battle of Bunker Hill, 190
Battle of Rhode Island, 189
Battle of Yorktown, 190
beacons, 6, 9. *See also* lighthouses

beavers, 7, 12, 16, 26, 33, 35, 39, 41–49
beaver dams, 17, 35, 41–48. *See also* dams
beaver populations, 39, 42–43
"Beavertail," 161–170, 174–175, 231
Beckwith, T., 211
Belknap, Jeremy, 43, 150
Bellomont, Lord, 102
Blackstone, Lord, 210
Blackstone Canal, 196–227
Blackstone Canal Company, 6, 9–10, 199, 204–214, 221–227
Blackstone River, 9, 57, 98, 112, 117, 118, 125, 196, 199, 203–212, 217, 219–221, 224–225
Blaskowitz, Charles, 173–181
Block Island, 11, 14, 28, 65, 78, 80, 85, 94, 106, 130, 162, 165, 172, 183, 231
Board of Trade, 85, 103–104, 168, 173, 177
borderlands, 3, 83, 98–107, 118, 123–128, 186, 190, 195, 229
Boston, 5, 12, 39, 53–56, 79, 82, 104–105, 110, 114, 130, 139, 141, 159, 161–163, 182, 189, 190, 197–200, 202, 205, 210, 211, 222–223
Boston Centinel, 202, 222
Boston Neck, 64, 96
Boston News-Letter, 165
Boston Patriot, 202
Boundary Commission, 125, 127
boundary disagreements, 86–87, 92–94, 118, 124–127
Bowen, Emanuel, 87
Bowen, John, 115
Bradford, William, 11, 15, 28, 38, 54, 111, 114–115
Braham, William De, 173
Brenton, Benjamin, 106
Brenton, Ebenezer, 80
Brenton, William, 65
Bridenbaugh, Carl, 51
Briggs, Joseph, 218
Brissot de Warville, Jacques-Pierre, 192–193
Bristol, England, 54
Bristol, Rhode Island, 73, 97, 110, 115, 118, 126, 130, 183–186, 193
Brown, George, 162

Brown, John, 162, 199–200
Brown, Moses, 193, 199, 209
Brown, Nicholas, 199, 225
Brown, Obadiah, 199
buccaneers, 103, 165
Bucklin, John, 209
Buffon, Comte de, 134–137, 140, 156–158
Bulger, Richard, 69
Bull, Henry, 74–75, 80
Burnaby, Andrew, 138

Cadillac, La Mothe, 65
Callender, John, 137–138
canals, 5, 9–10, 196–227
canoes, 32, 34, 56, 59, 99–100, 229
Canonchet, 100
Canonicus, 56, 58, 99
Cape Cod, 228
Carr, Caleb, 102
Carr, George, 64
Carr, Peleg, 162
Carrington, Edward, 225
Carson, Rachel, 233, 238
Cartesian philosophy, 154
Cartier, Jacques, 4, 13
cartographers, 20, 88, 234, 236. *See also* mapmaking
cattle, 7, 46, 50, 52, 65–71, 76, 78, 130–131, 148, 180, 183, 223
Chandler, William, 90–91, 94–98, 107, 127, 180
Charles I, King, 82
Charles II, King, 102, 120, 123
Charles River, 112, 190
Chase, Benjamin, 116
Cheese, Peter, 119
Cheese, Thomas, 119
Chesapeake Bay, 4, 190
Chronological History of New England, in the Form of Annals, 114
chronometer, 181
Church, Thomas, 108
clams, 6–7, 25–26, 69
Clarke, John, 78
classification system, 153–155
climate: changes in, 140–147; observations of, 132–135; understanding, 132–140

coastal communities: boundary changes for, 10, 86–87; boundary disagreements of, 92–94, 118; progress for, 9–10, 228–238
coastal space: improving, 3, 5, 161–195; nature of, 172–173; securing, 165–173; wartime and, 161–195
coastline: changes in, 6–9, 17–18, 195, 234, 236; imagination of 129, 173, 195, 231; industrialization of, 196–227; division of, 6
Coddington, William, 59
Coggeshall, John, 59
Coginiquant, 61
Colden, Cadwallader, 91, 96, 98, 107, 127
Comer, John, 141–142
compass, 55, 98, 124, 181
Conanicut Island, 22, 25, 62, 65, 96, 108, 161, 172, 174, 183, 187. *See also* Jamestown
Constitution: Rhode Island ratification of US, 195; Rhode Island crisis (Dorr Rebellion), 224
Cook, John, 116
Cooke, Nicholas, 182
Cornell, Thomas, Jr., 69
Cotton, John, 54
Cranston, Samuel, 103, 104, 165, 168
Cranston, Town of, 218, 238
Cree, 33
Crusoe, Robinson, 90
cultural complexities, 3–4
cultural exchange, 5
Culverwell, Thomas, 65–66

dams: beaver dams, 17, 35, 42–48; building, 5, 207–214; mill dams, 206–210, 219
Defoe, Daniel, 89, 125
Denham, Sir John, 2, 10
Des Barres, Joseph Frederick Wallet, 173, 177, 180, 194
d'Estaing, Comte Jean-Baptiste, 187
diplomacy, 60, 100, 122, 229
disease, 62, 99, 119, 123, 136, 190
Dodge, Jacob, 213
Doughty, Francis, 60–61
Douglass, William, 68
Dudingston, William, 177

Dudley, Joseph, 67
Dumman, William, 165
dunes, 31, 70, 94, 202
dung, 50, 70–76, 80. *See also* manure
Dunn, David, 214
Dutch Island, 35.
Dutch West India Company, 15, 34–35, 39
Duties of Neutral Princes towards Belligerent Princes, 170
Dwight, Timothy, 217
Dyck, Gysbert op, 60

earthquakes, 142
ecological complexities, 3–4
ecological revolution, 47–48
Elekens, Jacques, 34–35
Endecott, John, 55
Enlightenment, 132–135, 146, 153–154, 157, 230
environmental change, 10, 17, 79–80, 115, 139, 157, 195–200
estuaries: changes in, 3–4, 17–18, 46–48; definition of, 22; descriptions of, 31, 35; ecology of, 83–84; flushing time of, 23–24, 48; mythic view of, 31–33; shellfish in, 25–26
Exclusive Economic Zone, 234

Faden, William, 177, 180, 194
Farnsworth, William, 216
"Fate of the Mouse, The," 155
Fenner, James, 196
fertilizers, 70–73
First Memoir, 154
fish, 4, 24–25, 48, 206–208. *See also* shellfish
Fish Act, 208
fisheries, 67, 78, 206, 208, 231, 234
Fishery Conservation and Management Act, 234
Fishery Management Council, 234
Fleet, Henry, 14
flood plains, 236–238
fluvial systems, 46
Fogland, 185
food chain, 25, 48

Fort George, 168–173
forts, building, 6, 9, 160, 164
François-Alexandre-Frédéric, Duke, 193–194
Franklin, Abel, 162
Franklin, Benjamin, 145, 149
Freeborn, Gideon, 147–148
"freedom of the seas," 169
French: presence on America's Atlantic coast, 9, 13, 17, 27, 65, 104, 138, 147, 165, 173, 190–191, 194; presence in Narragansett Bay, 187–190, 194
French and Indian War, 138, 164, 173
freshwater sources, 2, 6, 10, 17, 22–23, 48, 86, 139, 199
From Here to Eternity, 234
frontier, 3, 39, 229
fur trade, 12, 15–17, 27–28, 35, 38–39

Gallop, John, 11–13
garden metaphor, 137–138
Garden of New England, 137
Gardner, Henry, 66, 106
Gardner, John, 68
Gardner, Robert, 141
geographic boundaries, 8, 112
geographic quicksilver, 82–83
George, King, 162, 164
George II, King, 168
Gibbs, Elisha, 116
Gilded Age, 232–233
glaciers, 25
goats, 61, 66, 69,
Goat Island, 167–168, 187
Gookin, Daniel, 15, 28, 114
Gorton, Samuel, 59
Greene, Christopher, 208
Greene, Fleet, 186–188
Greene, Griffin, 208
Greene, John, 208
Greene, Nathaniel, 208
Greene, Samuel, 145
Greene, William, 130, 132, 146
Grotius, Hugo, 169

Hall, William, 69

Harnet, Cornelius, 106
Harper's Monthly, 230
Harris, Charles, 165
Harris, Richard, 159–160
Harris, William, 65, 66, 68
Harrison, John, 181
Harrison, Joseph, 162, 175–176
Harrison, Peter, 173–176
Hazard, George, 65–66
Hazard, Robert, 64
Hazard, Tom, 64, 70
Helme, James, 73–75, 91, 96, 98, 107, 127
Hiawatha, 33
Higden, Ranulph, 88
Histoire Naturelle, 134, 156
Historisch Verhael, 35
History of America, 136
History of New-England, 138
History of the Royal Society, 144
hogs, 25, 66–69, 148, 152,
Hog Island, 22, 97, 183
Holland, Samuel, 173
Hooke, Robert, 143–144
Hope Furnace Company, 206–208, 217–218
Hope Island, 22, 58, 96
Hope, Mount, 97, 110, 118, 125, 186
Hope Bay, Mount, *See* Mount Hope Bay
Hopkins, Rufus, 218
Hopkins, Stephen, 218
Hopkins, W., 111
horses, 13, 19, 50, 52, 66, 68, 70, 72, 95–96, 98, 143, 148, 180, 184, 191, 196, 229.
Hosmore, Stephen, 142
Howe, Lord, 181, 189–190
Hubbard, William, 49, 100
Hudson, Henry, 34
Hull, John, 62, 64, 66
hurricanes, 145, 189, 237. *See also* weather
Hutchinson, Anne, 59

ice age, 25
industrial revolution, 5, 9, 198
industrialization: canals and, 199–227; industrial revolution, 5, 9, 198; progress and, 9–10, 228–238

Iroquois, 16, 30, 33, 46
Ives, Thomas, 199, 225

James, Henry, 232–233
Jamestown, Rhode Island, 162, 184, 231
Jefferson, Thomas, 140, 192
Jenks, Eleazar, 209
Jenks, Stephen, 209
Jones, John Paul, 191
Josselyn, John, 66, 100

Kalm, Peter, 160
Kant, Immanuel, 132
Kensett, John Frederick, 231
Kerr, Deborah, 234
keystone species, 42, 47
Kidd, William "Captain," 102
King, Benjamin, 176
King George's War, 162–164
King Philip's War, 64, 82, 100–101, 108, 121–123
Kingstown, 64, 75, 90, 96, 105, 107, 172, 183

La Mettrie, Julien Offray de, 154–155
Lancaster, Burt, 234
Leclerc, Georges Louis, 134
Lepore, Jill, 101
Lewis, Martin W., 87
"Libel of English Policy," 88
lighthouses, 6, 9, 162–172, 175
lime, 8, 76, 133, 149, 151–152, 159–160, 172, 217, 221
Lincoln, Benjamin, 150
Lincoln, John W., 205, 211, 213, 221
Lincoln, William, 211, 224
Lindsey, Benjamin, 177
Linnaeus, Carl, 152–153, 160
Lippit, Christopher, 218
littoral spaces: environmental changes in, 10; improving, 3–5; inhabitants of, 3–6, 127; value of, 63–64
livestock: exporting, 50, 78; on islands, 58–59; loss of, 130–131; manure from, 43, 47, 70–76, 80; in Massachusetts Bay, 69, 72; in Narragansett Bay, 43, 47, 70–76, 80; number of, 68–72, 183; raising, 50–52, 58–59, 62–68, 75–81; stockyards, 64–75
lobster, 24, 148
Low, Edward, 165

Mackenzie, Frederick, 184–188, 191
MacSparran, James, 50–52, 80, 105, 106
"Mad Jack," 11–13, 15, 17, 58
Magnuson Act, 234
Malbone, Godfrey, 162
manure, 43, 47, 70–76, 80
mapmaking: cartographers and, 20, 88, 234, 236; navigational instruments and, 175–176, 181; topographical charts and, 174–177, 180
marchlands, 4. *See also* borderlands
Mare Clausum, 169
Mare Liberum, 169
marine animals, 4, 7, 32. *See also* aquatic creatures
marine chronometers, 181
maritime infrastructure, 51–52
marshes: creation of, 31; dams and, 43; decline of, 227, 237; draining, 75–76, 135–136, 140, 237; life in, 95; nurturing, 238; for refuge, 101–102, 105, 128; waterways through, 22, 85, 112, 115
Massachusetts Bay: jurisdiction of, 123–126; livestock in, 69, 72; Native Americans and, 15, 110, 118–122; settlements in, 12, 55–59; shoreline of, 87–89
Massachusetts Spy and Worcester County Advertiser, 202, 205, 210, 219–220
Massachusetts Yeoman, 212
Massasoit, 56, 110–111, 124
Matheson, James, 206
Maushop, 31
Maverick, Samuel, 66–68
meadows, 5, 7–8, 45–46, 57–58, 60–62, 66–67, 70, 75–76, 81–83, 94–95, 123, 138, 213, 214, 216
Megapolensis, Reverend, 46–47
Mellen, Reverend, 76
Melville, Robert, 174, 177

Mercator, Gerardus, 181
Merchant of Venice, The, 88
merchant ships, 7, 106, 165, 168
meridian, 126, 145
Meteaûhock, 28. *See also* wampum
Miantonomi, 56, 82, 99
Middleborough, Massachusetts, 110, 118
military preparations, 181–195
mill dams, 206–210, 219. *See also* dams
Millbury, Massachusetts, 203, 212–213
mineral kingdom, 152–153
Miranda, Francisco, 192
Mohegans, 16, 28
Monamet Bay, 114
Morse, Jedidiah, 192
Morton, Thomas, 12, 19, 24, 25, 46, 48
Moshassuck River, 215, 227, 236
Mott, Adam, 68
Mount Hope Bay, 108, 116, 127
mowhacheis, 28, 30. *See also* wampum
mudflats, 4, 18, 25, 60, 69, 128, 147, 167
Mumford, Thomas, 62
mussels, 25–26, 153, 249
Mystic River (Connecticut), 28
Mystic River (Massachusetts), 190
Mystic, Connecticut, 15, 288, 290
mythic tales, 31–34

Narragansett Bay: borders of, 98–107; boundaries of, 86–87, 92–94; changes in, 3–9, 46–49, 80–81, 227–238; defining, 86–87; descriptions of, 18–22, 35, 38, 95–96, 130–131; first settlement of, 6; flushing time of, 23–24, 48; geographic quicksilver of, 82–129; livestock in, 43, 47, 70–76, 80; location of, 85–86, 108–116, 162; mapping, 173–195; maritime identity of, 51–52, 164, 222; maritime militarization of, 173–195; nature of, 130–160; placing, 108–123; pollution of, 7–8; progress of, 9–10, 228–238; securing, 165–173; settlements in, 3–13, 53–59, 63–70, 86–90; shoreline of, 70, 89, 131, 230–234, 237; size of, 22; survey of, 91, 96–97; in transition, 130–160; water flow through, 22–23; watershed of, 6, 9; winter in, 53, 130–138, 219
Narragansett Country, 2, 11, 30, 52, 61–66, 70, 85, 94–95, 101, 104, 111
Narragansett Indians, 1, 12–15, 27–31, 58–62, 99, 111, 122, 161
National Aegis, 204, 213, 214, 216, 220
Native Americans: changing patterns of, 26–27; conflicts with, 12–15; property of, 54–55; trading with, 6–7, 12, 15–16, 27–28. *See also specific tribes*
natural theology, 134–135
nature: Enlightenment and, 132–135, 146, 153–154, 157, 230; as God's work, 134–135, 154; of Narragansett Bay, 130–160; of space, 172–173
New England: desiccation of, 11–49; early history of, 3–9; population growth of, 6, 26, 71–72, 161, 197–198, 218, 224. *See also* Massachusetts Bay; Narragansett Bay
New England Fishery Management Council, 234
New England's Prospect, 112
New Netherland Company, 34–35
New York Times, 234
New-England Courant, 166
New-England Weekly Journal, 155
Newport, Rhode Island, ix, 9, 18, 60, 64–65, 67, 70–71, 79, 103–104, 106–107, 116, 130, 141–142, 144, 149, 161–163, 165–168, 170–177, 182–189, 191–193, 197–198, 230–233
Newport Mercury, 106
Newtonian philosophy, 154
Niagara Falls, 44
Niantics, 16
Nipmucks, 122
Norton, Walter, 12
Notes on the State of Virginia, 140

O'Connor, Patrick, 216
Oldham, John (Mad Jack), 11–13, 15, 17, 58
Oyster Act, 160
oyster cultivation, 133, 150–152
oysters, classifying, 8, 152–160

Paine, Thomas, 102
Palairet, Jean, 138
Partridge, Richard, 103
Patience Island, ix, 22, 58, 96
Pawtucket: City of, 57, 205, 208; Falls, 87, 98, 126, 207–209
Pawcatuck River, 22, 28, 70, 77, 83, 91, 150, 206
Pawcatuck, Village of, 94
Pawtuxet River, 22, 57, 207–208, 238
Pawtuxet, Town of, 57, 96. *See also* Cranston
peage, 27. *See also* wampum
Peirce, Joseph, 43, 45–46
"Pequot Path, The," 61
Pequot War, 63
Pequots, 12–15, 22, 28, 61, 63
Pettaquamscutt: Pond, 64, 75, 82, 84, 102; Purchase, 62, 64; River, 62, 64, 70, 74–75, 82, 84, 96
Philip, King, 97, 110, 117–119, 121–125, 128. *See also* King Philip's War
pigs, 52, 68–69, 72–73. *See also* hogs
pirates, 5, 102–104, 165–168
Plimouth Township, 109, 111, 115
Plymouth Colony, 11, 38, 56, 60, 66, 85, 90, 99, 109–112, 117–118
Pocanocket, 110–118, 124–125
Point Judith, 64, 66, 80, 96, 111, 116, 125, 162–163, 207
Polychronicon, 88
"polyps," 153–155
ponds: building, 43–45, 212–213, 221; draining, 45–46; feeder ponds, 212, 220; holding ponds, 199, 217, 221, 225; salt ponds, 22, 30, 70–72, 76–80, 95, 104–105, 111
population growth, 6, 26, 71–72, 161, 197–198, 218, 224
Porter, John, 62
portolan charts, 181
Portsmouth, Rhode Island, 59–62, 64–66, 68–69, 79, 116, 119, 162, 184–186, 189, 231, 250
prehistoric objects, 27
Prescott, Richard, 185
Prince, Thomas, 110–112, 114–115

privateers, 5, 103, 106, 172, 190–191
progress, 9–10, 228–238. *See also* industrialization
Providence, City of, 9, 36, 39, 56–59, 65, 71, 79, 86–87, 96–97, 99, 110, 112, 118, 126, 130, 151–152, 159, 176–177, 180, 182–183, 193, 196–211, 214–217, 220–226, 235–237
Providence American, 212
Providence Evening Press, 225
Providence Gazette, 211, 220
Providence Journal, 221
Providence Patriot, 214
Providence River, ix, 120, 126, 177, 185
Providence Sunday Journal, 225, 227
provisioning trade, 7, 67–68
Prudence Island, ix, 22, 58, 96, 147–149, 174, 183–185, 251
Pynchon, William, 39

quadrant, 175–176
quahogs, 7, 17, 25–26, 30, 49, 131
Queen Anne's War, 82, 103, 166
Quequaganuet, 61–62
Quesney, François, 153

railroads, 223–224
Ramusio, Giovanni Battista, 20
Randall, William, 225
Randolph, Edward, 65
Rasieres, Isaack de, 38
Réaumur, René Ferchault de, 154
Redwood, Abraham, 176
Regnum animale, 152–153
Regnum lapideum, 152–153
Regnum vegetabile, 152–153
Rehoboth, 115, 118–120, 123
Reid, William, 176
Rhode Island American, 220
Rice, Daniel, 214
rivers: biogeochemical makeup of, 17–18; changes to, 46–48; descriptions of, 35
Robertson, James, 190
Robertson, William, 136–137, 190
Rochambeau, Comte de, 189

Rogers, Robert, 138
Ryther, John, 73

Sakonnet: Point, 31, 98, 163; River, 89, 108, 110–111, 116, 119–120, 126, 185; Sakonnet, Town of, 98, 108, 116, 120–121
Salem, Town of, 53–55
saltwater, 1–10, 17, 26, 86, 89, 211
Sanderson, John, 217
Sassamon, John, 122
Saunders, Samuel, 224
Sawamsett, 110, 117–118. *See also* Assawampset, Sawampset, Sowams, Sowamsett
Sawampset, 110
Schröter, Johann Friedrich, 41
Scituate, Rhode Island, 64, 206, 218
Scultob, 61
sea: changes in, 3–4; inspiration from, 231–234; mythic view of, 32–33; progress of, 228–238; pull of, 228–238
Sea Around Us, The, 233, 234
sea levels, rising, 10, 25–26
sea polyps, 153–155
seawater, 1–10, 17, 26, 86, 89, 211
Seekonk, 56, 83, 87, 96–97, 114–116, 118, 120,
sedimentation, 17, 46–48, 77
Segreve, John, 214
Selden, John, 169
Sergeant, William, 207, 209–210
Serle, Ambrose, 190
Seton, Ernest Thompson, 42
settlements, 3–13, 53–59, 63–70, 86–90
sewan, 27, 38. *See also* wampum
Sewell, Samuel, 64
Shakespeare, William, 88, 89
Shearmn, Philip, 70
Sheffield, Edmond, 78
sheep,7, 52, 64–68, 72, 122, 130, 180, 183–184, 192
shell beads, 6–7, 12–16, 27–35. *See also* wampum
shellfish, 6, 25, 148–154. *See also* fish, quahog, oyster
Siege of Rhode Island, 189

Siege of Yorktown, 190
Simon, John, 110
Slater, Samuel, 208–209
slavery, 5, 64, 67–68, 104–107, 123, 128, 172
Smith, Adam, 4, 5, 10, 153
Smith, Alfred, 230
Smith, Caleb, 175
Smith, Richard, 39, 60–62
Smith, Stephen, 205, 211
Smithfield, 159–160
Solgard, Captain, 165
Somerset, 110
Sowams, 115. *See also* Assawampset, Sawampset, Sawamsett, Sowamsett
Sowamsett, 39, 118, 120, 123. *See also* Assawampset, Sawampset, Sawamsett, Sowams
Spain: Spanish presence on America's Atlantic coast, 13, 34, 102–103, 162, 191–192
Spelman, Henry, 13–14
Spencer, Joseph, 78
Sprat, Thomas, 138, 144
Squanto, 111
Squinimo, Benjamin, 110
Stafford, Joseph, 146
Standish, Myles, 12
Staple Act, 67–68
Stiles, Ezra, 31, 82–84, 142, 182–184, 191
Stilgoe, John, 163
stockyards, 64–75
Stone, John, 12
Story, Joseph, 209–210
Stowell, Samuel, 213
Strabo, 228–229
Stuart-Wortley, Emmeline, 217
suckáuhock, 30. *See also* wampum
Sullivan, John, 188
"super storms," 10, 237. *See also* weather
swamps, 33, 45, 75–76, 100–105, 128, 190. *See also* marshes
Swansea, 110, 118–123
sweetwater, 1–10
Systema naturae, 152–153

Takamunna, 121

Taunton, Town of, 6061, 101, 121, 126–127
Taunton River, 22, 101, 108, 110, 112, 116, 125, 126–127, 177, 211
Thames River, 2, 167
Thompson, Benjamin, 101
Thompson, David, 43
Thoreau, Henry David, 228–229, 238.
Tillinghast, Samuel, 130, 133, 142–145
Tillman, Benjamin, 231
Timucuan Indians, 13
Titticut, Samuel, 118–119
Titus, Joseph, 118
Tiverton, 108, 116, 119, 126, 185
Topographical Chart of the Bay of Narragansett in the Province of New England, 177
topographical charts, 174–177, 180
trade: fur trade, 12, 15–17, 27–28, 35, 38–39; monopoly on, 34–35; wampum trade, 6–7, 12, 15–17, 27–28, 34, 39
Trembley, Abraham, 154
Trip, John, 69
Tyler v. Wilkinson, 209

Under the Sea-Wind, 233
Uxbridge, Massachusetts, 212–214

Van Bynkershoerk, Cornelius, 170
Van der Donck, Adriaen, 35, 45–47
Van Leeuwenhoeck, Antony, 153
Van Wassenaer, Nicolaes, 35, 38, 45
vegetable kingdom, 152–153
venison, 28
Verrazano, Giovanni da, 4, 17–20, 22, 26, 60

Wallace, James, 182
Wampanoags, 31, 56, 62, 86, 99, 110, 122, 127
wampum: definition of, 7; myths about, 31–34; spiritual symbol of, 30; stringing, 28–30; trading, 6–7, 12, 15–17, 27–28, 34, 39; value of, 16–17, 30, 49. *See also* shell beads
"wampum revolution," 47

wampompeag, 28. *See also* wampum
wampompeage, 27. *See also* wampum
Wanton, Gideon, 162
Wanton, Joseph, 182
War for Independence, 9, 164, 188–194
war preparations, 181–195
war ships, 181–195
Ward, Governor, 172
Warren, 110, 126, 186, 193
Warren River, 97
wartime, 161–195
Warwick, 25, 59, 60, 69, 79, 96, 130–131, 177, 238
Washington, George, 183
Watch Hill, Town of, 94, 231
Waterhouse, Benjamin, 192–193
waterscape, 83–84, 105, 115, 128, 229
watershed: biogeochemical makeup of, 17–18; impact on, 47–48, 70–72, 76, 206, 209, 212; reconfiguration of, 6, 9; size of, 22, 47
Watson, Job, 183
weather: atmospheric changes, 140–147; barometers for, 145; humidity, 135–136, 144; hurricanes, 145, 189, 237; instruments for, 143, 145; observations of, 132–147; predicting, 128, 139, 144–145; storms, 10, 130–134, 140–142, 145, 237–238; thermometers for, 145; winter weather, 53, 130–138, 219
Weld, Isaac, 217
Welland, John, 165
Wesson, Abel, 213
Wesson, John, 213
Westerly, Town of, 65, 77–80.
"wet fragmentation," 135
Whale, Theophilus, 82–86, 102
Whalley, Edward, 82–83
whelk, 7, 17, 28, 49
Whittredge, Thomas Worthington, 231
Wickcom, Thomas, 105
Wilbore, Samuel, 62
Willet, 82
Williams, Roger, 1, 10, 22, 27–28, 30–32, 39, 42, 48, 53–63, 76, 83, 99, 122
Williamson, Hugh, 139–140
Wilson, Samuel, 62
Windsor, Thomas, 104

Wing, Elisha, 108–109
Winslow, Edward, 19, 56, 60, 111, 114–115
Winthrop, John, 12, 53, 55–59, 65, 76
Wood, William, 15, 24–28, 30, 32, 48, 69, 99, 112–114, 117
Woodley, Robert, 216

Worcester, Town of, 9, 196–197, 199–203, 205–206, 211–212, 214–217, 221–225
World Atlas, 87
Wright, Benjamin, 200, 203

zeewan, 27, 35. *See also* wampum